IET CONTROL ENGINEERING SERIES 84

Nonlinear and Adaptive Control Systems

Other volumes in this series:

Nonlinear and Adaptive Control Systems

Zhengtao Ding

The Institution of Engineering and Technology

Published by The Institution of Engineering and Technology, London, United Kingdom

The Institution of Engineering and Technology is registered as a Charity in England & Wales (no. 211014) and Scotland (no. SC038698).

The Institution of Engineering and Technology
Michael Faraday House
Six Hills Way, Stevenage
Herts, SG1 2AY, United Kingdom

www.theiet.org

British Library Cataloguing in Publication Data
A catalogue record for this product is available from the British Library

ISBN 978-1-84919-574-4 (hardback)
ISBN 978-1-84919-575-1 (PDF)

Typeset in India by MPS Limited
Printed in the UK by CPI Group (UK) Ltd, Croydon

To my family, Yinghong, Guangyue and Xiang

Contents

Preface

This book is intended for the use as a textbook at MSc and senior undergraduate level in control engineering and related disciplines such as electrical, mechanical, chemical and aerospace engineering and applied mathematics. It can also be used as a reference book by control engineers in industry and research students in automation and control. It is largely, although not entirely, based on the course unit bearing the same name as the book title that I have been teaching for several years for the MSc course at Control Systems Centre, School of Electrical and Electronic Engineering, The University of Manchester. The beginning chapters cover fundamental concepts in nonlinear control at moderate mathematical level suitable for students with a first degree in engineering disciplines. Simple examples are used to illustrate important concepts, such as the difference between exponential stability and asymptotic stability. Some advanced and recent stability concepts such as input-to-state stability are also included, mainly as an introduction at a less-demanding mathematical level compared with their normal descriptions in the existing books, to research students who may encounter those concepts in literature. Most of the theorems in the beginning chapters are introduced with the proofs, and some of the theorems are simplified with less general scopes, but without loss of rigour. The later chapters cover several topics which are closely related to my own research activities, such as nonlinear observer design and asymptotic disturbance rejection of nonlinear systems. They are included to demonstrate the applications of fundamental concepts in nonlinear and adaptive control to MSc and research students, and to bridge the gap between a normal textbook treatment of control concepts and that of research articles published in academic journals. They can also be used as references for the students who are working on the related topics. At the end of the book, applications to less traditional areas such as control of circadian rhythms are also shown, to encourage readers to explore new applied areas of nonlinear and adaptive control.

This book aims at a unified treatment of adaptive and nonlinear control. It is well known that the dynamics of an adaptive control system for a linear dynamic system with unknown parameters are nonlinear. The analysis of such adaptive systems requires similar techniques to the analysis for nonlinear systems. Some more recent control design techniques such as backstepping relies on Lyapunov functions to establish the stability, and they can be directly extended to adaptive control of nonlinear systems. These techniques further reduce the traditional gap between adaptive control and nonlinear control. Therefore, it is now natural to treat adaptive control as a part of nonlinear control systems. The foundation for linear adaptive control and nonlinear adaptive control is the positive real lemma, which is related to passive systems in nonlinear control and Lyapunov analysis. It is decided to use the positive real lemma and related results in adaptive control and nonlinear control as the main theme of the book, together with Lyapunov analysis. Other important results such as circle criterion and backstepping are introduced as extensions and further developments from this main theme.

For a course unit of 15 credits on nonlinear and adaptive control at the Control Systems Centre, I normally cover Chapters 1–4, 6 and 7, and most of the contents of Chapter 5, and about half of the materials in Chapter 9. Most of the topics covered in Chapters 8, 10 and 11 have been used as MSc dissertation projects and some of them as PhD projects. The contents may also be used for an introductory course on nonlinear control systems, by including Chapters 1–5, 8 and the first half of Chapter 9, and possibly Chapter 6. For a course on adaptive control of nonlinear systems, an instructor may include Chapters 1, 2, 4, 5, 7 and 9. Chapter 8 may be used alone as a brief introduction course to nonlinear observer design. Some results shown in Chapters 8, 10 and 11 are recently published, and can be used as references for the latest developments in related areas.

Nonlinear and adaptive control is still a very active research area in automation and control, with many new theoretic results and applications continuing to merge. I hope that the publication of this work will have a good impact, however small, on students' interests to the subject. I have been benefited from my students, both undergraduate and MSc students, through my teaching and other interactions with them, in particular, their questions to ask me to explain many of the topics covered in this book with simple languages and examples. My research collaborators and PhD students have contributed to several topics covered in the book through joint journal publications, whose names may be found in the references cited at the end of the book. I would like to thank all the researchers in the area who contributed to the topics covered in the book, who are the very people that make this subject fascinating.

Chapter 1

Introduction to nonlinear and adaptive systems

Nonlinearity is ubiquitous, and almost all the systems are nonlinear systems. Many of them can be approximated by linear dynamic systems, and significant amount of analysis and control design tools can then be applied. However, there are intrinsic nonlinear behaviours which cannot be described using linear systems, and analysis and control are necessarily based on nonlinear systems. Even for a linear system, if there are uncertainties, nonlinear control strategies such as adaptive control may have to be used. In the last two decades, there have been significant developments in nonlinear system analysis and control design. Some of them are covered in this book. In this chapter, we will discuss typical nonlinearities and nonlinear behaviours, and introduce some basic concepts for nonlinear system analysis and control.

1.1 Nonlinear functions and nonlinearities

A dynamic system has its origin from dynamics in classic mechanics. The behaviour of a dynamic system is often specified by differential equations. Variables in a dynamic system are referred to as states, and they can be used to determine the status of a system. Without external influences, the state variables are sufficient to determine the future status for a dynamic system. For a dynamic system described by continuous differential equations, state variables cannot be changed instantly, and this reflects the physical reality. Many physical and engineering systems can be modelled as dynamic systems, using ordinary differential equations. Application areas of dynamic systems have expanded rapidly to other areas such as biological systems, financial systems, etc. Analysis of the behaviours of dynamic systems is essential to the understanding of various applications in many science and engineering disciplines. The behaviour of a dynamic system may be altered by exerting external influences, and quite often this kind of influences is based on knowledge of the current state. In this sense, the dynamic system is controlled to achieve certain behaviours.

State variables are denoted by a vector in an appropriate dimension for convenience, and the dynamic systems are described by first-order differential equations of the state vector. A linear dynamic system is described by

$$
\begin{aligned}
\dot{x} &= Ax + Bu \\
y &= Cx + Du,
\end{aligned}
\tag{1.1}
$$

where $x \in \mathbb{R}^n$ is the state, $u \in R^m$ is the external influence, which is referred to as the input, and $y \in \mathbb{R}^s$ is a vector that contains the variables for measurement, which is referred to as the output, and $A \in \mathbb{R}^{n \times n}$, $B \in \mathbb{R}^{n \times m}$, $C \in \mathbb{R}^{s \times n}$ and $D \in \mathbb{R}^{s \times m}$ are matrices that may depend on time. If the system matrices A, B, C and D are constant matrices, the system (1.1) is a linear time-invariant system. For a given input, the state and the output can be computed using the system equation (1.1). More importantly, the superposition principle holds for linear dynamic systems.

Nonlinear dynamic systems are the dynamic systems that contain at least one nonlinear component, or in other words, the functions in the differential equations contain nonlinear functions. For example we consider a Single-Input-Single-Output (SISO) system with input saturation

$$\dot{x} = Ax + B\sigma(u)$$
$$y = Cx + Du, \tag{1.2}$$

where $\sigma : \mathbb{R} \to \mathbb{R}$ is a saturation function defined as

$$\sigma(u) = \begin{cases} -1 & \text{for } u < -1, \\ u & \text{otherwise}, \\ 1 & \text{for } u > 1. \end{cases} \tag{1.3}$$

The only difference between the systems (1.2) and (1.1) is the saturation function σ. It is clear that the saturation function σ is a nonlinear function, and therefore this system is a nonlinear system. Indeed, it can be seen that the superposition principle does not apply, because after the input saturation, any increase in the input amplitude does not change the system response at all.

A general nonlinear system is often described by

$$\dot{x} = f(x, u, t)$$
$$y = h(x, u, t), \tag{1.4}$$

where $x \in \mathbb{R}^n$, $u \in \mathbb{R}^m$ and $y \in \mathbb{R}^s$ are the system state, input and output respectively, and $f : \mathbb{R}^n \times \mathbb{R}^m \times \mathbb{R} \to \mathbb{R}^n$ and $h : \mathbb{R}^n \times \mathbb{R}^m \times \mathbb{R} \to \mathbb{R}^s$ are nonlinear functions.

Nonlinearities of a dynamic system are described by nonlinear functions. We may roughly classify nonlinear functions in nonlinear dynamic systems into two types. The first type of nonlinear functions are analytical functions such as polynomials, sinusoidal functions and exponential functions, or composition of these functions. The derivatives of these functions exist, and their Taylor series can be used to obtain good approximations at any points. These nonlinearities may arise from physical modelling of actual systems, such as nonlinear springs and nonlinear resistors, or due to nonlinear control design, such as nonlinear damping and parameter adaptation law for adaptive control. There are nonlinear control methods such as backstepping which requires the existence of derivatives up to certain orders.

Other nonlinearities may be described by piecewise linear functions, but with a finite number of points where the derivatives do not exist or the functions are not even continuous. The saturation function mentioned earlier is a piecewise linear function that is continuous, but not smooth at the two joint points. A switch, or an idea relay, can be modelled using a signum function which is not continuous. There may also be nonlinearities which are multi-valued, such as relay with hysteresis, which is described in Chapter 3 in detail. A nonlinear element with multi-valued nonlinearity returns a single value at any instance, depending on the history of the input. In this sense, the multi-valued nonlinearities have memory. Other single-valued nonlinearities are memoryless. The nonlinearities described by piecewise linear functions are also called hard nonlinearities, and the common hard nonlinearities include saturation, relay, dead zone, relay with hysteresis, backlash, etc. One useful way to study hard nonlinearities is by describing functions. Hard nonlinearities are also used in control design, for example signum function is used in siding mode control, and saturation functions are often introduced for control inputs to curb the peaking phenomenon for the semi-global stability. In the following example, we show a nonlinearity that arises from adaptive control of a linear system.

Example 1.1. Consider a first-order linear system

$$\dot{x} = ax + u,$$

where a is an unknown parameter. How to design a control system to ensure the stability of the system? If a range $a^- < a < a^+$ is known, we can design a control law as

$$u = -cx - a^+x$$

with $c > 0$, which results in the closed-loop system

$$\dot{x} = -cx + (a - a^+)x.$$

Adaptive control can be used in the case of completely unknown a,

$$u = -cx - \hat{a}x$$
$$\dot{\hat{a}} = x^2.$$

If we let $\tilde{a} = a - \hat{a}$, the closed-loop system is described by

$$\dot{x} = -cx + \tilde{a}x$$
$$\dot{\tilde{a}} = -x^2.$$

This adaptive system is nonlinear, even though the original uncertain system is linear. This adaptive system is stable, and it does need the stability theory introduced later in Chapter 7 of the book.

1.2   Common nonlinear systems behaviours

Many nonlinear systems may be approximated by linearised systems around operating points, and their behaviours in close neighbourhoods can be predicted from the linear dynamics. This is the justification of applying linear system theory to control study, as almost all the practical systems are nonlinear systems.

There are many nonlinear features that do not exist in linear dynamic systems, and therefore linearised dynamic models cannot describe the behaviours associated with these nonlinear features. We will discuss some of those nonlinear features in the following text.

Multiple equilibrium points are common for nonlinear systems, unlike the linear system. For example

$$\dot{x} = -x + x^2.$$

This system has two equilibria at $x = 0$ and $x = 1$. The behaviours around these equilibrium points are very different, and they cannot be described by a single linearised model.

Limit cycles are a phenomenon that periodic solutions exist and attract nearby trajectories in positive or negative time. Closed curve solutions may exist for linear systems, such as solution to harmonic oscillators. But they are not attractive to nearby trajectories and not robust to any disturbances. Heart beats of human body can be modelled as limit cycles of nonlinear systems.

High-order harmonics and subharmonics occur in the system output when subject to a harmonic input. For linear systems, if the input is a harmonic function, the output is a harmonic function, with the same frequency, but different amplitude and phase. For nonlinear systems, the output may even have harmonic functions with fractional frequency of the input or multiples of the input frequency. This phenomenon is common in power distribution networks.

Finite time escape can happen in a nonlinear system, i.e., the system state tends to infinity at a finite time. This will never happen for linear systems. Even for an unstable linear system, the system state can only grow at an exponential rate. The finite time escape can cause a problem in nonlinear system design, as a trajectory may not exist.

Finite time convergence to an equilibrium point can happen to nonlinear systems. Indeed, we can design nonlinear systems in this way to achieve fast convergence. This, again, cannot happen for linear systems, as the convergence rate can only be exponential, i.e., a linear system can only converge to its equilibrium asymptotically.

Chaos can only happen in nonlinear dynamic systems. For some class of nonlinear systems, the trajectories are bounded, but not converge to any equilibrium or limit cycles. They may have quasi-periodic solutions, and the behaviour is very difficult to predict.

There are other nonlinear behaviours such as bifurcation, etc., which cannot happen in linear systems. Some of the nonlinear behaviours are covered in detail in this book, such as limit cycles and high-order harmonics. Limit cycles and chaos are discussed in Chapter 2, and limit cycles also appear in other problems considered in this book. High-order harmonics are discussed in disturbance rejection. When the disturbance is a harmonic signal, the internal model for disturbance rejection has to consider the high-order harmonics generated due to nonlinearities.

1.3 Stability and control of nonlinear systems

Nonlinear system behaviours are much more complex than those of linear systems. The analytical tools for linear systems often cannot be applied to nonlinear systems. For linear systems, the stability of a system can be decided by eigenvalues of system matrix A, and obviously for nonlinear systems, this is not the case. For some nonlinear systems, a linearised model cannot be used to determine the stability even in a very small neighbourhood of the operating point. For example for a first-order system

$$\dot{x} = x^3,$$

linearised model around the origin is $\dot{x} = 0$. This linearised model is critically stable, but the system is unstable. In fact, if we consider another system

$$\dot{x} = -x^3,$$

which is stable, but the linearised model around the origin is still $\dot{x} = 0$. Frequency domain methods also cannot be directly applied to analysing input–output relationship for nonlinear systems, as we cannot define a transfer function for a general nonlinear system.

Of course, some basic concept may still be applicable to nonlinear systems, such as controllability, but these systems are often in different formulation and use different mathematical tools. Frequency response method can be applied to analysing a nonlinear system with one nonlinear component based on approximation in frequency domain using describing function method. High gain control and zero dynamics for nonlinear systems originate from their counterparts in linear systems.

One important concept that we need to address is stability. As eigenvalues are no longer a suitable method for nonlinear systems, stability concepts and methods to check stability for nonlinear systems are necessary. Among various definition, Lyapunov stability is perhaps the most fundamental one. It can be checked by using a Lyapunov function for stability analysis. Some of the other stability concepts such as input-to-state stability may also be interpreted using Lyapunov functions. Lyapunov functions can also provide valid information in control design. Lyapunov stability will be the main stability concept used in this book.

Compared with the stability concepts of nonlinear systems, control design methods for nonlinear systems are even more diversified. Unlike control design for linear systems, there is a lack of systematic design methods for nonlinear systems. Most of the design methods can only apply to specific classes of nonlinear systems. Because of this, nonlinear control is often more challenging and interesting. It is impossible for the author to cover all the major areas of nonlinear control design and analysis, partially to author's knowledge base, and partially due to the space constraint.

People often start with linearisation of a nonlinear system. If the control design based on a linearised model works, then there is no need to worry about nonlinear control design. Linearised models depend on operating points, and a switching strategy might be needed to move from one operating point to the others. Gain scheduling and linear parameter variation (LPV) methods are also closely related to linearisation around operating points.

Linearisation can also be achieved for certain class of nonlinear systems through a nonlinear state transformation and feedback. This linearisation is very much different from linearisation around operating points. As shown later in Chapter 6, a number of geometric conditions must be satisfied for the existence of such a nonlinear transformation. The linearisation obtained in this way works globally in the state space, not just at one operating point. Once the linearised model is obtained, further control design can be carried out using design methods for linear systems.

Nonlinear functions can be approximated using artificial neuron networks, and fuzzy systems and control methods have been developed using these approximation methods. The stability analysis of such systems is often similar to Lyapunov function-based design method and adaptive control. We will not cover them in this book. Other nonlinear control design methods such as band–band control and sliding mode control are also not covered.

In the last two decades, there were developments for some more systematic control design methods, such as backstepping and forwarding. They require the system to have certain structures so that these iterative control designs can be carried out. Among them, backstepping method is perhaps the most popular one. As shown in Chapter 9 in the book, it requires the system state space function in a sort of lower-triangular form so that at each step a virtual control input can be designed. Significant amount of coverage of this topic can be found in this book. Forwarding control design can be interpreted as a counterpart of backstepping in principle, but it is not covered in this book.

When there are parametric uncertainties, adaptive control can be introduced to tackle the uncertainty. As shown in a simple example earlier, an adaptive control system is nonlinear, even for a linear system. Adaptive technique can also be introduced together with other nonlinear control design methods, such as backstepping method. In such a case, people often give it a name, adaptive backstepping. Adaptive control for linear systems and adaptive backstepping for nonlinear systems are covered in details in Chapter 7 and Chapter 9 in this book.

Similar to linear control system design, nonlinear control design methods can also be grouped as state-feedback control design and output-feedback control design. The difference is that the separation principle is not valid for nonlinear control design in general, that is if we replace the state in the control input by its estimate, we would not be able to guarantee the stability of the closed-loop system using state estimate. Often state estimation must be integrated in the control design, such as observer backstepping method.

State estimation is an important topic for nonlinear systems on its own. Over the last three decades, various observer design methods have been introduced. Some of them may have their counterparts in control design. Design methods are developed for different nonlinearities. One of them is for systems with Lipschitz nonlinearity, as shown in Chapter 8. A very neat nonlinear observer design is the observer design with output injection, which can be applied to a class of nonlinear systems whose nonlinearities are only of the system output.

In recent years, the concept of semi-global stability is getting more popular. Semi-global stability is not as good as global stability, but the domain of attraction can be as big as you can specify. The relaxation in the global domain of attraction does give control design more freedom in choosing control laws. One common strategy is to use high gain control together with saturation. We will not cover it in this book, but the design methods in semi-global stability can be easily followed once a reader is familiar with the control design and analysis methods introduced in this book.

Chapter 2
State space models

The nonlinear systems under consideration in this book are described by differential equations. In the same way as for linear systems, we have system state variables, inputs and outputs. In this chapter, we will provide basic definitions for state space models of nonlinear systems, and tools for preliminary analysis, including linearisation around operating points. Typical nonlinear behaviours such as limit cycles and chaos will also be discussed with examples.

2.1 Nonlinear systems and linearisation around an operating point

A system is called a dynamic system if its behaviours depend on its history. States of a system are the variables that represent the information of history. At any time instance, the current state value decides the system's future behaviours. In this sense, the directives of the state variables decide the system's behaviours, and hence we describe a dynamic system by a set of first-order differential equations in vector form as

$$\dot{x} = f(x, u, t), \quad x(0) = x_0, \tag{2.1}$$

where $x \in \mathbb{R}^n$ is the state of the system, $f : \mathbb{R}^n \to \mathbb{R}^n$ is a continuous function and $u \in \mathbb{R}^m$ denotes the external influence, which is usually referred to as the input to the system.

For the differential equation (2.1) to exist as a unique solution for a given initial condition, we need to impose a restriction on the nonlinear function f that f must be Lipschitz with respect to the variable x. The definition of Lipschitz condition is given below.

Definition 2.1. A function $f : \mathbb{R}^n \times \mathbb{R}^m \times \mathbb{R} \to \mathbb{R}^n$ is Lipschitz with a Lipschitz constant γ if for any vectors $x, \hat{x} \in D_x \subset \mathbb{R}^n$, and $u \in D_u \subset \mathbb{R}^m$ and $t \in I_t \subset \mathbb{R}$, with D_x, D_u being the regions of interest and I_t being an time interval,

$$\|f(x, u, t) - f(\hat{x}, u, t)\| \leq \gamma \|x - \hat{x}\|, \tag{2.2}$$

with $\gamma > 0$.

Note that Lipschitz condition implies continuity with respect to x. The existence and uniqueness of a solution for (2.1) are guaranteed by the function f being Lipschitz and being continuous with respect to t.

Remark 2.1. The continuity of f with respect to t and state variable x might be stronger than we have in real applications. For example, a step function is not continuous in time. In the case that there are finite number of discontinuities in a given interval, we can solve the equation of a solution in each of the continuous region, and join them together. There are situations of discontinuity with state variable, such as an ideal relay. In such a case, the uniqueness of the solution can be an issue. Further discussion on this is beyond the scope of this book. In the systems considered in the book, we would assume that there would be no problem with the uniqueness of a solution.   ◁

The system state contains the whole information of the behaviour. However, for a particular application, only a subset of the state variables or a function of state variables is of interest, which can be denoted as $y = h(x, u, t)$ with $h : \mathbb{R}^n \times \mathbb{R}^m \times \mathbb{R} \to \mathbb{R}^s$, normally with $s < n$. We often refer to y as the output of the system. To write them together with the system dynamics, we have

$$\dot{x} = f(x, u, t), \quad x(0) = x_0$$
$$y = h(x, u, t).$$

In this book, we mainly deal with time-invariant systems. Hence, we can drop the variable t in f and h and write the system as

$$\dot{x} = f(x, u), \quad x(0) = x_0$$
$$y = h(x, u), \tag{2.3}$$

where $x \in \mathbb{R}^n$ is the state of the system, $y \in \mathbb{R}^s$ and $u \in \mathbb{R}^m$ are the output and the input of the system respectively, and $f : \mathbb{R}^n \times \mathbb{R}^m \to \mathbb{R}^n$ and $h : \mathbb{R}^n \times \mathbb{R}^m \to \mathbb{R}^s$ are continuous functions.

Nonlinear system dynamics are much more complex than linear systems in general. However, when the state variables are subject to small variations, we would expect the behaviours for small variations to be similar to linear systems, based on the fact that

$$f(x + \delta x, u + \delta u) \approx f(x, u) + \frac{\partial f}{\partial x}(x, u)\delta x + \frac{\partial f}{\partial u}(x, u)\delta u,$$

when δx and δu are very small.

An operating point at (x_e, u_e) is taken with $x = x_e$ and $u = u_e$ being constants such that $f(x_e, u_e) = 0$. A linearised model around the operation point can then be obtained. Let

$$\bar{x} = x - x_e,$$
$$\bar{u} = u - u_e,$$
$$\bar{y} = h(x, u) - h(x_e, u_e),$$

then the linearised model is given by

$$\dot{\bar{x}} = A\bar{x} + B\bar{u}$$
$$\bar{y} = C\bar{x} + D\bar{u}, \tag{2.4}$$

where $A \in \mathbb{R}^{n \times n}, B \in \mathbb{R}^{n \times m}, C \in \mathbb{R}^{s \times n}, D \in \mathbb{R}^{s \times m}$ matrices with elements $a_{i,j}, b_{i,j}, c_{i,j}$ and $d_{i,j}$ respectively shown by, assuming that f and h are differentiable,

$$a_{i,j} = \frac{\partial f_i}{\partial x_j}(x_e, u_e),$$

$$b_{i,j} = \frac{\partial f_i}{\partial u_j}(x_e, u_e),$$

$$c_{i,j} = \frac{\partial h_i}{\partial x_j}(x_e, u_e),$$

$$d_{i,j} = \frac{\partial h_i}{\partial u_j}(x_e, u_e).$$

Remark 2.2. For a practical system, a control input can keep the state in an equilibrium point, i.e., at a point such that $\dot{x} = 0$, and therefore it is natural to look at the linearisation around this point. However, we can obtain linearised model at points that are not at equilibrium. If (x_e, u_e) is not an equilibrium point, we have $f(x_e, u_e) \neq 0$. We can carry out the linearisation in the same way, but the resultant linearised system is given by

$$\dot{\bar{x}} = A\bar{x} + B\bar{u} + d$$
$$\bar{y} = C\bar{x} + D\bar{u},$$

where $d = f(x_e, u_e)$ is a constant vector. ◁

2.2 Autonomous systems

For a system in (2.3), the external influence can only be exerted through the input u. If a system does not take any input, its future state only depends on the initial state. It means that the system is not influenced by external factors, and the behaviours are completely determined by the system state. Such a system is often referred to as an autonomous system. A definition is given below.

Definition 2.2. An autonomous system is a dynamic system whose behaviour does not explicitly depend on time.

In terms of differential equations, an autonomous system can be expressed as

$$\dot{x} = f(x), \tag{2.5}$$

where $x \in \mathbb{R}^n$ is the state of the system, and $f : \mathbb{R}^n \to \mathbb{R}^n$.

Remark 2.3. It is easy to see that for a system in (2.3), if the control input remains constant, then it is an autonomous system. We only need to re-define the function f as $f_a(x) := f(x, u_c)$ where u_c is a constant input. Even if the inputs are polynomials and sinusoidal functions of time, we can convert the system to the autonomous system by modelling the sinusoidal and polynomial functions as the state variables of a linear dynamic system, and integrate this system into the original system. The augmented system is then an autonomous system. ◁

Definition 2.3. For an autonomous system (2.5), a point $x_e \in \mathbb{R}^n$ is a singular point if $f(x_e) = 0$.

It is easy to see that singular points are equilibrium points. Singular points are more preferred for autonomous systems, especially for second-order systems.

Since autonomous system do not have external input, the set of all the trajectories provides a complete geometrical representation of the dynamic behaviour. This is often referred to as the *phase portrait*, especially for second-order systems in the format

$$\dot{x}_1 = x_2$$
$$\dot{x}_2 = \phi(x_1, x_2).$$

In the above system, if we interpret x_1 as the displacement, then x_2 is the velocity. The state variables often have clear physical meanings. Phase portraits can be obtained by a number of methods, including analysing the behaviours near the singular points. In fact, singular points might get the name from their positions in the phase portrait. In a phase portrait, the lines usually do not intercept each other due to the uniqueness of the solutions. However, they meet at the points where $f(x) = 0$, seemingly intercepting each other. Those points are singular in this sense.

2.3   Second-order nonlinear system behaviours

For a second-order system, we can write the system equation as

$$\dot{x}_1 = f_1(x_1, x_2)$$
$$\dot{x}_2 = f_2(x_1, x_2).$$

For an equilibrium point (x_{1e}, x_{2e}), the linearised model is given by

$$\begin{bmatrix} \dot{\bar{x}}_1 \\ \dot{\bar{x}}_2 \end{bmatrix} = A_e \begin{bmatrix} \bar{x}_1 \\ \bar{x}_2 \end{bmatrix}, \tag{2.6}$$

where

$$A_e = \begin{bmatrix} \dfrac{\partial f_1}{\partial x_1}(x_{1e}, x_{2e}) & \dfrac{\partial f_1}{\partial x_2}(x_{1e}, x_{2e}) \\ \dfrac{\partial f_2}{\partial x_1}(x_{1e}, x_{2e}) & \dfrac{\partial f_2}{\partial x_2}(x_{1e}, x_{2e}) \end{bmatrix}.$$

Therefore, the behaviour about this equilibrium or singular point is determined by the properties of matrix A_e. Based on the eigenvalues of A_e, we can classify the singular points in the following six different cases. We use λ_1 and λ_2 to denote the two eigenvalues of A_e.

Stable node, for $\lambda_1 < 0$, $\lambda_2 < 0$. This is a case when both eigenvalues are negative real numbers. The linearised model is stable, and a typical phase portrait around this singular point is shown below.

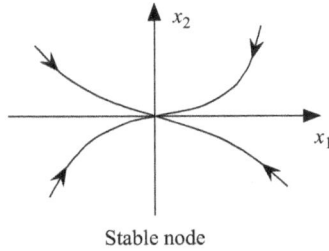

Stable node

Unstable node, for $\lambda_1 > 0$, $\lambda_2 > 0$. This singular point is unstable, and the trajectories diverge from the point, but not spiral around it.

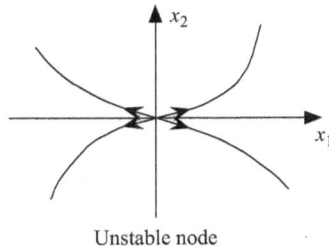

Unstable node

Saddle point, for $\lambda_1 < 0$, $\lambda_2 > 0$. With one positive and one negative eigenvalues, the hyperplane in three dimensions may look like a saddle. Some trajectories converge to the singular point, and others diverge, depending on the directions of approaching the point.

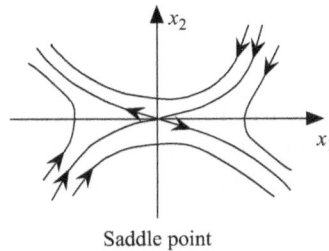

Saddle point

Stable focus, for $\lambda_{1,2} = \mu \pm j\nu$, $(\mu < 0)$. With a negative real part for a pair of conjugate poles, the singular point is stable. Trajectories converge to the singular point, spiralling around. In time domain, the solutions are similar to decayed sinusoidal functions.

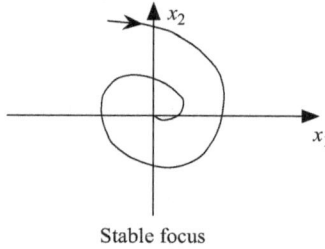

Stable focus

Unstable focus, for $\lambda_{1,2} = \mu \pm jv$, $(\mu > 0)$. The real part is positive, and therefore the singular point is unstable, with the trajectories spiralling out from the singular point.

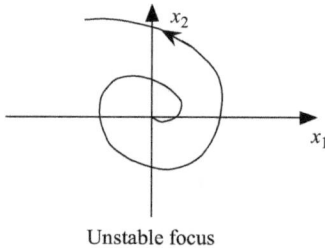

Unstable focus

Centre, for $\lambda_{1,2} = \pm jv$, $(v > 0)$. For the linearised model, when the real part is zero, the norm of the state is constant. In phase portrait, there are closed orbits around the singular point.

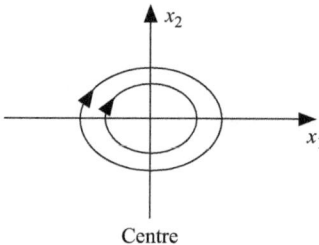

Centre

To draw a phase portrait, the analysis of singular points is the first step. Based on the classification of the singular points, the behaviours in neighbourhoods of these points are more or less determined. For other regions, we can calculate the directions of the movement from the directives.

At any point, the *slope of trajectory* can be computed by

$$\frac{dx_2}{dx_1} = \frac{f_2(x_1, x_2)}{f_1(x_1, x_2)}.$$

With enough points in the plane, we should be able to sketch phase portraits connecting the points in the directions determined by the slopes.

Indeed, we can even obtain curves with constant slopes, which are named as *isoclines*. An isocline is a curve on which $(f_2(x_1, x_2)/f_1(x_1, x_2))$ is constant. This, again, can be useful in sketching a phase portrait for a second-order nonlinear system.

It should be noted that modern computer simulation can provide accurate solutions to many nonlinear differential equations. For this reason, we will not go to further details of drawing phase portraits based on calculating slope of trajectories and isoclines.

Example 2.1. Consider a second-order nonlinear system

$$\dot{x}_1 = x_2$$
$$\dot{x}_2 = -x_2 - 2x_1 + x_1^2.$$

Setting

$$0 = x_2,$$
$$0 = -x_2 - 2x_1 + x_1^2,$$

we obtain two singular points $(0, 0)$ and $(-2, 0)$.

Linearised system matrix for the singular point $(0, 0)$ is obtained as

$$A = \begin{bmatrix} 0 & 1 \\ -1 & -2 \end{bmatrix},$$

and the eigenvalues are obtained as $\lambda_1 = \lambda_2 = -1$. Hence, this singular point is a stable node.

For the singular point $(-2, 0)$, the linearised system matrix is obtained as

$$A = \begin{bmatrix} 0 & 1 \\ 2 & -1 \end{bmatrix},$$

and the eigenvalues are $\lambda_1 = -2$ and $\lambda_2 = 1$. Hence, this singular point is a saddle point. It is useful to obtain the corresponding eigenvectors to determine which direction is converging and which is diverging. The eigenvalues v_1 and v_2, for $\lambda_1 = -2$ and $\lambda_2 = 1$, are obtained as

$$v_1 = \begin{bmatrix} 1 \\ -2 \end{bmatrix}, \quad v_2 = \begin{bmatrix} 1 \\ 1 \end{bmatrix}$$

This suggests that along the direction of v_1, relative to the singular point, the state converges to the singular point, while along v_2, the state diverges. We can clearly see from Figure 2.1 that there is a stable region near the singular point $(0, 0)$. However, in the neighbourhood of $(-2, 1)$ one part of it is stable, and the other part is unstable.

◁

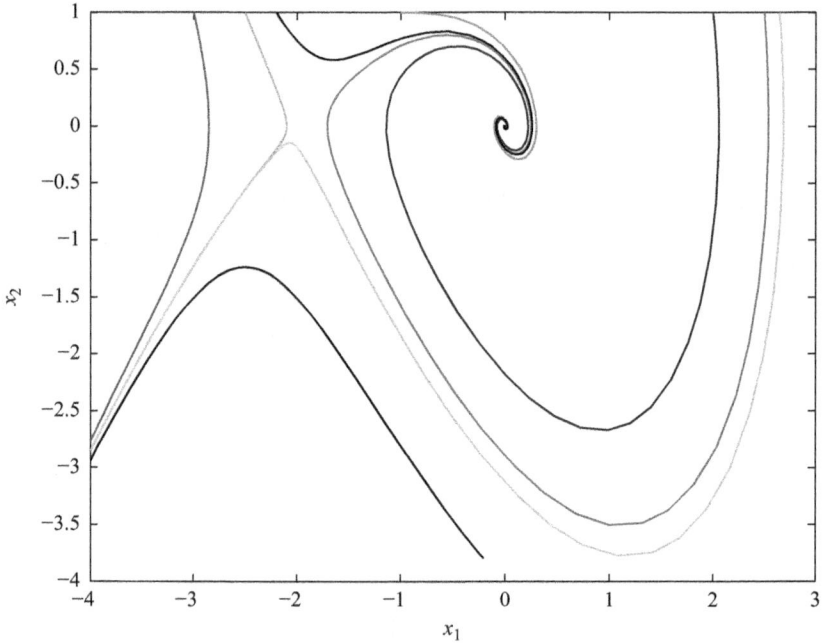

Figure 2.1   Phase portrait of Example 2.1

Example 2.2. Consider the swing equation of a synchronous machine

$$H\ddot{\delta} = P_m - P_e \sin(\delta), \tag{2.7}$$

where H is the inertia, δ is the rotor angle, P_m is the mechanical power and P_e is the maximum electrical power generated. We may view P_m as the input and $P_e \sin\delta$ as the output. For the convenience of presentation, we take $H = 1$, $P_m = 1$ and $P_e = 2$.

The state space model is obtained by letting $x_1 = \delta$ and $x_2 = \dot{\delta}$ as

$$\dot{x}_1 = x_2$$
$$\dot{x}_2 = 1 - 2\sin(x_1).$$

For the singular points, we obtain

$$x_{1e} = \frac{1}{6}\pi \text{ or } \frac{5}{6}\pi,$$
$$x_{2e} = 0.$$

Note that there are an infinite number of singular points, as $x_{1e} = 2k\pi + \frac{1}{6}\pi$ and $x_{1e} = 2k\pi + \frac{5}{6}\pi$ are also solutions for any integer value of k.

Let us concentrate on the analysis of the two singular points $(\frac{1}{6}\pi, 0)$ and $(\frac{5}{6}\pi, 0)$. The linearised system matrix is obtained as

$$A = \begin{bmatrix} 0 & 1 \\ -2\cos(x_{1e}) & 0 \end{bmatrix}.$$

For $(\frac{1}{6}\pi, 0)$, the eigenvalues are $\lambda_{1,2} = \pm 3^{1/4}j$, and therefore this singular point it a centre.

For $(\frac{5}{6}\pi, 0)$, the eigenvalues are $\lambda_1 = -3^{1/4}$ and $\lambda_2 = 3^{1/4}$. Hence, this singular point it a saddle point. The eigenvalues v_1 and v_2, for $\lambda_1 = -3^{1/4}$ and $\lambda_2 = 3^{1/4}$, are obtained as

$$v_1 = \begin{bmatrix} 1 \\ -3^{1/4} \end{bmatrix}, \quad v_2 = \begin{bmatrix} 1 \\ 3^{1/4} \end{bmatrix}.$$

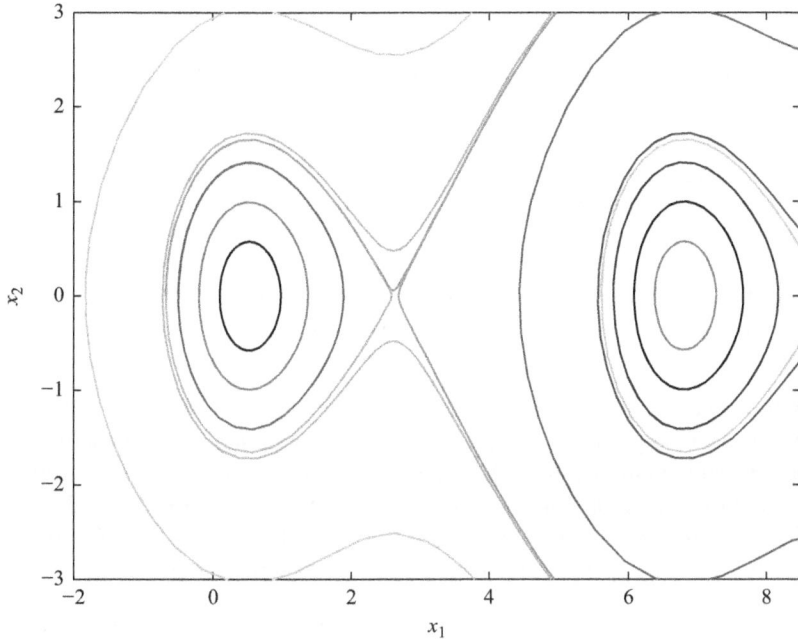

Figure 2.2   Phase portrait of Example 2.2

Figure 2.2 shows a phase portrait obtained from computer simulation. The centre at $(\frac{1}{6}\pi, 0)$ and the saddle point at $(\frac{5}{6}\pi, 0)$ are clearly shown in the figure. The directions of the flow can be determined from the eigenvectors of the saddle point. For example the trajectories start from the points around $(5, -3)$ and move upwards and to the left, along the direction pointed by the eigenvector v_1 towards the saddle point. Along the direction pointed by v_2, the trajectories depart from the saddle point. ◁

2.4   Limit cycles and strange attractors

Some trajectories appear as closed curves in phase portrait. For autonomous systems, they represent periodic solutions, as the solutions only depend on the states, and they are referred to as cycles. For linear systems, periodic solutions appear in harmonic oscillators. For example a harmonic oscillator described by

$$\dot{x}_1 = x_2$$
$$\dot{x}_2 = -x_1$$

has solutions as sinusoidal functions, and the amplitude is determined by the initial values. It is easy to see that the function $V = x_1^2 + x_2^2$ remains a constant, as we have

$$\dot{V} = 2x_1(-x_2) + 2x_2x_1 = 0.$$

Therefore, the solutions of this oscillator are circles around the origin. For a given initial state, the radius does not change. When two initial points are very close, their solutions will be very close, but they will not converge to one cycle, no matter how close the initial values. Also the solution for this harmonic oscillator is not robust, as any small disturbance will destroy the cycle.

There are cycles in the phase portrait shown in Figure 2.2. Even though they are solutions to a nonlinear dynamic system, they are similar to the cycles obtained from the harmonic oscillator in the sense that cycles depend on the initial values and they do not converge or attract to each other, no matter how close the two cycles are.

The cycles discussed above are not limit cycles. For limit cycles we have the following definition.

Definition 2.4. A closed curve solution, or in other word, a cycle, of an autonomous system is a limit cycle, if some non-periodic solutions converge to the cycle as $t \to \infty$ or $t \to -\infty$.

A limit cycle is stable if nearby trajectories converge to it asymptotically, unstable if move away. One property of a limit cycle is that amplitude of the oscillation may not depend on the initial values. A limit cycle may be attractive to the nearby region. One of the most famous ones is van der Pol oscillator. This oscillator does have a physical meaning. It can be viewed as a mathematical model of an RLC circuit, with the resistor being possible to take negative values in certain regions.

Example 2.3. One form of van der Pol oscillator is described in the following differential equation:

$$\ddot{y} - \epsilon(1 - y^2)\dot{y} + y = 0, \tag{2.8}$$

where ϵ is a positive real constant.

If we take $x_1 = y$ and $x_2 = \dot{y}$, we obtain the state space equation

$$\dot{x}_1 = x_2$$
$$\dot{x}_2 = -x_1 + \epsilon(1 - x_1^2)x_2.$$

From this state space realisation, it can be seen that when $\epsilon = 0$, van der Pol oscillator is the same as a harmonic oscillator. For ϵ with small values, one would expect that it behaves like a harmonic oscillator.

A more revealing state transformation for ϵ with big values is given by

$$x_1 = y$$
$$x_2 = \frac{1}{\epsilon}\dot{y} + f(y),$$

where $f(y) = y^3/3 - y$. Under the above transformation, we have the system as

$$\dot{x}_1 = \epsilon(x_2 - f(x_1))$$
$$\dot{x}_2 = -\frac{1}{\epsilon}x_1. \tag{2.9}$$

It can be obtained that

$$\frac{dx_2}{dx_1}(x_2 - f(x_1)) = \frac{x_1}{\epsilon^2} \tag{2.10}$$

This equation suggests that as $\epsilon \to \infty$, we have $\frac{dx_2}{dx_1} = 0$ or $x_2 = f(x_1)$. This can be seen from the phase portrait for very big values of ϵ in Figure 2.4.

Let us stick with the state space model (2.9). The only singular point is at the origin $(0, 0)$. The linearised system matrix at the origin is obtained as

$$A = \begin{bmatrix} \epsilon & 1 \\ -\frac{1}{\epsilon} & 0 \end{bmatrix}.$$

From the eigenvalues of A, we can see that this singular point is either an unstable node or an unstable focus, depending on the value of ϵ. Phase portrait of van der Pol oscillator with $\epsilon = 1$ is shown in Figure 2.3 for two trajectories, one with initial condition outside the limit cycle and one from inside. The broken line shows $x_2 = f(x_1)$. Figure 2.4 shows the phase portrait with $\epsilon = 10$. It is clear from Figure 2.4 that the trajectory sticks with the line $x_2 = f(x_1)$ along the outside and then moves almost horizontally to the other side, as predicted in the analysis earlier. ◁

Limit cycles also exist in high-order nonlinear systems. As seen later in Chapter 11, circadian rhythms can also be modelled as limit cycles of nonlinear dynamic systems. For second-order autonomous systems, limit cycles are very typical trajectories. The following theorem, Poincare–Bendixson theorem, describes the features of trajectories of the second-order systems, from which a condition on the existence of a limit cycle can be drawn.

Theorem 2.1. *If a trajectory of the second-order autonomous system remains in a finite region, then one of the following is true:*

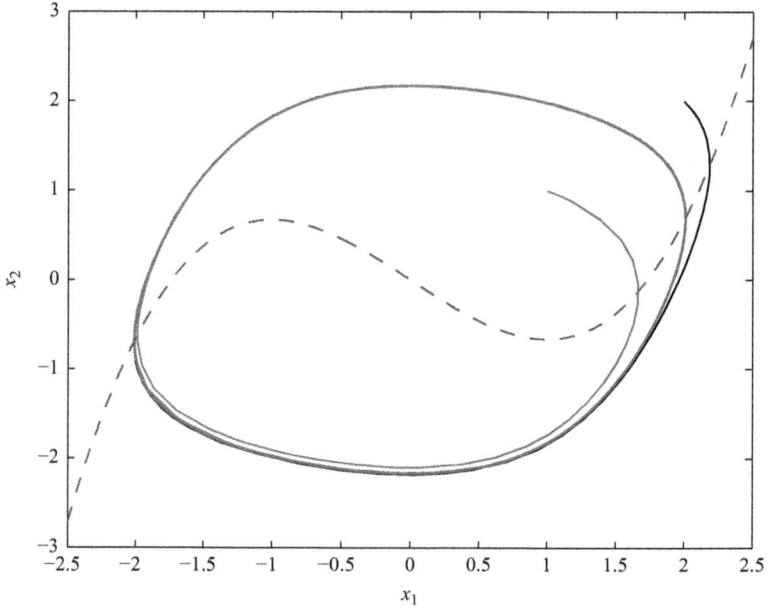

Figure 2.3   Phase portrait of van der Pol oscillator with $\epsilon = 1$

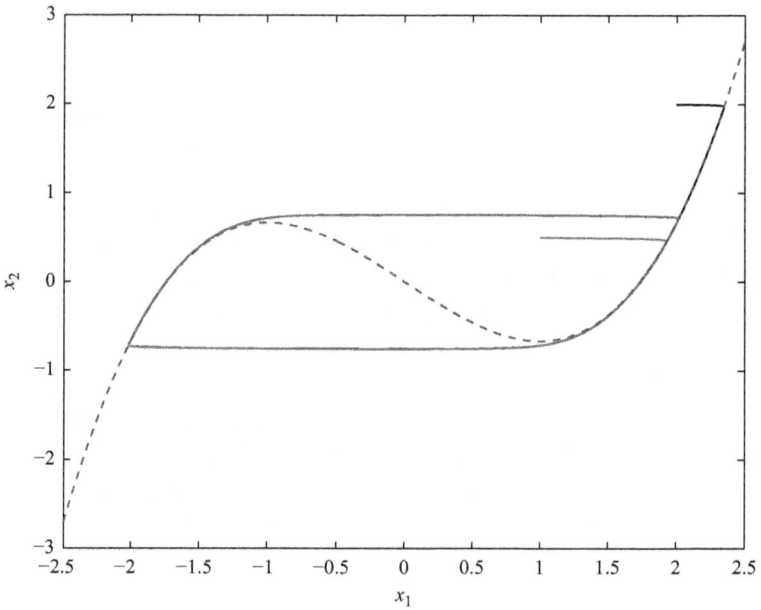

Figure 2.4   Phase portrait of van der Pol oscillator with $\epsilon = 10$

- *The trajectory goes to an equilibrium point.*
- *The trajectory tends to an asymptotically stable limit cycle.*
- *The trajectory is itself a cycle.*

For high-order nonlinear systems, there are more complicated features if the trajectories remain in a bounded region. For the asymptotic behaviours of dynamic systems, we define positive limit sets.

Definition 2.5. Positive limit set of a trajectory is the set of all the points for which the trajectory converges to, as $t \to \infty$.

Positive limit sets are also referred to as ω-limit sets, as ω is the last letter of Greek letters. Similarly, we can define negative limit sets, and they are called α-limit sets accordingly. Stable limit cycles are positive limit sets, so do stable equilibrium points. The dimension for ω-limit sets is zero or one, depending on singular points or limit cycles.

Strange limit sets are those limit sets which may or may not be asymptotically attractive to the neighbouring trajectories. The trajectories they contain may be locally divergent from each other, within the attracting set. Their dimensions might be fractional. Such structures are associated with the quasi-random behaviour of solutions called *chaos*.

Example 2.4. *The Lorenz attractor.* This is one of the most widely studied examples of strange behaviour in ordinary differential equations, which is originated from studies of turbulent convection by Lorenz. The equation is in the form

$$\dot{x}_1 = \sigma(x_2 - x_1)$$
$$\dot{x}_2 = (1 + \lambda - x_3)x_1 - x_2 \qquad\qquad (2.11)$$
$$\dot{x}_3 = x_1 x_2 - bx_3,$$

where σ, λ and b are positive constants. There are three equilibrium points $(0, 0, 0)$, $(\sqrt{b\lambda}, \sqrt{b\lambda}, \lambda)$ and $(-\sqrt{b\lambda}, -\sqrt{b\lambda}, \lambda)$. The linearised system matrix around the origin is obtained as

$$A = \begin{bmatrix} -\sigma & \sigma & 0 \\ \lambda + 1 & -1 & 0 \\ 0 & 0 & -b \end{bmatrix},$$

and its eigenvalues are obtained as $\lambda_{1,2} = -(\sigma - 1) \pm \sqrt{(\sigma - 1)^2 + 4\sigma\lambda}/2$ and $\lambda_3 = -b$. Since the first eigenvalue is positive, this equilibrium is unstable. It can be shown that the other equilibrium points are unstable when the parameters satisfies

$$\sigma > b + 1,$$
$$\lambda > \frac{(\sigma + 1)(\sigma + b + 1)}{\sigma - b - 1}.$$

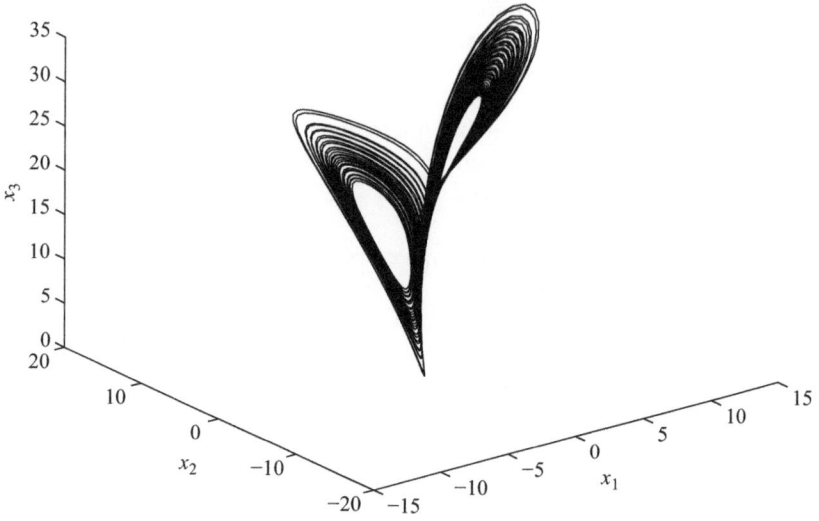

Figure 2.5   A typical trajectory of Lorenz attractor

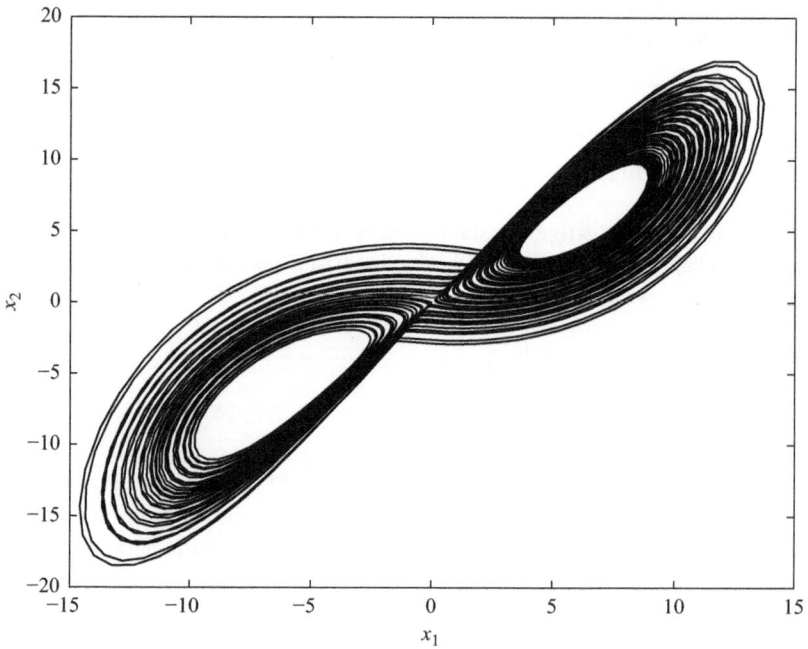

Figure 2.6   Projection to (x_1, x_2)

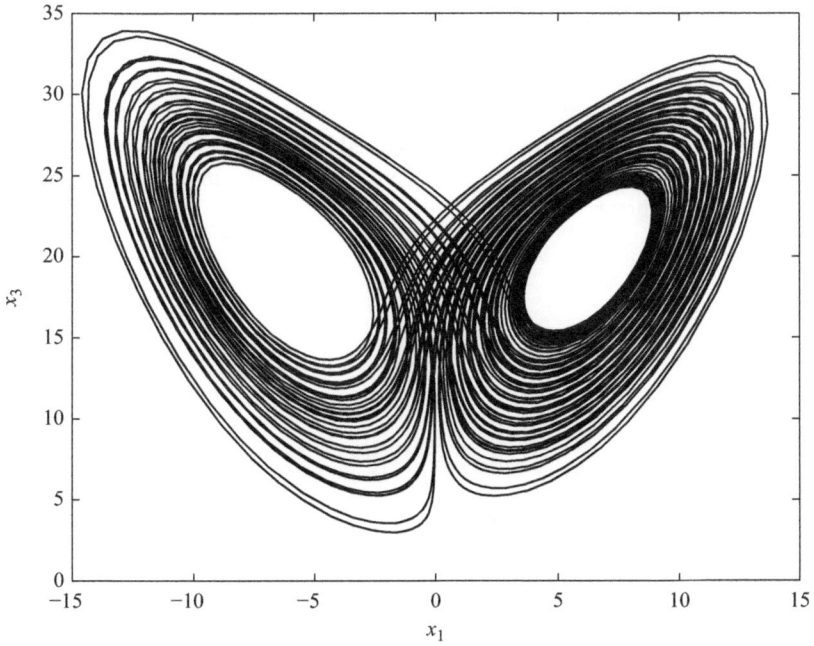

Figure 2.7   Projection to (x_1, x_3)

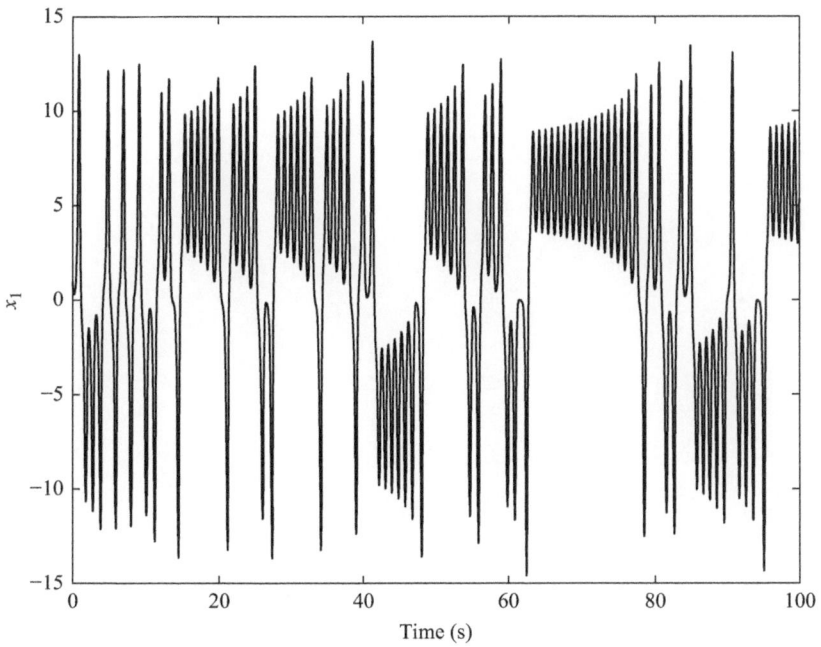

Figure 2.8   Time trajectory of x_1

It can be established that the trajectories converge to a bounded region specified by

$$(\lambda + 1)x_1^2 + \sigma x_2^2 + \sigma(x_3 - 2(\lambda + 1))^2 \leq C$$

for a positive constant C. When all the three equilibria are unstable, the behaviour of Lorenz system is chaotic. A trajectory is plotted in Figure 2.5 for $b = 2$, $\sigma = 10$ and $\lambda = 20$. Projections to (x_1, x_2) and (x_1, x_3) are shown in Figures 2.6 and 2.7. The time trajectory of x_1 is shown in Figure 2.8.   ◁

Chapter 3

Describing functions

In classical control, frequency response is a powerful tool for analysis and control design of linear dynamic systems. It provides graphical presentation of system dynamics and often can reflect certain physical features of engineering systems. The basic concept of frequency response is that for a linear system, if the input is a sinusoidal function, the steady-state response will still be a sinusoidal function, but with a different amplitude and a different phase. The ratio of the input and output amplitudes and the difference in the phase angles are determined by the system dynamics. When there is a nonlinear element in a control loop, frequency response methods cannot be directly applied. When a nonlinear element is a static component, i.e., the input and output relationship can be described by an algebraic function, its output to any periodic function will be a periodic function, with the same period as the input signal. Hence, the output of a static nonlinear element is a periodic function when the input is a sinusoidal function. It is well known that any periodic function with piece-wise continuity has its Fourier series which consists of sinusoidal functions with the same period or frequency as the input with a constant bias, and other sinusoidal functions with high multiple frequencies. If we take the term with the fundamental frequency, i.e., the same frequency as the input, as an approximation, the performance of the entire dynamic system may be analysed using frequency response techniques. Describing functions are the frequency response functions of nonlinear components with their fundamental frequency terms as their approximate outputs. In this sense, describing functions are first-order approximation in frequency domain. It can also be viewed as a linearisation method in frequency domain for nonlinear components.

Describing function analysis remains as an important tool for analysis of nonlinear systems with static components despite several more recent developments in nonlinear control and design. It is relatively easy to use, and closely related to frequency response analysis of linear systems. It is often used to predict the existence of limit cycles in a nonlinear system, and it can also be used for prediction of subharmonics and jump phenomena of nonlinear systems. In this chapter, we will present basic concept of describing functions, calculation of describing functions of common nonlinear elements and how to use describing functions to predict the existence of limit cycles.

3.1   Fundamentals

For a nonlinear component described by a nonlinear function $f : \mathbb{R} \to \mathbb{R}$, its

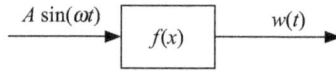

$$A \sin(\omega t) \quad \boxed{f(x)} \quad w(t)$$

output

$$w(t) = f(A \sin(\omega t))$$

to a sinusoidal input $A \sin(\omega t)$ is a periodical function, although it may not be sinusoidal in general. Assuming that the function f is piecewise-continuous, $w(t)$ is a piecewise-continuous periodic function with the same period as the input signal. A piecewise periodical function can be expanded in Fourier series

$$w(t) = \frac{a_0}{2} + \sum_{n=1}^{\infty} (a_n \cos(n\omega t) + b_n \sin(n\omega t)), \tag{3.1}$$

where

$$a_0 = \frac{1}{\pi} \int_{-\pi}^{\pi} w(t) d(\omega t)$$

$$a_n = \frac{1}{\pi} \int_{-\pi}^{\pi} w(t) \cos(n\omega t) d(\omega t)$$

$$b_n = \frac{1}{\pi} \int_{-\pi}^{\pi} w(t) \sin(n\omega t) d(\omega t).$$

Remark 3.1. For a piecewise-continuous function $w(t)$, the Fourier series on the right-hand side of (3.1) converges to $w(t)$ at any continuous point, and to the average of two values obtained by taking limits from both sides at a dis-continuous point. If we truncate the series up to order k,

$$w_k(t) = \frac{a_0}{2} + \sum_{n=1}^{k} (a_n \cos(n\omega t) + b_n \sin(n\omega t)),$$

where w_k is the best approximation in least squares, i.e., in L_2.   ◁

Taking the approximation to the first order, we have

$$w_1 = \frac{a_0}{2} + a_1 \cos(\omega t) + b_1 \sin(\omega t).$$

If $a_0 = 0$, which can be guaranteed by setting the nonlinear function f to an odd function, we have the approximation

$$w_1 = a_1 \cos(\omega t) + b_1 \sin(\omega t) \tag{3.2}$$

which is an approximation at the fundamental frequency. The above discussion shows that for a nonlinear component described by the nonlinear function f, the approximation at the fundamental frequency, i.e., the frequency of the input signal, to an input signal $A \sin(\omega t)$, is a sinusoidal function in (3.2) with the Fourier coefficients a_1 and b_1 shown in (3.1). Hence, we can analyse the frequency response of this nonlinear component.

We can rewrite w_1 in (3.2) as

$$w_1 = M \sin(\omega t + \phi), \tag{3.3}$$

where

$$M(A, \omega) = \sqrt{a_1^2 + b_1^2},$$

$$\phi(A, \omega) = \arctan(a_1/b_1).$$

In complex expression, we have

$$w_1 = M e^{j(\omega t + \phi)} = (b_1 + ja_1)e^{j\omega t}.$$

The describing function is defined, similar to frequency response, as the complex ratio of the fundamental component of the nonlinear element against the input by

$$N(A, \omega) = \frac{M e^{j\omega t + \phi}}{A e^{j\omega t}} = \frac{b_1 + ja_1}{A}. \tag{3.4}$$

Remark 3.2. A clear difference between the describing function of a nonlinear element and the frequency response of a linear system is that the describing function depends on the input amplitude. This reflects the nonlinear nature of the describing function. ◁

Remark 3.3. If f is a single-valued odd function, i.e., $f(-x) = f(x)$, we have

$$a_1 = \frac{1}{\pi} \int_{-\pi}^{\pi} f(A \sin(\omega t)) \cos(\omega t) d(\omega t)$$

$$= \frac{1}{\pi} \int_{-\pi}^{0} f(A \sin(\omega t) \cos(\omega t) d(\omega t) + \frac{1}{\pi} \int_{0}^{\pi} f(A \sin(\omega t) \cos(\omega t) d(\omega t)$$

$$= \frac{1}{\pi} \int_{0}^{\pi} f(A \sin(-\omega t) \cos(-\omega t) d(\omega t) + \frac{1}{\pi} \int_{0}^{\pi} f(A \sin(\omega t) \cos(\omega t) d(\omega t)$$

$$= 0.$$

If $a_1 = 0$, the describing function is a real value. ◁

Example 3.1. The characteristics of a hardening spring are given by

$$f(x) = x + \frac{x^3}{2}.$$

Given the input $A \sin(\omega t)$, the output is

$$w(t) = f(A \sin(\omega t))$$

$$= A \sin(\omega t) + \frac{A^3}{2} \sin^3(\omega t).$$

Since f is an odd function, we have $a_1 = 0$. The coefficient b_1 is given by

$$b_1 = \frac{1}{\pi} \int_{-\pi}^{\pi} \left[A \sin(\omega t) + \frac{A^3}{2} \sin^3(\omega t) \right] \sin(\omega t) d(\omega t)$$

$$= \frac{4}{\pi} \int_0^{\pi/2} \left[A \sin^2(\omega t) + \frac{A^3}{2} \sin^4(\omega t) \right] d(\omega t).$$

Using the integral identity

$$\int_0^{\pi/2} \sin^n(\omega t) d(\omega t) = \frac{n-1}{n} \int_0^{\pi/2} \sin^{n-2}(\omega t) d(\omega t) \qquad \text{for } n > 2,$$

we have

$$b_1 = A + \frac{3}{8} A^3.$$

Therefore, the describing function is

$$N(A, \omega) = N(A) = \frac{b_1}{A} = 1 + \frac{3}{8} A^2.$$

Alternatively, we can also use the identity

$$\sin(3\omega t) = 3 \sin(\omega t) - 4 \sin^3(\omega t)$$

to obtain

$$w(t) = A \sin(\omega t) + \frac{A^3}{2} \sin^3(\omega t)$$

$$= A \sin(\omega t) + \frac{A^3}{2} \left(\frac{3}{4} \sin(\omega t) - \frac{1}{4} \sin(3\omega t) \right)$$

$$= \left(A + \frac{3}{8} A^3 \right) \sin(\omega t) - \frac{1}{8} A^3 \sin(3\omega t).$$

Hence, we obtain $b_1 = A + \frac{3}{8} A^3$ from the first term.   ◁

Through the above discussion, describing functions are well defined for nonlinear components whose input–output relationship can be well defined by piecewise-continuous functions. These functions are time-invariant, i.e., the properties of nonlinear elements do not vary with time. This is in line with the assumption for frequency response analysis, which can only be applied to time-invariant linear systems. We treat describing functions as the approximations at the fundamental frequencies, and therefore in our analysis, we require $a_0 = 0$ which is guaranteed by odd functions for the nonlinear components. With the describing function of a nonlinear component, we can then apply analysis in frequency responses for the entire system. For the convenience of this kind of analysis, we often assume that the nonlinear component for which the describing function is used to approximate its behaviours is the only

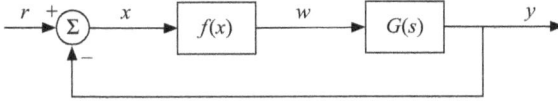

Figure 3.1   Block diagram for describing function analysis

nonlinear component in the system, as shown in Figure 3.1. Hence, in the remaining part of this chapter, we use the following assumptions for describing function analysis:

- There is only a single nonlinear component in the entire system.
- The nonlinear component is time-invariant.
- The nonlinearity is odd.

3.2   Describing functions for common nonlinear components

In this section, we will calculate the describing functions of common nonlinear elements in a number of examples.

Example 3.2. *Saturation.* A saturation function shown in Figure 3.2 is described by

$$f(x) = \begin{cases} kx, & \text{for } |x| < a, \\ \text{sign}(x)ka, & \text{otherwise.} \end{cases} \tag{3.5}$$

The output to the input $A \sin(\omega t)$, for $A > a$, is symmetric over quarters of a period, and in the first quarter,

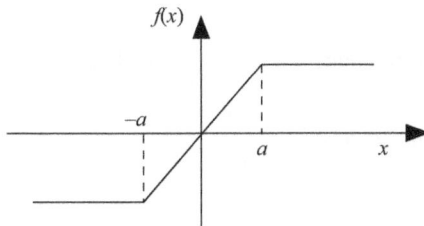

Figure 3.2   Saturation

$$w(t) = \begin{cases} kA \sin(\omega t), & 0 \le \omega t \le \gamma, \\ ka, & \gamma < \omega t \le \pi/2, \end{cases} \tag{3.6}$$

where $\gamma = \sin^{-1}(a/A)$. The function is odd, hence we have $a_1 = 0$, and the symmetry of $w_1(t)$ implies that

$$b_1 = \frac{4}{\pi} \int_0^{\pi/2} w_1 \sin(\omega t) d(\omega t)$$

$$= \frac{4}{\pi} \int_0^{\gamma} kA \sin^2(\omega t) d(\omega t) + \frac{4}{\pi} \int_{\gamma}^{\pi/2} ka \sin(\omega t) d(\omega t)$$

$$= \frac{2kA}{\pi} \left(\gamma - \frac{1}{2} \sin(2\gamma) \right) + \frac{4ka}{\pi} \cos(\gamma)$$

$$= \frac{2kA}{\pi} \left(\gamma - \frac{a}{A} \cos(\gamma) \right) + \frac{4ka}{\pi} \cos(\gamma)$$

$$= \frac{2kA}{\pi} \left(\gamma + \frac{a}{A} \cos(\gamma) \right)$$

$$= \frac{2kA}{\pi} \left(\gamma + \frac{a}{A} \sqrt{1 - \frac{a^2}{A^2}} \right).$$

Note that we have used $\sin \gamma = \frac{a}{A}$ and $\cos \gamma = \sqrt{1 - \frac{a^2}{A^2}}$. Therefore, the describing function is given by

$$N(A) = \frac{b_1}{A} = \frac{2k}{\pi} \left(\sin^{-1} \frac{a}{A} + \frac{a}{A} \sqrt{1 - \frac{a^2}{A^2}} \right). \tag{3.7}$$

◁

Example 3.3. *Ideal relay.* The output from the ideal relay shown in Figure 3.3 (signum function) is described by, with $M > 0$,

$$w(t) = \begin{cases} -M, & -\pi \le \omega t < 0, \\ M, & 0 \le \omega t < \pi. \end{cases} \tag{3.8}$$

It is again an odd function, hence we have $a_1 = 0$. The coefficient b_1 is given by

$$b_1 = \frac{2}{\pi} \int_0^{\pi} M \sin(\omega t) d(\omega t) = \frac{4M}{\pi}$$

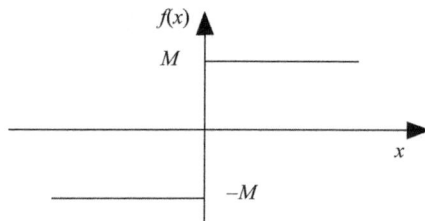

Figure 3.3   Ideal relay

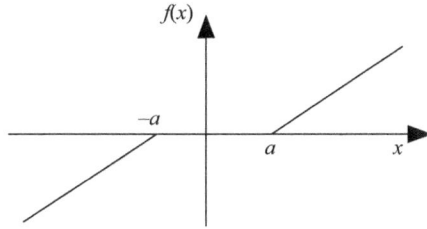

Figure 3.4   Dead zone

and therefore the describing function is given by

$$N(A) = \frac{4M}{\pi A}.$$
(3.9)

◁

Example 3.4. *Dead zone.* A dead zone is a complement to saturation. A dead zone shown in Figure 3.4 can be described by a nonlinear function

$$f(x) = \begin{cases} k(x - a), & \text{for } x > a, \\ 0, & \text{for } |x| < a, \\ k(x + a), & \text{for } x < -a. \end{cases}$$
(3.10)

The output to the input $A \sin(\omega t)$, for $A > a$, is symmetric over quarters of a period, and in the first quarter,

$$w(x) = \begin{cases} 0, & 0 \leq \omega t \leq \gamma, \\ k(A \sin(\omega t) - a), & \gamma < \omega t \leq \pi/2, \end{cases}$$
(3.11)

where $\gamma = \sin^{-1}(a/A)$. The function is odd, hence we have $a_1 = 0$, and the symmetry of $w(t)$ implies that

$$\begin{aligned} b_1 &= \frac{4}{\pi} \int_0^{\pi/2} w(t) \sin(\omega t) d(\omega t) \\ &= \frac{4}{\pi} \int_\gamma^{\pi/2} k(A \sin(\omega t) - a) d(\omega t) \\ &= \frac{2kA}{\pi} \left(\left(\frac{\pi}{2} - \gamma \right) + \frac{1}{2} \sin(2\gamma) \right) - \frac{4ka}{\pi} \cos(\gamma) \\ &= kA - \frac{2kA}{\pi} \left(\gamma + \frac{a}{A} \cos(\gamma) \right) \\ &= kA - \frac{2kA}{\pi} \left(\gamma + \frac{a}{A} \sqrt{1 - \frac{a^2}{A^2}} \right). \end{aligned}$$

Similar to the calculation of the describing function for saturation, we have used $\sin \gamma = \frac{a}{A}$ and $\cos \gamma = \sqrt{1 - \frac{a^2}{A^2}}$. The describing function for a dead zone is given by

$$N(A) = \frac{b_1}{A} = k - \frac{2k}{\pi}\left(\sin^{-1}\frac{a}{A} + \frac{a}{A}\sqrt{1 - \frac{a^2}{A^2}}\right). \tag{3.12}$$

◁

Remark 3.4. The dead-zone function shown in (3.10) complements the saturation function shown in (3.5) in the sense that if we use f_s and f_d to denote the saturation function and dead-zone function, we have $f_s + f_d = k$ for the describing functions shown in (3.7) and (3.12), the same relationship holds. ◁

Example 3.5. *Relay with hysteresis.* Consider a case when there is a delay in the ideal relay as shown in Figure 3.5. The nonlinear function for relay with hysteresis can be described by

$$f(x) = \begin{cases} M, & \text{for } x \geq a, \\ -M, & \text{for } |x| < a, \quad \dot{x} > 0, \\ M, & \text{for } |x| < a, \quad \dot{x} < 0, \\ -M, & \text{for } x \leq -a. \end{cases} \tag{3.13}$$

When this nonlinear component takes $A\sin(\omega t)$ as the input with $A > a$, the output $w(t)$ is given by

$$w(t) = \begin{cases} M, & \text{for } -\pi \leq \omega t < (\pi - \gamma), \\ -M, & \text{for } -(\pi - \gamma) \leq \omega t < \gamma, \\ M, & \text{for } \gamma \leq \omega t < \pi, \end{cases} \tag{3.14}$$

where $\gamma = \sin^{-1}(\frac{a}{A})$. In this case, we still have $a_0 = 0$, but not a_1. For a_1 we have

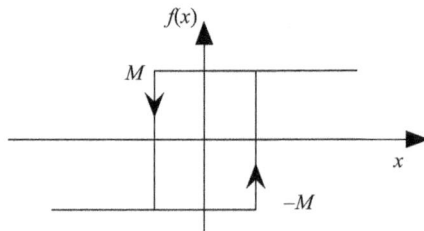

Figure 3.5   Relay with hysteresis

$$a_1 = \frac{1}{\pi} \int_{-\pi}^{-(\pi-\gamma)} M \cos(\omega t) d(\omega t) + \frac{1}{\pi} \int_{-(\pi-\gamma)}^{\gamma} -M \cos(\omega t) d(\omega t)$$

$$+ \frac{1}{\pi} \int_{\gamma}^{\pi} M \cos(\omega t) d(\omega t)$$

$$= -\frac{4M}{\pi} \sin(\gamma)$$

$$= -\frac{4M}{\pi} \frac{a}{A}.$$

Similarly, we have

$$b_1 = \frac{1}{\pi} \int_{-\pi}^{-(\pi-\gamma)} M \sin(\omega t) d(\omega t) + \frac{1}{\pi} \int_{-(\pi-\gamma)}^{\gamma} -M \sin(\omega t) d(\omega t)$$

$$+ \frac{1}{\pi} \int_{\gamma}^{\pi} M \sin(\omega t) d(\omega t)$$

$$= \frac{4M}{\pi} \cos(\gamma)$$

$$= \frac{4M}{\pi} \sqrt{1 - \frac{a^2}{A^2}}.$$

From

$$N(A, \omega) = \frac{b_1 + ja_1}{A},$$

we have

$$N(A) = \frac{4M}{\pi A} \left(\sqrt{1 - \frac{a^2}{A^2}} - j\frac{a}{A} \right). \tag{3.15}$$

Using the identity $\cos(\gamma) + j\sin(\gamma) = e^{jr}$, we can rewrite the describing function as

$$N(A) = \frac{4M}{\pi A} e^{-j \arcsin(a/A)}.$$

◁

Remark 3.5. Comparing the describing function of the relay with hysteresis with that of ideal relay in (3.9), the describing functions indicate that there is a delay in the relay with hysteresis by $\arcsin(a/A)$ in terms of phase angle. There is indeed a delay of $\gamma = \arcsin(a/A)$ in the time response $w(t)$ shown in (3.14) with that of the ideal relay. In fact, we could use this fact to obtain the describing function for the relay with hysteresis. ◁

3.3   Describing function analysis of nonlinear systems

One of the most important applications of describing functions is to predict the existence of a limit cycle in a closed-loop system that contains a nonlinear component with a linear transfer function, as shown in Figure 3.1. Consider a system with a linear transfer function $G(s)$ and a nonlinear element with describing function $N(A, \omega)$ in the forward path, under unit feedback. The input–output relations of the system component by setting $r = 0$ can be described by

$$w = N(A, \omega)x$$

$$y = G(j\omega)w$$

$$x = -y,$$

with y as the output and x as the input to the nonlinear component. From the above equations, it can be obtained that

$$y = G(j\omega)N(A, \omega)(-y),$$

and it can be arranged as

$$(G(j\omega)N(A, \omega) + 1)y = 0.$$

If there exists a limit cycle, then $y \neq 0$, which implies that

$$G(j\omega)N(A, \omega) + 1 = 0, \tag{3.16}$$

or

$$G(j\omega) = -\frac{1}{N(A, \omega)}. \tag{3.17}$$

Therefore, the amplitude A and frequency ω of the limit cycle must satisfy the above equation. Equation (3.17) is difficult to solve in general. Graphic solutions can be found by plotting $G(j\omega)$ and $-1/N(A, \omega)$ on the same graph to see if they intersect each other. The intersection points are the solutions, from which the amplitude and frequency of the oscillation can be obtained.

Remark 3.6. The above discussion is based on the assumption that the oscillation, or limit cycle, can be well approximated by a sinusoidal function, and the nonlinear component is well approximated by its describing function. The describing function analysis is an approximate method in nature.   ◁

Only a stable limit cycle may exist in real applications. When we say stable limit cycle, we mean that if the state deviates a little from the limit cycle, it should come back. With the amplitude as an example, if A is perturbed from its steady condition, say with a very small increase in the amplitude, for a stable limit cycle, the system will decay to its steady condition.

As describing functions are first-order approximations in the frequency domain, stability criteria in the frequency domain may be used for the stability analysis of limit cycles. Nyquist criterion can be extended to give the conditions for stability of limit cycles.

Recall the case for a linear system with the forward transfer function $G(s)$ with unit feedback. The characteristic equation is given by

$$G(s) + 1 = 0, \text{ or } G(s) = -1.$$

The Nyquist criterion determines stability of the closed-loop system from the number of encirclements of the Nyquist plot around point -1, or $(-1, 0)$ in the complex plain.

In the case that there is a control gain K in the forward transfer function, the characteristic equation is given by

$$KG(s) + 1 = 0, \text{ or } G(s) = \frac{-1}{K}.$$

In this case, the Nyquist criterion can be extended to determine the stability of the closed loop by counting the encirclements of the Nyquist plot around $(-1/K, 0)$ in the complex plain in the same way as around $(-1, 0)$. The Nyquist criterion for non-unity forward path gain K is also referred to as the extended Nyquist criterion. The same argument holds when k is a complex number.

We can apply the extended Nyquist criterion to determine the stability of a limit cycle. When the condition specified in (3.17) is satisfied for some (A_0, ω_0), A_0 and ω_0 are the amplitude and frequency of the limit cycle respectively, and $N(A_0, \omega_0)$ is a complex number. We can use the extended Nyquist criterion to determine the stability of the limit cycle with the amplitude A_0 and frequency ω_0 by considering a perturbation of A around A_0.

To simplify our discussion, let us assume that $G(s)$ is stable and minimum phase. It is known from the Nyquist criterion that the closed-loop system with constant gain K is stable if the Nyquist plot does not encircle $(-1/K, 0)$. Let us consider a perturbation in A to A^+ with $A^+ > A_0$. In such a case, $-1/N(A^+, \omega_0)$ is a complex number in general. If the Nyquist plot does not encircle the point $-1/N(A^+, \omega_0)$, we conclude that the closed-loop system is stable with the complex gain $-1/N(A^+, \omega_0)$. Therefore, in a stable closed-loop system, the oscillation amplitude decays, which makes A^+ return to A_0. This implies that the limit cycle (A_0, ω_0) is stable. Alternatively, if the Nyquist plot encircles the point $-1/N(A^+, \omega_0)$, we conclude that the closed-loop system is unstable with the complex gain $-1/N(A^+, \omega_0)$. In such a case, the oscillation amplitude may grow even further, and does not return to A_0. Therefore, the limit cycle is unstable.

Similar arguments can be made for the perturbation to a smaller amplitude. For an $A^- < A_0$, if the Nyquist plot does encircle the point $-1/N(A^-, \omega_0)$, the limit cycle is stable. If the Nyquist plot does not encircle the point $-1/N(A^-, \omega_0)$, the limit cycle is unstable.

When we plot $-1/N(A, \omega_0)$ in the complex plane with A as a variable, we obtain a line with direction of the increment of A. Based on the discussion above, the way

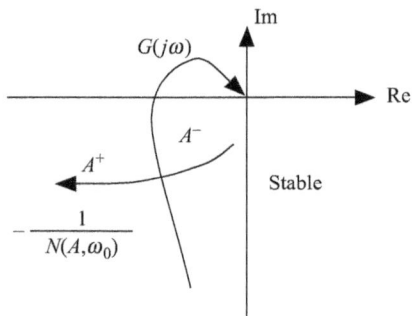

Figure 3.6   Digram for stable limit cycle

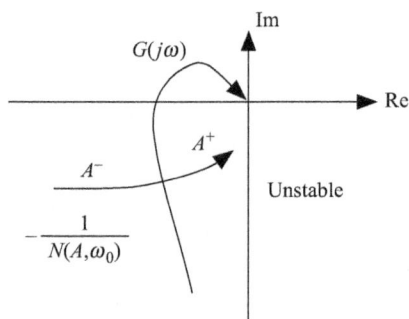

Figure 3.7   Digram for unstable limit cycle

of the line for $-1/N(A, \omega_0)$ intersects with the Nyquist plot determines the stability of the limit cycle. Typical Nyquist plots of stable minimum-phase systems are shown in Figures 3.6 and 3.7 for stable and unstable limit cycles with nonlinear elements respectively.

We can summarise the above discussion for the stability criterion of limit cycles using describing function.

Theorem 3.1. *Consider a unity-feedback system with the forward path with stable minimum phase transfer function $G(s)$ and a nonlinear component with the describing function $N(A, \omega)$, and suppose that the plots, $-1/N$ and $G(j\omega)$ intersect at the point with $A = A_0$ and $\omega = \omega_0$. The limit cycle at (A_0, ω_0) is stable if the plot of $-1/N(A, \omega_0)$ crosses the Nyquist plot from the inside of the encirclement to the outside of the encirclement as A increases. The limit cycle at (A_0, ω_0) is unstable if the plot of $-1/N(A, \omega_0)$ crosses the Nyquist plot from the outside of the encirclement to the inside of the encirclement as A increases.*

Remark 3.7. Theorem 3.1 requires the transfer function to be stable and minimum phase, for the simplicity of the presentation. This theorem can be easily extended to the case when $G(s)$ is unstable or has unstable zeros by using corresponding stability conditions based on the Nyquist criterion. For example if $G(s)$ is stable and has one unstable zero, then the stability criterion for the limit cycle will be opposite to the condition stated in the theorem, i.e., the limit cycle is stable if the plot of $-1/N(A, \omega_0)$ crosses the Nyquist plot from the outside of the encirclement to the inside of the encirclement as A increases.   ◁

Example 3.6. Consider a linear transfer function $G(s) = \dfrac{K}{s(s+1)(s+2)}$ with K a positive constant and an ideal relay in a closed loop, as shown in Figure 3.8. We will determine if there exists a limit cycle and analyse the stability of the limit cycle.

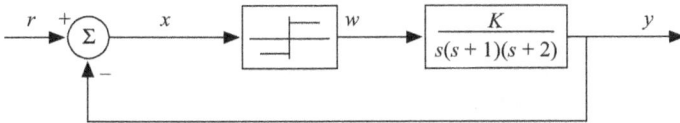

Figure 3.8   Closed-loop system for Example 3.6

For the ideal relay, we have $N = \dfrac{4M}{\pi A}$. For the transfer function, we can obtain that

$$G(j\omega) = \frac{K}{j\omega(j\omega+1)(j\omega+2)}$$
$$= K\frac{-3\omega^2 - j\omega(2 - \omega^2)}{(-3\omega^2)^2 + \omega^2(2 - \omega^2)^2}.$$

From

$$G(j\omega) = -\frac{1}{N},$$

we obtain two equations for real and imaginary parts respectively as

$$\Im(G(j\omega)) = 0,$$
$$\Re(G(j\omega)) = -\frac{\pi A}{4M}.$$

From the equation of the imaginary part, we have

$$K\frac{-\omega(2 - \omega^2)}{(-3\omega^2)^2 + \omega^2(2 - \omega^2)^2} = 0,$$

which gives $\omega = \sqrt{2}$. From the equation of the real part, we have

$$K\frac{-3\omega^2}{(-3\omega^2)^2 + \omega^2(2 - \omega^2)^2} = -\frac{\pi A}{4M},$$

which gives $A = 2KM/3\pi$.

Hence, we have shown that there exists a limit cycle with amplitude and frequency at $(A, \omega) = (2KM/3\pi, \sqrt{2})$.

The plot of $-(1/N(A)) = -(\pi A/4M)$ overlaps with the negative side of the real axis. As A increases from 0, $-(1/N(A))$ moves from the origin towards left. Therefore, as A increases, $-(1/N(A))$ moves from inside of the encirclement of the Nyquist plot to outside of the encirclement, and the limit cycle is stable, based on Theorem 3.1.

A simulation result for $K = M = 1$ is shown in Figure 3.9 with the amplitude $A = 0.22$ and period $T = 4.5$ s, not far from the values $A = 0.2212$ and $T = 4.4429$, predicted from the describing function analysis.   ◁

Example 3.7. In this example, we consider a van der Pol oscillator described by

$$\ddot{y} + \epsilon(3y^2 - 1)\dot{y} + y = 0. \tag{3.18}$$

We will use describing function analysis to predict the existence of a limit cycle, and compare the predicted amplitudes and periods for different ϵ values with the simulated ones.

To use the describing analysis, we need to formulate the system in the format of one linear transfer function and a nonlinear element. Rearranging (3.18), we have

$$\ddot{y} - \epsilon\dot{y} + y = -\epsilon\frac{d}{dt}y^3.$$

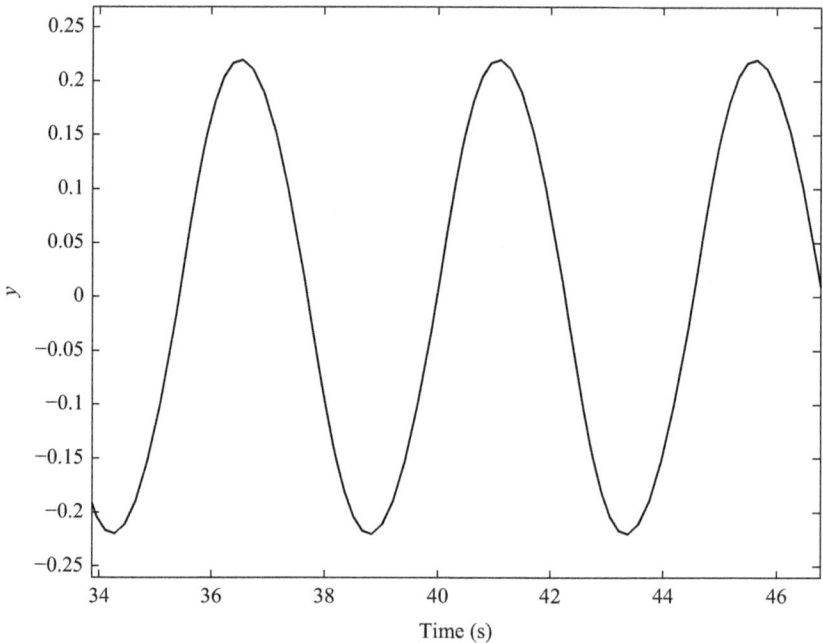

Figure 3.9   The simulated output for Example 3.6

Hence, the system (3.18) can be described by a closed-loop system with a nonlinear component

$$f(x) = x^3$$

and a linear transfer function

$$G(s) = \frac{\epsilon s}{s^2 - \epsilon s + 1}.$$

Using the identity

$$\sin(3\omega t) = 3\sin(\omega t) - 4\sin^3(\omega t)$$

in a similar way as in Example 3.1, we obtain the describing function for $f(x) = x^3$ as

$$N(A) = \frac{3}{4}A^2.$$

Setting

$$\Im G(j\omega) = -\Im\left(\frac{1}{N}\right) = 0,$$

we have

$$\frac{\epsilon\omega(1 - \omega^2)}{(1 - \omega^2)^2 + \epsilon^2\omega^2} = 0,$$

which gives $\omega = 1$. From the equation for the real part, we obtain

$$\frac{-\epsilon^2\omega^2}{(1 - \omega^2)^2 + \epsilon^2\omega^2} = -\frac{4}{3A^2}$$

which gives $A = 2\sqrt{3}/3$.

The linear part of the transfer function has one unstable pole. We need to take this into consideration for the stability of the limit cycle. As A increases, $-1/N(A)$ moves from the left to the right along the negative part of the real axis, basically from the outside of the encirclement of the Nyquist plot to the inside of the encirclement. This suggests that the limit cycle is stable, as there is an unstable pole in the linear transfer function. The simulation results for $\epsilon = 1$ and $\epsilon = 30$ are shown in Figures 3.10 and 3.11. In both cases, the amplitudes are very close to the predicted one from the describing function analysis. For the period, the simulation result for $\epsilon = 1$ in Figure 3.10 is very close to 2π, but the period for $\epsilon = 30$ is much better than 2π. This suggests that the describing function analysis gives a better approximation for the case of $\epsilon = 1$ than $\epsilon = 30$. In fact, for a small value of ϵ the oscillation is very similar to a sinusoidal function. With a big value of ϵ, the wave form is very different from a sinusoidal function, and therefore the describing function method cannot provide a good approximation. ◁

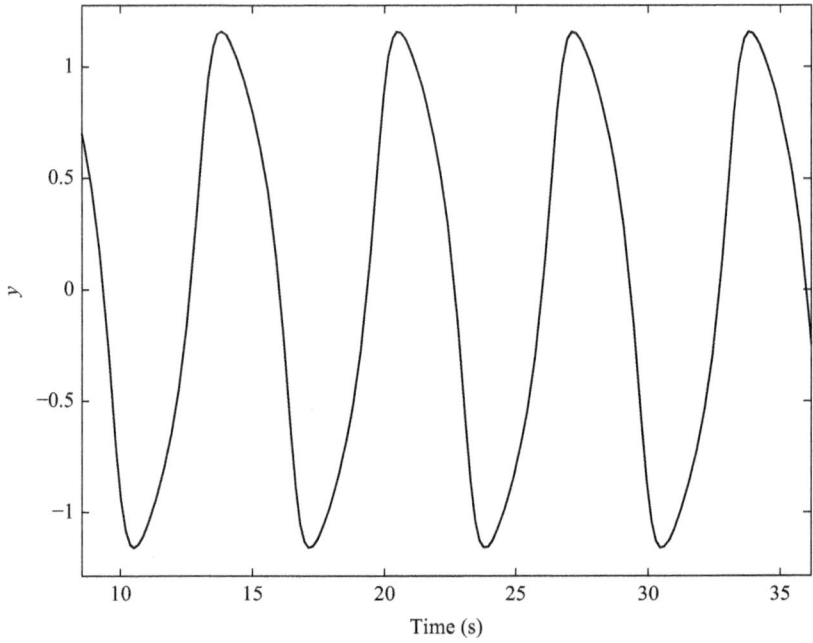

Figure 3.10   The simulated output for Example 3.7 with $\epsilon = 1$

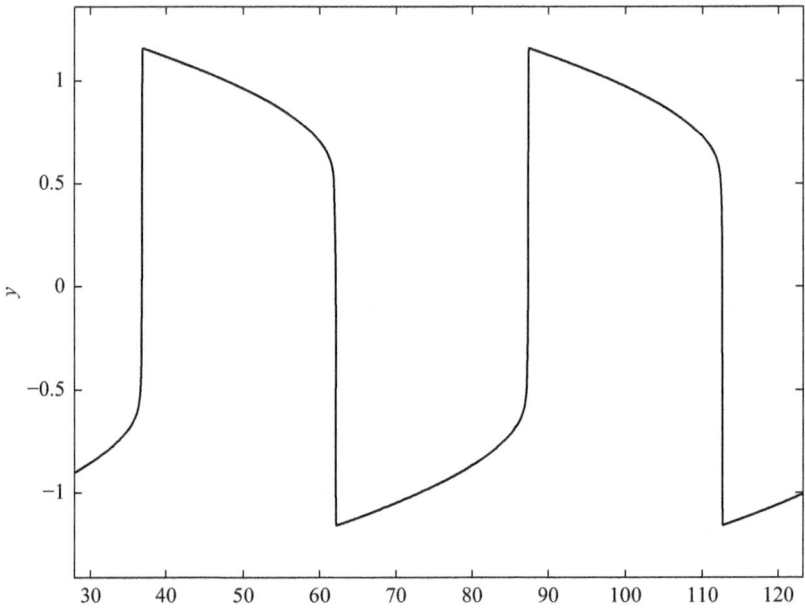

Figure 3.11   The simulated output for Example 3.7 with $\epsilon = 30$

Chapter 4
Stability theory

For control systems, design, one important objective is to ensure the stability of the closed-loop system. For a linear system, the stability can be evaluated in time domain or frequency domain, by checking the eigenvalues of the system matrix or the poles of the transfer function. For nonlinear systems, the dynamics of the system cannot be described by equations in linear state space or transfer functions in general. We need more general definitions about the stability of nonlinear systems. In this chapter, we will introduce basic concepts of stability theorems based on Lyapunov functions.

4.1 Basic definitions

Consider a nonlinear system

$$\dot{x} = f(x), \tag{4.1}$$

where $x \in \mathcal{D} \subset \mathbb{R}^n$ is the state of the system, and $f : \mathcal{D} \subset \mathbb{R}^n \longrightarrow \mathbb{R}^n$ is a continuous function, with $x = 0$ as an equilibrium point, that is $f(0) = 0$, and with $x = 0$ as an interior point of $\mathcal{D}$. Here we use $\mathcal{D}$ to denote a domain around the equilibrium $x = 0$. This domain can be interpreted as a set with 0 as its interior point, or it can also be simplified as $\mathcal{D} = \{x \mid \|x\| < r\}$ for some positive r. In the remaining part of this chapter, we will use $\mathcal{D}$ in this way.

When we say the stability of the system, we refer to the behaviour of the system around the equilibrium point. Here we assume that $x = 0$ is an equilibrium point without loss of generality. In case that the system has an equilibrium point at x_0, we can always define a state transformation with $x - x_0$ as the new state, to shift the equilibrium point to the origin.

The system in (4.1) is referred to as an autonomous system, as it does not depend on the signals other than the system state. For nonlinear control systems, we can write

$$\dot{x} = f(x, u), \tag{4.2}$$

where $u \in \mathbb{R}^m$ is the control input. With u as an external signal, the system (4.2) is not autonomous. However, for such a system, if we design a feedback control law $u = g(x)$ with $g : \mathbb{R}^n \longrightarrow \mathbb{R}^m$ as a continuous function, the closed-loop system

$$\dot{x} = f(x, g(x))$$

becomes an autonomous system.

In this chapter, we will present basic definitions and results for stability of autonomous systems. As discussed above, control systems can be converted to autonomous systems by state feedback control laws.

There are many different definitions of stability for dynamics systems. Often different definitions are needed for different purposes, and many of them are actually the same when the system is linear. Among different definitions, the most fundamental one is the Lyapunov stability.

Definition 4.1 (Lyapunov stability). For the system (4.1), the equilibrium point $x = 0$ is said to be Lyapunov stable if for any given positive real number R, there exists a positive real number r to ensure that $\|x(t)\| < R$ for all $t \geq 0$ if $\|x(0)\| < r$. Otherwise the equilibrium point is unstable.

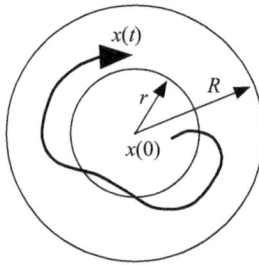

Figure 4.1   Lyapunov stability

The definition of Lyapunov stability concerns with the behaviours of a dynamic system with respect to the initial state. If a system is Lyapunov stable, we can impose a restriction on the initial state of the system to make sure that the state variables stay in a certain region. For the two positive numbers in the definition, R and r, the definition did not explicitly require $R \geq r$. However, if we set $r > R$, from the continuity of the solution, we cannot ensure $\|x(t)\| < R$ for t close to 0, because of $\|x(0)\| > R$. Therefore, when using this definition, we need $r \leq R$.

Example 4.1. Consider a linear system

$$\dot{x} = Ax, \quad x(0) = x_0,$$

where

$$A = \begin{bmatrix} 0 & \omega \\ -\omega & 0 \end{bmatrix}$$

with $\omega > 0$. For this linear system, we can explicitly solve the differential equation to obtain

$$x(t) = \begin{bmatrix} \cos \omega t & \sin \omega t \\ -\sin \omega t & \cos \omega t \end{bmatrix} x_0.$$

It is easy to check that we have $\|x(t)\| = \|x_0\|$. Hence, to ensure that $\|x(t)\| \leq R$, we only need to set $r = R$, i.e., if $\|x_0\| \leq R$, we have $\|x(t)\| \leq R$ for all $t > 0$. ◁

Note that for the system in Example 4.1, the system matrix has two eigenvalues on the imaginary axis, and this kind of systems is referred to as critically stable in many undergraduate texts. As shown in the example, this system is Lyapunov stable. It can also be shown that for a linear system, if all the eigenvalues of the system matrix A are in the closed left half of the complex plane, and the eigenvalues on the imaginary axis are simple, the system is Lyapunov stable. However, if the system matrix has multiple poles on the imaginary axis, the system is not Lyapunov stable. For example let $\dot{x}_1 = x_2$, and $\dot{x}_2 = 0$ with $x_1(0) = x_{1,0}, x_2(0) = x_{2,0}$. It is easy to obtain that $x_1(t) = x_{1,0} + x_{2,0}t$ and $x_2(t) = x_{2,0}$. If we want $\|x(t)\| \leq R$, there does not exist a positive r for $\|x(0)\| \leq r$ to guarantee $\|x(t)\| \leq R$. Therefore, this system is not Lyapunov stable.

For linear systems, when a system is stable, the solution will converge to the equilibrium point. This is not required by Lyapunov stability. For more general dynamic systems, we have the following definition concerning with the convergence to the equilibrium.

Definition 4.2 (Asymptotic stability). For the system (4.1), the equilibrium point $x = 0$ is asymptotically stable if it is stable (Lyapunov) and furthermore $\lim_{t \to \infty} x(t) = 0$.

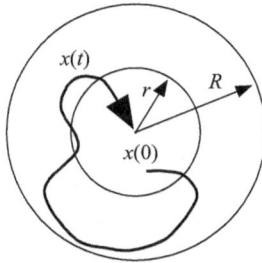

Figure 4.2 Asymptotic stability

Linear systems with poles in the open left half of the complex plane are asymptotically stable. The asymptotic stability only requires that a solution converges the equilibrium point, but it does not specify the rate of convergence. In the following definition, we specify a stability property with an exponential rate of convergence.

Definition 4.3 (Exponential stability). For the system (4.1), the equilibrium point $x = 0$ is exponentially stable if there exist two positive real numbers a and λ such that the following inequality holds:

$$\|x(t)\| < a\|x(0)\|e^{-\lambda t} \tag{4.3}$$

for $t > 0$ in some neighbourhood $\mathcal{D} \subset \mathbb{R}^n$ containing the equilibrium point.

For linear systems, the stability properties are relatively simple. If a linear system is asymptotically stable, it can be shown that it is exponentially stable. Of course, for nonlinear systems, we may have a system that is asymptotically stable, but not exponentially stable.

Example 4.2. Consider a nonlinear system

$$\dot{x} = -x^3, \quad x(0) = x_0 > 0,$$

where $x \in \mathbb{R}$. Let us solve this differential equation. From the system equation we have

$$-\frac{dx}{x^3} = dt,$$

which gives

$$\frac{1}{x^2(t)} - \frac{1}{x_0^2} = 2t$$

and

$$x(t) = \frac{x_0}{\sqrt{1 + 2x_0^2 t}}.$$

It is easy to see that $x(t)$ decreases as t increases, and also $\lim_{t\to\infty} x(t) = 0$. Therefore, this system is asymptotically stable. However, this system is not exponentially stable, as there does not exist a pair of a and γ to satisfy

$$\frac{x_0}{\sqrt{1 + 2x_0^2 t}} \le a x_0 e^{-\gamma t}.$$

Indeed, if there exist such constants a and γ, we have

$$\sqrt{1 + 2x_0^2 t}\, e^{-\gamma t} \ge \frac{1}{a},$$

which is not satisfied for any choices of a and γ, because the left-hand side converges to zero. Hence, the system considered in this example is asymptotically stable, but not exponentially stable. ◁

Lyapunov, asymptotic and exponential, stabilities are defined around equilibrium points. If the properties hold for any initial points in the entire state space, they are referred to as global stability properties. In the following definitions, we give their global versions.

Definition 4.4 (Globally asymptotic stability). If the asymptotic stability defined in Definition 4.2 holds for any initial state in $\mathbb{R}^n$, the equilibrium point is said to be globally asymptotically stable.

Definition 4.5 (Globally exponential stability). If the exponential stability defined in Definition 4.3 holds for any initial state in $\mathbb{R}^n$, the equilibrium point is said to be globally exponentially stable.

The stability property discussed in Example 4.2 is globally and asymptotically stable. In the two examples shown in this section, the stability properties are checked based on the actual solutions of the systems. In general, explicit solutions of nonlinear systems are difficult to obtain, and it is expected to check the stability properties of nonlinear systems without knowing the solutions. In the later part of this chapter, we will show a number of results to establish stability properties without their solutions.

4.2 Linearisation and local stability

In this section, we introduce a result for checking the stability of nonlinear systems based on its linearised model.

Theorem 4.1 (Lyapunov's linearisation method). *For a linearised model, there are three cases*:

- *If the linearised system has all the system's poles in the open left half of the complex plane, the equilibrium point is asymptotically stable for the actual nonlinear system.*
- *If the linearised system has poles in the open right half of the complex plane, then the equilibrium point is unstable.*
- *If the linearised system has poles on the imaginary axis, then the stability of the original system cannot be concluded using the linearised model.*

We do not show a proof of this theorem here. It is clear that this theorem can be applied to check local stabilities of nonlinear systems around equilibrium points. For the case that the linearised model has poles on the imaginary axis, this theorem cannot give conclusive result about the stability. This is not a surprise, because stable and unstable systems can have the same linearised model. For example the systems $\dot{x} = -x^3$ and $\dot{x} = x^3$ have the same linearised model at $x = 0$, that is $\dot{x} = 0$, which is marginally stable. However, as we have seen in Example 4.2, the system $\dot{x} = -x^3$ is asymptotically stable, and it is not difficult to see that $\dot{x} = x^3$ is unstable. For both the stable and unstable cases of linearised models, the linearised model approximates the original system better when the domain around the equilibrium point gets smaller. Hence, the linearised model is expected to reflect on the stability behaviours around the equilibrium point.

Example 4.3. Consider a nonlinear system

$$\dot{x}_1 = x_2 + x_1 - x_1^3,$$
$$\dot{x}_2 = -x_1.$$

It can be seen that $x = (0, 0)$ is an equilibrium point of the system. The linearised model around $x = (0, 0)$ is given by

$$\dot{x} = Ax$$

where

$$A = \begin{bmatrix} 1 & 1 \\ -1 & 0 \end{bmatrix}$$

The linearised system is unstable as $\lambda(A) = \frac{1 \pm \sqrt{3}j}{2}$. Indeed, this nonlinear system is a van der Pols system, and the origin is unstable. Any trajectories that start from initial point close to the origin and within the limit cycle will spiral out, and converge to the limit cycle.   ◁

4.3   Lyapunov's direct method

Lyapunov's linearisation method can only be used to check local stabilities, and also there is a limitation in the case of marginal stability. Fortunately, there is a direct method to check the stability of dynamic systems. This method is based on Lyapunov functions. We need a few definitions before we can show some of the results on stability based on Lyapunov functions.

Definition 4.6 (Positive definite function). A function $V(x) : \mathcal{D} \subset \mathbb{R}^n \to \mathbb{R}$ is said to be locally positive definite if $V(x) > 0$ for $x \in \mathcal{D}$ except at $x = 0$ where $V(0) = 0$. If $\mathcal{D} = \mathbb{R}^n$, i.e., the above property holds for the entire state space, $V(x)$ is said to be globally positive definite.

There are many examples of positive definite functions, such as $x^T P x$ for P being a positive definite matrix, or even $\|x\|$.

Definition 4.7 (Lyapunov function). If in $\mathcal{D} \subset \mathbb{R}^n$ containing the equilibrium point $x = 0$, the function $V(x)$ is positive definite and has continuous partial derivatives, and if its time derivative along any state trajectory of system (4.1) is non-positive, i.e.,

$$\dot{V}(x) \leq 0 \tag{4.4}$$

then $V(x)$ is a Lyapunov function.

Stability analysis based on a Lyapunov function is probably the most commonly used method to establish the stability of nonlinear dynamic systems. A fundamental theorem on Lyapunov function is given below.

Theorem 4.2 (Lyapunov theorem for local stability). *Consider the system (4.1). If in $\mathcal{D} \subset \mathbb{R}^n$ containing the equilibrium point $x = 0$, there exists a function $V(x)$: $\mathcal{D} \subset \mathbb{R}^n \to \mathbb{R}$ with continuous first-order derivatives such that*

- *$V(x)$ is positive definite in $\mathcal{D}$*
- *$\dot{V}(x)$ is non-positive definite in $\mathcal{D}$*

then the equilibrium point $x = 0$ is stable. Furthermore, if $\dot{V}(x)$ is negative definite, i.e., $-\dot{V}(x)$ is positive definite in $\mathcal{D}$, then the stability is asymptotic.

Proof. We need to find a value for r such that when $\|x(0)\| < r$, we have $\|x(t)\| < R$. Define

$$B_R := \{x | \|x\| \le R\} \subset \mathcal{D},$$

and let

$$a = \min_{\|x\|=R} V(x).$$

Since $V(x)$ is positive definite, we have $a > 0$. We then define the level set within B_R

$$\Omega_c := \{x \in B_R | V(x) < c\},$$

where c is a positive real constant and $c < a$. The existence of such a positive real constant c is guaranteed by the continuity and positive definiteness of V. From the definition of Ω_c, $x \in \Omega_c$ implies that $\|x\| < R$. Since $\dot{V} \le 0$, we have $V(x(t)) \le V(x(0))$. Hence, for any $x(0) \in \Omega_c$, we have

$$V(x(t)) \le V(x(0)) < c,$$

which implies

$$\|x(t)\| < R.$$

Since Ω_c contains 0 as an interior point, and V is a continuous function, there must exist a positive real r such that

$$B_r := \{x | \|x\| < r\} \subset \Omega_c.$$

Hence, we have

$$B_r \subset \Omega_c \subset B_R.$$

Therefore, for any $x(0) \in B_r$, we have

$$V(x(t)) \leq V(x(0)) < c,$$

and $\|x(t)\| < R$. We have established that if $\dot{V}$ is non-positive, the system is Lyapunov stable.

Next, we will establish the asymptotic stability from the negative definiteness of $\dot{V}$. For any initial point in $\mathcal{D}$, $V(x(t))$ monotonically decreases with time t. Therefore, there must be a lower limit such that

$$\lim_{t \to \infty} V(x(t)) = \beta \geq 0.$$

The asymptotic stability can be established if we can show that $\beta = 0$. We can prove it by seeking a contradiction. Suppose $\beta > 0$. Let

$$\alpha = \min_{x \in \mathcal{D} - \Omega_\beta} (-\dot{V}(x)),$$

where $\Omega_\beta := \{x \in \mathcal{D} | V(x) < \beta\}$. Since $\dot{V}$ is negative definite, we have $\alpha > 0$. From the definition of α, we have

$$V(x(t)) \leq V(x(0)) - \alpha t.$$

The right-hand side turns to negative when t is big enough, which is a contradiction. Therefore, we can conclude that $\lim_{t \to \infty} V(x(t)) = 0$, which implies $\lim_{t \to \infty} x(t) = 0$. $\qquad\square$

Example 4.4. A pendulum can be described by

$$\ddot{\theta} + \dot{\theta} + \sin \theta = 0,$$

where θ is the angle. If we let $x_1 = \theta$ and $x_2 = \dot{\theta}$, we re-write the dynamic system as

$$\dot{x}_1 = x_2$$
$$\dot{x}_2 = -\sin x_1 - x_2.$$

Consider the scalar function

$$V(x) = (1 - \cos x_1) + \frac{x_2^2}{2}. \tag{4.5}$$

The first term $(1 - \cos x_1)$ in (4.5) can be viewed as the potential energy and the second term $\frac{x_2^2}{2}$ as the kinetic energy. This function is positive definite in the domain $\mathcal{D} = \{|x_1| \leq \pi, x_2 \in \mathbb{R}\}$. A direct evaluation gives

$$\dot{V}(x) = -\sin x_1 \dot{x}_1 + x_2 \dot{x}_2$$
$$= -x_2^2.$$

Hence, the system is stable at $x = 0$. However, we cannot conclude the asymptotic stability of the system from Theorem 4.2. This system is in fact asymptotically stable by using more advanced stability theorem such as invariant set theorem, which is not covered in this book. ◁

When establishing global stability using Lyapunov functions, we need the function $V(x)$ to be unbounded as x tends to infinity. This may sound strange. The reason behind this point is that we need the property that if $V(x)$ is bounded, then x is bounded, in order to conclude the boundedness of x from the boundedness of $V(x)$. This property is defined in the following function as the radial unboundedness of V.

Definition 4.8 (Radially unbounded function). A positive definite function $V(x) : \mathbb{R}^n \to \mathbb{R}$ is said to be radially unbounded if $V(x) \to \infty$ as $\|x\| \to \infty$.

Theorem 4.3 (Lyapunov theorem for global stability). *For the system (4.1) with $\mathcal{D} = \mathbb{R}^n$, if there exists a function $V(x) : \mathbb{R}^n \to \mathbb{R}$ with continuous first order derivatives such that*

- $V(x)$ *is positive definite*
- $\dot{V}(x)$ *is negative definite*
- $V(x)$ *is radially unbounded*

then the equilibrium point $x = 0$ is globally asymptotically stable.

Proof. The proof is similar to the proof of Theorem 4.2, except that for any given point in $\mathbb{R}^n$, we need to show that there is a level set defined by

$$\Omega_c = \{x \in \mathbb{R}^n | V(x) < c\}$$

to contain it.

Indeed, since the function V is radially unbounded, for any point in B_r with any positive real r, there exists a positive real constant c such that $B_r \subset \Omega_c$. It is clear that the level set Ω_c is invariant for any c, that is, for any trajectory that starts in Ω_c remains in Ω_c. The rest of the proof follows the same argument as in the proof of Theorem 4.2. □

Example 4.5. Consider the nonlinear system

$$\dot{x} = -x^3, \quad x(0) = x_0 > 0,$$

where $x \in \mathbb{R}$. In Example 4.2, we have shown that the equilibrium point $x = 0$ is asymptotically stable by checking the solution of the differential equation. In this example, we use a Lyapunov function.

Let

$$V = \frac{1}{2}x^2$$

and it is easy to see that this function is globally positive definite. Its derivative is given by

$$\dot{V} = -x^4$$

which is negative definite. Hence, from Theorem 4.3, we conclude $x = 0$ is asymptotically stable.   ◁

To conclude this section, we have another result for exponential stability.

Theorem 4.4 (Exponential stability). *For the system (4.1), if there exists a function $V(x) : \mathcal{D} \subset \mathbb{R}^n \to \mathbb{R}$ with continuous first-order derivatives such that*

$$a_1 \|x\|^b \leq V(x) \leq a_2 \|x\|^b, \tag{4.6}$$

$$\frac{\partial V}{\partial x} f(x) \leq -a_3 \|x\|^b, \tag{4.7}$$

where a_1, a_2, a_3 and b are positive real constants, the equilibrium point $x = 0$ is exponentially stable. Furthermore, the conditions hold for the entire state space, then the equilibrium point $x = 0$ is globally exponentially stable.

The proof of this theorem is relatively simple, and we are going to show it here. We need a technical lemma, which is also needed later for stability analysis of robust adaptive control systems.

Lemma 4.5 (Comparison lemma). *Let $g, V : [0, \infty) \to \mathbb{R}$. Then*

$$\dot{V}(t) \leq -aV(t) + g(t), \quad \forall t \geq 0 \tag{4.8}$$

implies that

$$V(t) \leq e^{-at} V(0) + \int_0^t e^{-\alpha(t-\tau)} g(\tau) d\tau, \quad \forall t \geq 0 \tag{4.9}$$

for any finite constant a.

Proof. From the derivative of Ve^{at}, we have

$$\frac{d}{dt}(Ve^{at}) = \dot{V}e^{at} + aVe^{at}.$$

Substituting $\dot{V}$ from (4.8) in the above equation, we have

$$\frac{d}{dt}(Ve^{at}) \leq e^{at}g(t). \tag{4.10}$$

Integrating (4.10), we have

$$V(t)e^{at} \leq V(0) + \int_0^t e^{a\tau}g(\tau)d\tau. \tag{4.11}$$

Multiplying both sides of (4.11) by $e^{-a\tau}$ gives (4.8). This completes the proof. $\square$

Now we are ready to prove Theorem 4.4.

Proof. From (4.6) and (4.7), we have

$$\dot{V} \leq -\frac{a_3}{a_2}V.$$

Applying the comparison lemma (Lemma 4.5), we have

$$V(t) \leq V(0)e^{-(a_3/a_2)t}. \tag{4.12}$$

Then from (4.6) and (4.12), we have

$$\begin{aligned}
\|x(t)\| &\leq \left(\frac{1}{a_1}V(t)\right)^{1/b} \\
&\leq \left(\frac{1}{a_1}V(0)\right)^{1/b} e^{-(a_3/a_2 b)t} \\
&\leq \left(\frac{a_2}{a_1}\right)^{1/b} \|x(0)\| e^{-(a_3/a_2 b)t}.
\end{aligned}$$

Hence, (4.3) is satisfied with $a = \left(\frac{a_2}{a_1}\right)^{1/b}$ and $\lambda = \frac{a_3}{a_2 b}$, and the equilibrium point is exponentially stable. $\square$

4.4 Lyapunov analysis of linear time-invariant systems

In this section, we will apply Lyapunov stability analysis to linear time-invariant (LTI) systems. Consider an LTI system

$$\dot{x} = Ax \tag{4.13}$$

where $x \in \mathbb{R}^n$ and $A \in \mathbb{R}^{n \times n}$, x is the state variable, and A is a constant matrix. From linear system theory, we know that this system is stable if all the eigenvalues of A are in the open left half of the complex plane. Such a matrix is referred to as a Hurwitz matrix. Here, we would like to carry out the stability analysis using a Lyapunov function. We can state the stability in the following theorem.

Theorem 4.6. *For the linear system shown in (4.13), the equilibrium $x = 0$ is globally and exponentially stable if and only if there exist positive definite matrices P and Q such that*

$$A^T P + PA = -Q \tag{4.14}$$

holds.

Proof. For sufficiency, let

$$V(x) = x^T P x, \tag{4.15}$$

and then the direct evaluation gives

$$\dot{V} = x^T A^T P x + x^T PAx = -x^T Q x. \tag{4.16}$$

Let us use $\lambda_{max}(\cdot)$ and $\lambda_{min}(\cdot)$ to denote maximum and minimum eigenvalues of a positive definite matrix. From (4.15), we have

$$\lambda_{min}(P)\|x\|^2 \leq V(x) \leq \lambda_{max}(P)\|x\|^2. \tag{4.17}$$

From (4.17) and (4.16), we obtain

$$\begin{aligned}
\dot{V} &\leq -\lambda_{min}(Q)\|x\|^2 \\
&\leq -\frac{\lambda_{min}(Q)}{\lambda_{max}(P)}\|x\|^2.
\end{aligned} \tag{4.18}$$

Now we can apply Theorem 4.4 with (4.17) and (4.18) to conclude that the equilibrium point is globally and exponentially stable. Furthermore, we can identify $a_1 = \lambda_{min}(P)$, $a_2 = \lambda_{max}(P)$, $a_3 = \lambda_{min}(Q)$ and $b = 2$. Following the proof of Theorem 4.4, we have

$$\|x(t)\| \leq \sqrt{\frac{\lambda_{max}(P)}{\lambda_{min}(P)}}\|x(0)\|e^{-\frac{\lambda_{min}(Q)}{2\lambda_{max}(P)}t}. \tag{4.19}$$

For the necessary part, we have

$$\|x(t)\| \leq a\|x(0)\|e^{-\lambda t}$$

for some positive real constants a and λ, which implies $\lim_{t\to\infty} x(t) = 0$. Since

$$x(t) = e^{At}x(0),$$

we can conclude $\lim_{t\to\infty} e^{At} = 0$. In such a case, for a positive definite matrix Q, we can write

$$\int_0^\infty d[\exp(A^T t)Q\exp(At)] = -Q. \tag{4.20}$$

For the left-hand side, we can obtain

$$\int_0^\infty d[\exp(A^T t)Q\exp(At)]$$

$$= A^T \int_0^\infty \exp(A^T t)Q\exp(At)t + \int_0^\infty \exp(A^T t)Q\exp(At)tA$$

Let

$$P = \int_0^\infty \exp(A^T t)Q\exp(At)t$$

and if we can show that P is positive definite, then we obtain (4.14), and hence complete the proof. Indeed, for any $z \in \mathbb{R}^n \neq 0$, we have

$$z^T P z = \int_0^\infty z^T \exp(A^T t)Q\exp(At)z dt.$$

Since Q is positive definite, and e^{At} is non-singular for any t, we have $z^T P z > 0$, and therefore P is positive definite. $\qquad\square$

Chapter 5
Advanced stability theory

Lyapunov direct method provides a tool to check the stability of a nonlinear system if a Lyapunov function can be found. For linear systems, a Lyapunov function can always be constructed if the system is asymptotically stable. In many nonlinear systems, a part of the system may be linear, such as linear systems with memoryless nonlinear components and linear systems with adaptive control laws. For such a system, a Lyapunov function for the linear part may be very useful in the construction for the Lyapunov function for the entire nonlinear system. In this chapter, we will introduce one specific class of linear systems, strict positive real systems, for which, an important result, Kalman–Yakubovich lemma, is often used to guarantee a choice of the Lyapunov function for stability analysis of several types of nonlinear systems. The application of Kalman–Yakubovich lemma to analysis of adaptive control systems will be shown in later chapters, while in this chapter, this lemma is used for stability analysis of systems containing memoryless nonlinear components and the related circle criterion. In Section 5.3 of this chapter, input-to-state stability (ISS) is briefly introduced.

5.1 Positive real systems

Consider a first-order system

$$\dot{y} = -ay + u,$$

where $a > 0$ is a constant, and y and $u \in \mathbb{R}$ are the output and input respectively. This is perhaps the simplest dynamic system we could possibly have. The performance of such a system is desirable in control design. One of the characteristics for such a system is that its transfer function $\frac{1}{s+a}$ is with positive real part if $s = \sigma + j\omega$ with $\sigma > 0$. If such a property holds for other rational transfer functions, they are referred to as positive real transfer functions. For this, we have the following definition.

Definition 5.1. A rational transfer function $G(s)$ is positive real if

- $G(s)$ is real for real s
- $\Re(G(s)) \geq 0$ for $s = \sigma + j\omega$ with $\sigma \geq 0$

For analysis of adaptive control systems, strictly positive real systems are more widely used than the positive real systems.

Definition 5.2. A proper rational transfer function $G(s)$ is strictly positive real if there exists a positive real constant ϵ such that $G(s - \epsilon)$ is positive real.

Example 5.1. For the transfer function $G(s) = \frac{1}{s+a}$, with $a > 0$, we have, for $s = \sigma + j\omega$,

$$G(s) = \frac{1}{a + \sigma + j\omega} = \frac{a + \sigma - j\omega}{(a+\sigma)^2 + \omega^2}$$

and

$$\Re(G(s)) = \frac{a + \sigma}{(a+\sigma)^2 + \omega^2} > 0.$$

Hence, $G(s) = \frac{1}{s+a}$ is positive real. Furthermore, for any $\epsilon \in (0, a)$, we have

$$\Re(G(s - \epsilon)) = \frac{a - \epsilon + \sigma}{(a - \epsilon + \sigma)^2 + \omega^2} > 0$$

and therefore $G(s) = \frac{1}{s+a}$ is also strictly positive real.   ◁

Definition 5.1 shows that a positive real transfer function maps the closed right half of the complex plane to itself. Based on complex analysis, we can obtain the following result.

Proposition 5.1. *A proper rational transfer function $G(s)$ is positive real if*

- *all the poles of $G(s)$ are in the closed left half of the complex plane*
- *any poles on the imaginary axis are simple and their residues are non-negative*
- *for all $\omega \in \mathbb{R}$, $\Re(G(j\omega)) \geq 0$ when $j\omega$ is not a pole of $G(s)$*

It can be seen that $G(s) = \frac{1}{s+a}$, with $a < 0$, is not positive real. If $G(s)$ is positive real, we must have $\Re(G(j\omega)) \geq 0$. Similarly, other necessary conditions can be obtained for a transfer function $G(s)$ to be positive real. We can state those conditions in an opposite way.

Proposition 5.2. *A transfer function $G(s)$ cannot be positive real if one of the following conditions is satisfied:*

- *The relative degree of $G(s)$ is greater than 1.*
- *$G(s)$ is unstable.*
- *$G(s)$ is non-minimum phase (i.e., with unstable zero).*
- *The Nyquist plot of $G(j\omega)$ enters the left half of the complex plane.*

Based on this proposition, the transfer functions $G_1 = \dfrac{s-1}{s^2+as+b}$, $G_2 = \dfrac{s+1}{s^2-s+1}$ and $G_3 = \dfrac{1}{s^2+as+b}$ are not positive real, for any real numbers a and b, because they are non-minimum phase, unstable, and with relative degree 2 respectively. It can also be shown that $G(s) = \dfrac{s+4}{s^2+3s+2}$ is not positive real as $G(j\omega) < 0$ for $\omega > 2\sqrt{2}$.

One difference between strictly positive real transfer functions and positive real transfer functions arises due to the poles on imaginary axis.

Example 5.2. Consider $G(s) = \dfrac{1}{s}$. For $s = \sigma + j\omega$, we have?

$$\Re(G(s)) = \Re\left(\frac{1}{\sigma+j\omega}\right) = \frac{\sigma}{\sigma^2+\omega^2}.$$

Therefore, $G(s) = \frac{1}{s}$ is positive real. However, $G(s) = \frac{1}{s}$ is not strictly positive real.

◁

For the stability analysis later in the book, we only need the result on strictly positive real transfer functions.

Lemma 5.3. *A proper rational transfer function $G(s)$ is strictly positive real if and only if*

- *$G(s)$ is Hurwitz, i.e., all the poles of $G(s)$ are in the open left half of the complex plane.*
- *The real part of $G(s)$ is strictly positive along the $j\omega$ axis, i.e.,*

  $$\forall \omega \geq 0, \quad \Re(G(j\omega)) > 0,$$

- $\lim_{s\to\infty} G(s) > 0$, *or in case of* $\lim_{s\to\infty} G(s) = 0$, $\lim_{\omega\to\infty} \omega^2 \Re(G(j\omega)) > 0$.

Proof. We show the proof for sufficiency here, and omit the necessity, as it is more involved. For sufficiency, we only need to show that there exists a positive real constant ϵ such that $G(s - \epsilon)$ is positive real.

Since $G(s)$ is Hurwitz, there must exist a positive real constant $\bar{\delta}$ such that for $\delta \in (0, \bar{\delta}]$, $G(s - \delta)$ is Hurwitz. Suppose (A, b, c^T, d) is a minimum state space realisation for $G(s)$, i.e.,

$$G(s) = c^T(sI - A)^{-1}b + d.$$

We have

$$
\begin{aligned}
G(s - \delta) &= c^T(sI - \delta I - A)^{-1}b + d \\
&= c^T(sI - A)^{-1}((sI - \delta I - A) + \delta I)(sI - \delta I - A)^{-1}b + d \\
&= G(s) + \delta E(s),
\end{aligned}
\tag{5.1}
$$

where

$$E(s) = c^T(sI - A)^{-1}(sI - \delta I - A)^{-1}b.$$

$E(s)$ is Hurwitz, and strictly proper. Therefore, we have

$$\Re(E(j\omega)) < r_1, \quad \forall \omega \in \mathbb{R}, \ \delta \in (0, \bar{\delta}] \tag{5.2}$$

for some positive real r_1 and the existence of $\lim_{\omega \to \infty} \omega^2 \Re(E(j\omega))$, which implies

$$\omega^2 \Re(E(j\omega)) < r_2, \quad \text{for } |\omega| > \omega_1, \ \delta \in (0, \bar{\delta}] \tag{5.3}$$

for some $\omega_1 > 0$.

If $\lim_{\omega \to \infty} \Re(G(j\omega)) > 0$, we have

$$\Re(G(j\omega)) > r_3, \quad \forall \omega \in \mathbb{R} \tag{5.4}$$

for some $r_3 > 0$. Hence, combining (5.2) and (5.4), we obtain, from (5.1), that

$$\Re(G(j\omega - \delta)) > r_3 - \delta r_1, \quad \forall \omega \in \mathbb{R}. \tag{5.5}$$

Then we have $\Re(G(j\omega - \delta)) > 0$, by setting $\delta < \frac{r_3}{r_1}$.

In the case that $\lim_{\omega \to \infty} \Re(G(j\omega)) = 0$, the condition

$$\lim_{\omega \to \infty} \omega^2 \Re(G(j\omega)) > 0$$

implies that

$$\omega^2 \Re(G(j\omega)) > r_4, \quad \text{for } |\omega| > \omega_2 \tag{5.6}$$

for some positive reals r_4 and ω_2. From (5.1), (5.3) and (5.6), we obtain that

$$\omega^2 \Re(G(j\omega - \delta)) > r_4 - \delta r_2, \quad \text{for } |\omega| > \omega_3 \tag{5.7}$$

where $\omega_3 = \max\{\omega_1, \omega_2\}$. From the second condition of the lemma, we have, for some positive real constant r_5,

$$\Re(G(j\omega)) > r_5, \quad \text{for } |\omega| \le \omega_3. \tag{5.8}$$

Then from (5.1), (5.2) and (5.8), we obtain that

$$\Re(G(j\omega - \delta)) > r_5 - \delta r_1, \quad \text{for } |\omega| \le \omega_3. \tag{5.9}$$

Combining the results in (5.7) and (5.9), we obtain that $\Re(G(j\omega - \delta)) > 0$ by setting $\delta = \min\{\frac{r_4}{r_2}, \frac{r_5}{r_1}\}$. Therefore, we have shown that there exists a positive real δ such that $G(s - \delta)$ is positive real. $\qquad \square$

The main purpose of introducing strictly positive real systems is for the following result, which characterises the systems using matrices in time domain.

Lemma 5.4 (Kalman–Yakubovich lemma). *Consider a dynamic system*

$$\dot{x} = Ax + bu$$
$$y = c^T x,$$

(5.10)

where $x \in \mathbb{R}^n$ is the state variable; y and $u \in \mathbb{R}$ are the output and input respectively; and A, b and c are constant matrices with proper dimensions, and (A, b, c^T) is controllable and observable. Its transfer function $G(s) = c^T(sI - A)^{-1}b$ is strictly positive real if and only if there exist positive definite matrices P and Q such that

$$A^T P + PA = -Q,$$
$$Pb = c.$$

(5.11)

Remark 5.1. We do not provide a proof here, because the technical details in the proof such as finding the positive definite P and the format of Q are beyond the scope of this book. In the subsequent applications for stability analysis, we only need to know the existence of P and Q, not their actual values for a given system. For example in the stability analysis for adaptive control systems in Chapter 7, we only need to make sure that the reference model is strictly positive real, which then implies the existence of P and Q to satisfy (5.11). ◁

5.2 Absolute stability and circle criterion

In this section, we will consider a dynamic system which consists of a linear part and a memoryless nonlinear component. Surely some engineering systems can be modelled in this format, such as linear systems with nonlinearity in sensors. Let us consider a closed-loop system

$$\dot{x} = Ax + bu$$
$$y = c^T x$$
$$u = -F(y)y,$$

(5.12)

where $x \in \mathbb{R}^n$ is the state variable; y and $u \in \mathbb{R}$ are the output and input respectively; and A, b and c are constant matrices with proper dimensions. The nonlinear component is in the feedback law. Similar systems have been considered earlier using describing functions for approximation to predict the existence of limit cycles. Nonlinear elements considered in this section are sector-bounded, i.e., the nonlinear feedback gain can be expressed as

$$\alpha < F(y) < \beta$$

(5.13)

for some constants α and β, as shown in Figure 5.2.

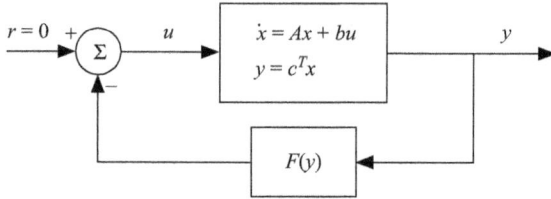

Figure 5.1   Block digram of system (5.12)

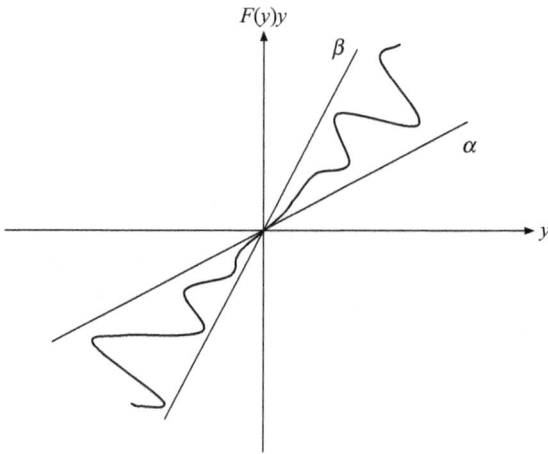

Figure 5.2   Sector-bounded nonlinear feedback gain

The absolute stability refers to the globally asymptotic stability of the equilibrium point at the origin for the system shown in (5.12) for a class of sector-bounded nonlinearities shown in (5.13). We will use Kalman–Yakubovich lemma for the stability analysis. If the transfer function for the linear part is strictly positive real, we can establish the stability by imposing a restriction on the nonlinear element.

Lemma 5.5. *For the system shown in (5.12), if the transfer function $c^T(sI - A)^{-1}b$ is strictly positive real, the system is absolutely stable for $F(y) > 0$.*

Proof. The proof is straightforward by invoking Kalman–Yakubovich lemma. Since the linear part of the system is strictly positive real, then there exist positive definite matrices P and Q such that (5.11) is satisfied.

Consider a Lyapunov function candidate

$$V = x^T P x.$$

Its derivative is given by

$$
\begin{aligned}
\dot{V} &= x^T(A^TP + PA)x + 2x^TPbu \\
&= -x^TQx - 2x^TPbF(y)y \\
&= -x^TQx - 2x^TcF(y)y \\
&= -x^TQx - 2F(y)y^2,
\end{aligned}
$$

where in obtaining the third line of equation, we used $Pb = c$ from the Kalman–Yakubovich lemma. Therefore, if $F(y) > 0$, we have

$$
\dot{V} \leq -x^TQx,
$$

and then system is exponentially stable by Theorem 4.4. □

Note that the conditions specified in Lemma 5.4 are the sufficient conditions.

With the result shown in Lemma 5.5, we are ready to consider the general case for $\alpha < F(y) < \beta$. Consider the function defined by

$$
\tilde{F} = \frac{F - \alpha}{\beta - F} \tag{5.14}
$$

and obviously we have $\tilde{F} > 0$. How to use this transformation for analysis of systems stability?

With $G(s) = c^T(sI - A)^{-1}b$, the characteristic equation of (5.12) can be written as

$$
G(s)F + 1 = 0. \tag{5.15}
$$

Manipulating (5.15) by adding and subtracting suitable terms, we have

$$
G(s)(F - \alpha) = -\alpha G - 1, \tag{5.16}
$$
$$
G(s)(\beta - F) = \beta G + 1. \tag{5.17}
$$

With (5.16) being divided by (5.17), we can obtain that

$$
\frac{1 + \beta G}{1 + \alpha G} \cdot \frac{F - \alpha}{\beta - F} + 1 = 0. \tag{5.18}
$$

Let

$$
\tilde{G} := \frac{1 + \beta G}{1 + \alpha G} \tag{5.19}
$$

and we can write (5.18) as

$$
\tilde{G}\tilde{F} + 1 = 0 \tag{5.20}
$$

which implies that the stability of the system (5.12) with the nonlinear gain shown in (5.13) is equivalent to the stability of the system with the forward transfer function $\tilde{G}$ and the feedback gain $\tilde{F}$. Based on Lemma 5.5 and (5.20), we can see that the system (5.12) is stable if $\tilde{G}$ is strictly positive real.

The expressions of $\tilde{F}$ in (5.14) and $\tilde{G}$ in (5.19) cannot deal with the case $\beta = \infty$. In such a case, we re-define

$$\tilde{F} = F - \alpha \tag{5.21}$$

which ensures that $\tilde{F} > 0$. With this $\tilde{F}$, we can obtain the manipulated characteristic equation as

$$\frac{G}{1 + \alpha G} \cdot (F - \alpha) + 1 = 0$$

which enables us to re-define

$$\tilde{G} := \frac{G}{1 + \alpha G}. \tag{5.22}$$

We summarise the results in the following theorem.

Theorem 5.6 (Circle criterion). *For the system (5.12) with the feedback gain satisfying the condition in (5.13), if the transfer function $\tilde{G}$ defined by*

$$\tilde{G}(s) = \frac{1 + \beta G(s)}{1 + \alpha G(s)}$$

or in case of $\beta = \infty$, by

$$\tilde{G}(s) = \frac{G(s)}{1 + \alpha G(s)}$$

is strictly positive real, with $G(s) = c^T(sI - A)^{-1}b$, the system is absolutely stable.

What is the condition of G if $\tilde{G}$ is strictly positive real? Let us assume that $\beta > \alpha > 0$. Other cases can be analysed similarly. From Lemma 5.3, we know that for $\tilde{G}$ to be strictly positive real, we need $\tilde{G}$ to be Hurwitz, and $\Re(\tilde{G}(j\omega)) > 0$, that is

$$\Re\left(\frac{1 + \beta G(j\omega)}{1 + \alpha G(j\omega)}\right) > 0, \quad \forall \omega \in \mathbb{R},$$

which is equivalent to

$$\Re\left(\frac{1/\beta + G(j\omega)}{1/\alpha + G(j\omega)}\right) > 0, \quad \forall \omega \in \mathbb{R}. \tag{5.23}$$

If $\frac{1}{\beta} + G(j\omega) = r_1 e^{j\theta_1}$ and $\frac{1}{\alpha} + G(j\omega) = r_2 e^{j\theta_2}$, the condition in (5.23) is satisfied by

$$-\frac{\pi}{2} < \theta_1 - \theta_2 < \frac{\pi}{2},$$

which is equivalent to the point $G(j\omega)$ that lies outside the circle centered at $(-\frac{1}{2}(1/\alpha + 1/\beta), 0)$ with radius of $\frac{1}{2}(1/\alpha - 1/\beta)$ in the complex plane. This circle intersects the real axis at $(-\frac{1}{\alpha}, 0)$ and $(-\frac{1}{\beta}, 0)$. Indeed, $\frac{1}{\beta} + G(j\omega)$ is represented as a vector from the point $(-\frac{1}{\beta}, 0)$ to $G(j\omega)$, and $\frac{1}{\alpha} + G(j\omega)$ as a vector from the point $(-\frac{1}{\alpha}, 0)$ to $G(j\omega)$. The angle between the two vectors will be less than $\frac{\pi}{2}$ when $G(j\omega)$ is outside the circle, as shown in Figure 5.3. Since the condition must hold for all $\omega \in \mathbb{R}$, the condition $\Re(\tilde{G}(j\omega)) > 0$ is equivalent to the Nyquist plot of $G(s)$ that lies outside the circle. The condition that $\tilde{G}$ is strictly positive real requires that the Nyquist plot of $G(s)$ does not intersect with the circle and encircles the circle counterclockwise the same number of times as the number of unstable poles of $G(s)$, as illustrated in Figure 5.4.

Alternatively, the circle can also be interpreted from complex mapping. From (5.19), it can be obtained that

$$G = \frac{\tilde{G} - 1}{\beta - \alpha \tilde{G}}. \tag{5.24}$$

The mapping shown in (5.24) is a bilinear transformation, and it maps a line to a line or circle. For the case of $\beta > \alpha > 0$, we have

$$G = -\frac{1}{\alpha} - \left(\frac{1}{\alpha} - \frac{1}{\beta}\right) \frac{\beta/\alpha}{\tilde{G} - \beta/\alpha}. \tag{5.25}$$

The function

$$\frac{\beta/\alpha}{\tilde{G} - \beta/\alpha}$$

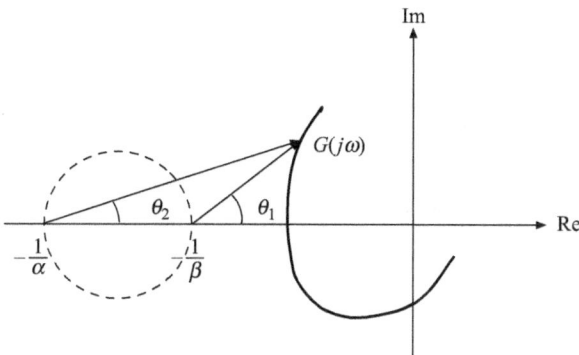

Figure 5.3   *Diagram of* $|\theta_1 - \theta_2| < \frac{\pi}{2}$

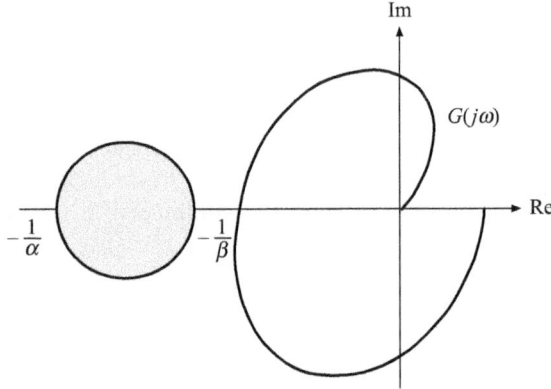

Figure 5.4   Circle criterion

maps the imaginary axis to a circle centred as $(-1/2, 0)$ with the radius $1/2$, i.e., the line from $(-1, 0)$ to $(0, 0)$ on the complex plane is the diameter of the circle. Then the function

$$-\left(\frac{1}{\alpha} - \frac{1}{\beta}\right) \frac{\beta/\alpha}{\tilde{G} - \beta/\alpha},$$

maps the imaginary axis to a circle with the diameter on the line from $(0, 0)$ to $(\frac{1}{\alpha} - \frac{1}{\beta}, 0)$ on the complex plane. Finally, it can be seen from (5.25) that this map of $\tilde{G}$ to G maps the imaginary axis to the circle with the diameter on the line from $(-\frac{1}{\alpha}, 0)$ to $(-\frac{1}{\beta}, 0)$, or in other words, the circle centered as $(-\frac{1}{2}(1/\alpha + 1/\beta), 0)$ with radius of $\frac{1}{2}(1/\alpha - 1/\beta)$. It can also be shown that the function maps the open left-hand complex plane to the domain inside the circle.

Indeed, we can evaluate the circle directly from (5.24). Let u and v denote the real and imaginary parts of the mapping of the imaginary axis, and we have

$$u = \Re \frac{j\omega - 1}{\beta - \alpha j\omega} = -\frac{\alpha + \beta\omega^2}{\alpha^2 + \beta^2\omega^2},$$

$$v = \Im \frac{j\omega - 1}{\beta - \alpha j\omega} = \frac{(\alpha - \beta)\omega}{\alpha^2 + \beta^2\omega^2}.$$

Denoting $\mu = \frac{\beta}{\alpha}\omega$, we obtain

$$u = -\frac{1/\alpha + (1/\beta)\mu^2}{1 + \mu^2} = -\frac{1}{2}\left(\frac{1}{\alpha} + \frac{1}{\beta}\right) + \frac{1}{2}\left(\frac{1}{\beta} - \frac{1}{\alpha}\right)\frac{1 - \mu^2}{1 + \mu^2},$$

$$v = \frac{(1/\beta - 1/\alpha)\mu}{1 + \mu^2} = \frac{1}{2}\left(\frac{1}{\beta} - \frac{1}{\alpha}\right)\frac{2\mu}{1 + \mu^2}.$$

It is now easy to see that

$$\left(u + \frac{1}{2}\left(\frac{1}{\alpha} + \frac{1}{\beta}\right)\right)^2 + v^2 = \left(\frac{1}{2}\left(\frac{1}{\alpha} - \frac{1}{\beta}\right)\right)^2.$$

which describes the circle that is discussed before.

5.3 Input-to-state stability and small gain theorem

In this section, we continue to consider the stability properties for systems with inputs. For linear systems, if a system is asymptotically stable, the state will remain bounded when the input is bounded. Does this property still hold for nonlinear systems?

Consider a nonlinear system

$$\dot{x} = f(x, u) \tag{5.26}$$

where $x \in \mathbb{R}^n$ is the state of the system, and $u \in \mathbb{R}^m$ is a bounded input, and $f : \mathbb{R}^n \times \mathbb{R}^m \longrightarrow \mathbb{R}^n$ is a continuous and locally Lipschitz function. If the autonomous system

$$\dot{x} = f(x, 0)$$

is asymptotically stable, will the state remain bounded for a bounded input signal u?

Example 5.3. Consider a nonlinear system

$$\dot{x} = -x + (1 + 2x)u, \quad x(0) = 0$$

where $x \in \mathbb{R}$. When $u = 0$, we have

$$\dot{x} = -x$$

which is asymptotically (exponentially) stable. However, the state of this system may not remain bounded for a bounded input. For example if we let $u = 1$, we have

$$\dot{x} = x + 1$$

of which the state variable x is unbounded. ◁

From the above example, it can be seen that even the corresponding autonomous system is asymptotically stable, the state may not remain bounded subject to a bounded input. We introduce a definition for systems with the property of bounded state with bounded input.

To show a definition of input-to-state stable (ISS), we need to use comparison functions, which are defined below.

Definition 5.3. A function $\gamma : [0, a) \rightarrow [0, \infty)$ is a class $\mathcal{K}$ function if γ is continuous, and strictly increasing with $\gamma(0) = 0$. If $a = \infty$ and $\lim_{r \rightarrow \infty} \gamma(r) = \infty$, the function is a class $\mathcal{K}_\infty$ function.

Definition 5.4. A function $\beta : [0, a) \times [0, \infty) \rightarrow [0, \infty)$ is a class $\mathcal{KL}$ function if it is continuous, for a fixed $t = t_0$, $\beta(\cdot, t_0)$ is a class $\mathcal{K}$ function, and for a fixed $x(0)$, $\lim_{t \rightarrow \infty} \beta(x(0), t) = 0$.

Definition 5.5. The system (5.26) is ISS if there exist a class $\mathcal{KL}$ function β and a class $\mathcal{K}$ function γ such that

$$\|x(t)\| \leq \beta(\|x(0)\|, t) + \gamma(\|u\|_\infty), \quad \forall t > 0. \tag{5.27}$$

There is an alternative definition, which may be convenient to use in some situations. Here, we state it as a proposition.

Proposition 5.7. *The system shown in (5.26) is ISS if and only if there exist a class $\mathcal{KL}$ function β and a class $\mathcal{K}$ function γ such that*

$$\|x(t)\| \leq \max\{\beta(\|x(0)\|, t), \gamma(\|u\|_\infty)\}, \quad \forall t > 0. \tag{5.28}$$

Proof. From (5.28), we have

$$\|x(t)\| \leq \beta(\|x(0)\|, t) + \gamma(\|u\|_\infty)$$

and hence the system is ISS. From (5.27), there exist a class $\mathcal{KL}$ function β_1 and a class $\mathcal{K}$ function γ_1

$$\begin{aligned} \|x(t)\| &\leq \beta_1(\|x(0)\|, t) + \gamma_1(\|u\|_\infty) \\ &\leq \max\{2\beta_1(\|x(0)\|, t), \ 2\gamma_1(\|u\|_\infty)\} \end{aligned}$$

and therefore the system satisfies (5.28) with $\beta = 2\beta_1$ and $\gamma = 2\gamma_1$. $\qquad\square$

Example 5.4. Consider a linear system

$$\dot{x} = Ax + Bu,$$

where $x \in \mathbb{R}^n$ and $u \in \mathbb{R}^m$ are the state and input respectively, and A and B are matrices with appropriate dimensions and A is Hurwitz. For this system, when $u = 0$, the system is asymptotically stable, as A is Hurwitz. From its solution

$$x(t) = e^{At}x(0) + \int_0^t e^{A(t-\tau)}Bu(\tau)d\tau,$$

we have

$$\|x(t)\| \le \|e^{At}\| \|x(0)\| + \int_0^t \|e^{A(t-\tau)}\| d\tau \|B\| \|u\|_\infty.$$

Since A is Hurwitz, there exist positive real constants a and λ such that $\|e^{A(t)}\| \le ae^{-\lambda t}$. Hence, we can obtain

$$\|x(t)\| \le ae^{-\lambda t} \|x(0)\| + \int_0^t ae^{-\lambda(t-\tau)} d\tau \|B\| \|u\|_\infty$$

$$\le ae^{-\lambda t} \|x(0)\| + \frac{a}{\lambda} \|B\| \|u\|_\infty$$

It is easy to see that the first term in the above expression is a $\mathcal{KL}$ function of t and $\|x(0)\|$ and the second term is a $\mathcal{K}$ function of $\|u\|_\infty$. Therefore, the linear system with Hurwitz system matrix A is ISS. ◁

Next, we show a result on establishing ISS property from a Lyapunov function.

Theorem 5.8. *For the system (5.26), if there exists a function $V(x) : \mathbb{R}^n \to \mathbb{R}$ with continuous first-order derivatives such that*

$$a_1 \|x\|^b \le V(x) \le a_2 \|x\|^b, \tag{5.29}$$

$$\frac{\partial V}{\partial x} f(x, u) \le -a_3 \|x\|^b, \quad \forall \|x\| \ge \rho(\|u\|) \tag{5.30}$$

where a_1, a_2, a_3 and b are positive real constants and ρ is a class $\mathcal{K}$ function, the system (5.26) is ISS.

Proof. From Theorem 4.4, we can see that the system is exponentially stable when $u = 0$, and $V(x)$ is a Lyapunov function for the autonomous system $\dot{x} = f(x, 0)$. To consider the case for non-zero input, let us find a level set based on V, defined by

$$\Omega_c := \{x | V(x) \le c\}$$

and determine the constant c such that for $x \notin \Omega_c$, we have $\|x\| > \rho(\|u\|_\infty)$. Indeed, let

$$c = a_2(\rho(\|u\|_\infty))^b.$$

In this case, from $V(x) > c$, we have

$$V(x) > a_2(\rho(\|u\|_\infty))^b,$$

which implies that

$$a_2 \|x\|^b > a_2(\rho(\|u\|_\infty))^b$$

and hence $\|x\| > \rho(\|u\|_\infty)$.

Therefore, for any x outside Ω_c, we can obtain from (5.29) and (5.30), in a similar way to the proof of Theorem 4.4,

$$\dot{V} \leq -\frac{a_3}{a_2}V$$

and then

$$\|x(t)\| \leq \left(\frac{a_2}{a_1}\right)^{1/b} \|x(0)\| e^{-\frac{a_3}{a_2 b}t}.$$

There will be a time, say t_1, at which the trajectory enters Ω_c.

We can show that Ω_c is invariant, i.e., the trajectory starts from Ω_c will remain in Ω_c, and this is shown by the fact that on the boundary of Ω_c, $\dot{V} \leq 0$. Therefore, after t_1, the trajectory will remain in Ω_c. Furthermore, for $x \in \Omega_c$, we have,

$$V(x) < a_2(\rho(\|u\|_\infty))^b$$

which implies that

$$\|x\| \leq \left(\frac{a_2}{a_1}\right)^{1/b} \rho(\|u\|_\infty).$$

Combining the cases for x outside and in Ω_c, we conclude that

$$\|x\| \leq \max\left\{\left(\frac{a_2}{a_1}\right)^{1/b}\|x(0)\|e^{-\frac{a_3}{a_2 b}t}, \left(\frac{a_2}{a_1}\right)^{1/b}\rho(\|u\|_\infty)\right\}.$$

Hence, the system is ISS with the gain function $\gamma(\cdot) = (\frac{a_2}{a_1})^{1/b}\rho(\cdot)$.

There is a more general result than Theorem 5.8 that requires $\dot{x} = f(x, 0)$ to be asymptotically stable, not necessarily exponential stable. The proof of that theorem is beyond the level of this text, and we include it here for completeness.

Theorem 5.9. *For the system (5.26), if there exists a function $V(x) : \mathbb{R}^n \to \mathbb{R}$ with continuous first-order derivatives such that*

$$\alpha_1(\|x\|) \leq V(x) \leq \alpha_2(\|x\|), \tag{5.31}$$

$$\frac{\partial V}{\partial x}f(x, u) \leq -\alpha_3(\|x\|), \quad \forall \|x\| \geq \rho(\|u\|), \tag{5.32}$$

where α_1, α_2, α_3 are $\mathcal{K}_\infty$ functions and ρ is a class $\mathcal{K}$ function, the system (5.26) is ISS with the gain function $\gamma(\cdot) = \alpha_1^{-1}(\alpha_2(\rho(\cdot)))$.

Note that class $\mathcal{K}_\infty$ functions are class $\mathcal{K}$ functions that satisfy the property

$$\lim_{r \to \infty} \alpha(r) = \infty.$$

There is a slightly different version of Theorem 5.9 which is shown below.

Corollary 5.10. *For the system (5.26), if there exists a function $V(x) : \mathbb{R}^n \to \mathbb{R}$ with continuous first-order derivatives such that*

$$\alpha_1(\|x\|) \leq V(x) \leq \alpha_2(\|x\|), \tag{5.33}$$

$$\frac{\partial V}{\partial x} f(x) \leq -\alpha(\|x\|) + \sigma(\|u\|), \tag{5.34}$$

where α_1, α_2, α are $\mathcal{K}_\infty$ functions and σ is a class $\mathcal{K}$ function, the system (5.26) is ISS.

The function which satisfies (5.31) and (5.32) or (5.33) and (5.34) is referred to as ISS-Lyapunov function. In fact, it can be shown that the existence of an ISS-Lyapunov function is also a necessary condition for the system to be ISS. Referring to (5.34), the gain functions α and σ characterise the ISS property of the system, and they are also referred to as an ISS pair. In other words, if we say a system is ISS with ISS pair (α, σ), we mean that there exists an ISS-Lyapunov function that satisfies (5.34).

Example 5.5. Consider a nonlinear system

$$\dot{x} = -x^3 + u,$$

where $x \in \mathbb{R}$ is the state, and u is input. The autonomous part $\dot{x} = -x^3$ is considered in Example 4.2, and it is asymptotically stable, but not exponentially stable. Consider an ISS-Lyapunov function candidate

$$V = \frac{1}{2}x^2.$$

Its derivative is given by

$$\begin{aligned}
\dot{V} &= -x^4 - xu \\
&\leq -\frac{1}{2}x^4 - \frac{1}{2}|x|(|x|^3 - 2|u|) \\
&\leq -\frac{1}{2}x^4, \quad \text{for}|x| \geq (2|u|)^{1/3}.
\end{aligned}$$

Hence the system is ISS with the gain function $\rho(|u|) = (2|u|)^{1/3}$, based on Theorem 5.9. Alternatively, using Young's inequality, we have

$$|x||u| \leq \frac{1}{4}|x|^4 + \frac{3}{4}|u|^{4/3}$$

which gives

$$\dot{V} \leq -\frac{3}{4}x^4 + \frac{3}{4}|u|^{4/3}.$$

Therefore, the system is ISS based on Corollary 5.10 with $\alpha(\cdot) = \frac{3}{4}(\cdot)^4$ and $\sigma(\cdot) = \frac{3}{4}(\cdot)^{4/3}$.   ◁

ISS property is useful in establishing the stability of interconnected systems. We include two results here to end this section.

Theorem 5.11. *If for the cascade connected system*

$$\dot{x}_1 = f_1(x_1, x_2), \tag{5.35}$$

$$\dot{x}_2 = f_1(x_2, u), \tag{5.36}$$

the subsystem (5.35) is ISS with x_2 as the input, and the subsystem (5.36) is ISS with u as input, the overall system with state $x = [x_1^T, x_2^T]^T$ and input u is ISS.

Theorem 5.12 (ISS small gain theorem). *If for the interconnected system*

$$\dot{x}_1 = f_1(x_1, x_2), \tag{5.37}$$

$$\dot{x}_2 = f_1(x_2, x_2, u), \tag{5.38}$$

the subsystem (5.37) is ISS with x_2 as the input with γ_1 as the ISS gain for x_2, and the subsystem (5.38) is ISS by viewing x_1 and u as the inputs, with the ISS input gain function γ_2 for x_1, the overall system with state $x = [x_1^T, x_2^T]^T$ and input u is ISS if

$$\gamma_1(\gamma_2(r)) < r, \quad \forall\, r > 0. \tag{5.39}$$

From Theorem 5.11, it can be seen that if the subsystem x_2 is globally asymptotically stable when $u = 0$, the overall system is globally asymptotically stable. Similarly, Theorem 5.12 can be used to establish the stability of the following system:

$$\dot{x}_1 = f_1(x_1, x_2)$$

$$\dot{x}_2 = f_1(x_2, x_2),$$

and the global and asymptotic stability of the entire system can be concluded if the gain condition shown in Theorem 5.12 is satisfied.

Example 5.6. Consider the second-order nonlinear system

$$\dot{x}_1 = -x_1^3 + x_2,$$

$$\dot{x}_2 = x_1 x_2^{2/3} - 3x_2.$$

From Example 5.5, we know that x_1-subsystem is ISS with the gain function $\gamma_1(\cdot) = (2\cdot)^{1/3}$. For the x_2-subsystem, we choose

$$V_2 = \frac{1}{2}x_2^2$$

and we have

$$
\begin{aligned}
\dot{V}_2 &= -3x_2^2 + x_1 x_2^{5/3} \\
&\leq -x_2^2 - |x_2|^{5/3}(2|x_2|^{1/3} - |x_1|) \\
&\leq -x_2^2, \quad \text{for } |x_2| > \left(\frac{|x_1|}{2}\right)^3
\end{aligned}
$$

Hence, the x_2-subsystem is ISS with the gain function $\gamma_2(\cdot) = (\frac{\cdot}{2})^3$. Now we have, for $r > 0$,

$$\gamma_1(\gamma_2(r)) = \left(2\left(\frac{r}{2}\right)^3\right)^{1/3} = \left(\frac{1}{4}\right)^{1/3} r < r.$$

Therefore, the system is globally asymptotically stable. ◁

5.4 Differential stability

Consider a nonlinear system

$$\dot{x} = f(x, u), \tag{5.40}$$

where $x \in \mathbb{R}^n$ is the state vector, $u \in \mathbb{R}^s$ is the input and $f : \mathbb{R}^n \times \mathbb{R}^s \to \mathbb{R}^n$ is a nonlinear smooth vector field with $f(0, u) = 0$.

Definition 5.6. This system (5.40) has differential stability if there exists a Lyapunov function $V(x)$ such that $V : \mathbb{R}^n \to \mathbb{R}$ for all $x, \hat{x} \in \mathbb{R}^n$, $u \in \mathbb{R}^s$, satisfies

$$
\begin{aligned}
\gamma_1(\|x\|) &\leq V(x) \leq \gamma_2(\|x\|), \\
\frac{\partial V(x - \hat{x})}{\partial x}(f(x, u) - f(\hat{x}, u)) &\leq -\gamma_3(\|x - \hat{x}\|), \\
c_1 \left\|\frac{\partial V(x)}{\partial x}\right\|^{c_2} &\leq \gamma_3(\|x\|),
\end{aligned}
\tag{5.41}
$$

where $\gamma_i, i = 1, 2, 3$, are $\mathcal{K}_\infty$ functions and $c_i, i = 1, 2$, are positive real constants with $c_2 > 1$.

Remark 5.2. The conditions specified in (5.41) are useful for observer design, in particular, for the stability analysis of the reduced-order observers in Chapter 8. A similar definition to differential stability is incremental stability. However, the conditions specified in (5.41) are not always satisfied by the systems with incremental stability. When $\hat{x} = 0$, i.e., in the case for one system only, the conditions specified in (5.41) are then similar to the properties of the nonlinear systems with exponential stability. The last condition in (5.41) is specified for interactions with other systems. This condition is similar to the conditions for the existence of changing the supply functions for inter-connection of ISS systems.   ◁

We include two illustrative examples below for the properties of differential stability.

Example 5.7. A linear system is differentially stable if the system is asymptotically stable. Consider

$$\dot{x} = Ax,$$

where $A \in \mathbb{R}^{n \times n}$. If the system is asymptotically stable, A must be Hurwitz. Therefore there exist positive definite matrices P and Q such that

$$A^T P + PA = -Q.$$

Let $V(x) = x^T Px$. In this case, the conditions (5.41) are satisfied with

$$\gamma_1(\|x\|) = \lambda_{min}(P)\|x\|^2,$$
$$\gamma_2(\|x\|) = \lambda_{max}(P)\|x\|^2,$$
$$\gamma_3(\|x\|) = \lambda_{min}(Q)\|x\|^2,$$
$$c_1 = \frac{\lambda_{min}(Q)}{4(\lambda_{max}(P))^2}, \quad c_2 = 2,$$

where $\lambda_{min}(\cdot)$ and $\lambda_{max}(\cdot)$ denote the minimum and maximum eigenvalues of a positive definite matrix.   ◁

The differential stability is closely related to observer design. Consider a system with input $u \in \mathbb{R}^s$

$$\dot{x} = Ax + Bu. \tag{5.42}$$

If $\dot{x} = Ax$ is differentially stable, an observer for (5.42) can be designed as

$$\dot{\hat{x}} = A\hat{x} + Bu, \quad \hat{x}(0) = 0. \tag{5.43}$$

It is easy to see that $x - \hat{x}$ converges to 0 exponentially.

However, for nonlinear systems, differential stability is not guaranteed by the asymptotic or even exponential stability of a system.

Example 5.8. We consider a first-order nonlinear system

$$\dot{x} = -x - 2\sin x.$$

Take $V = \frac{1}{2}x^2$, and we have

$$\dot{V} = -x^2 - 2x\sin x.$$

For $|x| \leq \pi$, we have $x\sin x \geq 0$, and therefore

$$\dot{V} \leq -x^2.$$

For $|x| > \pi$, we have

$$
\begin{aligned}
\dot{V} &\leq -x^2 - 2x\sin x \\
&\leq -x^2 + 2|x| \\
&= -\left(1 - \frac{2}{\pi}\right)x^2 - 2|x|\left(\frac{|x|}{\pi} - 1\right) \\
&\leq -\left(1 - \frac{2}{\pi}\right)x^2.
\end{aligned}
$$

Combining both the cases, we have

$$\dot{V} \leq -\left(1 - \frac{2}{\pi}\right)x^2.$$

Hence, the system is exponentially stable. But this system is not differentially stable. Indeed, let $e = x - \hat{x}$. We have

$$\dot{e} = -e - 2(\sin(x) - \sin(x + e)).$$

By linearising the system at $x = \pi$ and $\hat{x} = \pi$, and denoting the error at this point by e_l, we have

$$
\begin{aligned}
\dot{e}_l &= -e_l - 2\cos(x)|_{x=\pi}(x - \pi) + 2\cos(\hat{x})|_{\hat{x}=\pi}(\hat{x} - \pi) \\
&= e_l
\end{aligned}
$$

and the system is unstable in a neighbourhood of this point. ◁

Chapter 6

Feedback linearisation

Nonlinear systems can be linearised around operating points and the behaviours in the neighbourhoods of the operating points are then approximated by their linearised models. The domain for a locally linearised model can be fairly small, and this may result in that a number of linearised models are needed to cover an operating range of a system. In this chapter, we will introduce another method to obtain a linear model for nonlinear systems via feedback control design. The aim is to convert a nonlinear system to a linear one by state transformation and redefining the control input. The resultant linear model describes the system dynamics globally. Of course, there are certain conditions for the nonlinear systems to satisfy so that this feedback linearisation method can be applied.

6.1 Input–output linearisation

The basic idea for input–output linearisation is fairly straightforward. We use an example to demonstrate this.

Example 6.1. Consider the nonlinear system

$$\dot{x}_1 = x_2 + x_1^3$$
$$\dot{x}_2 = x_1^2 + u \qquad\qquad (6.1)$$
$$y = x_1.$$

For this system, taking the derivatives of y, we have

$$\dot{y} = x_2 + x_1^3,$$
$$\ddot{y} = 3x_1^2(x_2 + x_1^3) + x_1^2 + u.$$

Now, let us define

$$v = 3x_1^2(x_2 + x_1^3) + x_1^2 + u,$$

and we obtain

$$\ddot{y} = v.$$

Viewing v as the new control input, we see that the system is linearised. Indeed, let us introduce a state transformation

$$\xi_1 := y = x_1,$$
$$\xi_2 := \dot{y} = x_2 + x_1^3.$$

We then obtain a linear system

$$\dot{\xi}_1 = \xi_2$$
$$\dot{\xi}_2 = v.$$

We can design a state feedback law as

$$v = -a_1\xi_1 - a_2\xi_2$$

to stabilise the system with $a_1 > 0$ and $a_2 > 0$. The control input of the original system is given by

$$u = -3x_1^2(x_2 - x_1^3) - x_1^2 + v. \tag{6.2}$$

We say that the system (6.1) is linearised by the feedback control law (6.2). Notice that this linearisation works for the entire state space.   ◁

As shown in the previous example, we can keep taking the derivatives of the output y until the input u appears in the derivative, and then a feedback linearisation law can be introduced. The derivatives of the output also introduce a natural state transformation.

Consider a nonlinear system

$$\dot{x} = f(x) + g(x)u$$
$$y = h(x), \tag{6.3}$$

where $x \in \mathcal{D} \subset \mathbb{R}^n$ is the state of the system; y and $u \in \mathbb{R}$ are output and input respectively; and f and $g : \mathcal{D} \subset \mathbb{R}^n \rightarrow \mathbb{R}^n$ are smooth functions and $h : \mathcal{D} \subset \mathbb{R}^n \rightarrow \mathbb{R}$ is a smooth function.

Remark 6.1. The functions $f(x)$ and $g(x)$ are vectors for a given point x in the state space, and they are often referred to as vector fields. All the functions in (6.3) are required to be smooth in the sense that they have continuous derivatives up to certain orders when required. We use the smoothness of functions in the remaining part of the chapter in this way.   ◁

The input–output feedback linearisation problem is to design a feedback control law

$$u = \alpha(x) + \beta(x)v \tag{6.4}$$

with $\beta(x) \neq 0$ for $x \in \mathcal{D}$ such that the input–output dynamics of the system

$$\dot{x} = f(x) + g(x)\alpha(x) + g(x)\beta(x)v$$
$$y = h(x)$$

(6.5)

are described by

$$y^{(\rho)} = v$$

(6.6)

for $1 \le \rho \le n$.

For the system (6.3), the first-order derivative of the output y is given by

$$\dot{y} = \frac{\partial h(x)}{\partial x}(f(x) + g(x)u)$$
$$:= L_f h(x) + L_g h(x)u,$$

where the notations $L_f h$ and $L_g h(x)$ are Lie derivatives.

For any smooth function $f : \mathcal{D} \subset \mathbb{R}^n \to \mathbb{R}^n$ and a smooth function $h : \mathcal{D} \subset \mathbb{R}^n \to \mathbb{R}$, the Lie derivative $L_f h$, referred to as the derivative of h along f, is defined by

$$L_f h(x) = \frac{\partial h(x)}{\partial x} f(x).$$

This notation can be used iteratively, that is

$$L_f(L_f h(x)) = L_f^2 h(x),$$
$$L_f^k(h(x)) = L_f(L_f^{k-1}(h(x))),$$

where $k \ge 0$ is an integer.

The solution to this problem depends on the appearance of the control input in the derivatives of the output, which is described by the relative degree of the dynamic system.

Definition 6.1. The dynamic system (6.3) has relative degree ρ at a point x if the following conditions are satisfied:

$$L_g L_f^k h(x) = 0, \quad \text{for } k = 0, \dots, \rho - 2,$$
$$L_g L_f^{\rho-1} h(x) \ne 0.$$

Example 6.2. Consider the system (6.1). Comparing it with the format shown in (6.3), we have

$$f(x) = \begin{bmatrix} x_1^3 + x_2 \\ x_1^2 \end{bmatrix}, \quad g(x) = \begin{bmatrix} 0 \\ 1 \end{bmatrix}, \quad h(x) = x_1.$$

Direct evaluation gives

$$L_g h(x) = 0,$$
$$L_f L_g h(x) = 1.$$

Therefore, the relative degree of the system (6.1) is 2.   ◁

For SISO linear systems, the relative degree is the difference between the orders of the polynomials in the numerator and denominator of the transfer function. With the definition of the relative degree, we can present the input–output feedback linearisation using Lie derivatives.

Example 6.3. Consider the system (6.1) again, and continue from Example 6.2. With $L_g h(x) = 0$, we have

$$\dot{y} = L_f h(x)$$

where

$$L_f h(x) = x_1^3 + x_2.$$

Taking the derivative of $L_f h(x)$, we have

$$\ddot{y} = L_f^2 h(x) + L_g L_f h(x) u$$

where

$$L_f^2 h(x) = 3x_1^2(x_2 + x_1^3) + x_1^2,$$
$$L_g L_f h(x) = 1.$$

Therefore, we have

$$v = L_f^2 h(x) + L_g L_f h(x) u$$

or

$$u = -\frac{L_f^2 h(x)}{L_g L_f h(x)} + \frac{1}{L_g L_f h(x)} v,$$

which gives the same result as in Example 6.1.   ◁

The procedure shown in Example 6.3 works for systems with any relative degrees. Suppose that the relative degree for (6.3) is ρ, which implies that $L_g L_f^k h(x) = 0$ for $k = 0, \ldots, \rho - 2$. Therefore, we have the derivatives of y expressed by

$$y^{(k)} = L_f^k h(x), \quad \text{for } k = 0, \ldots, \rho - 1, \tag{6.7}$$
$$y^{(\rho)} = L_f^\rho h(x) + L_g L_f^{\rho-1} h(x) u. \tag{6.8}$$

If we define the input as

$$u = \frac{1}{L_g L_f^{\rho-1} h(x)} (-L_f^\rho h(x) + v)$$

it results in

$$y^{(\rho)} = v.$$

We can consider to use $\xi_i := y^{(i-1)} = L_f^{i-1} h$ as coordinates for the linearised input–output dynamics. The only remaining issue is to establish that $\frac{\partial \xi}{\partial x}$ has full rank. To do that, we need to introduce a few notations.

For any smooth functions $f, g : \mathcal{D} \subset \mathbb{R}^n \to \mathbb{R}^n$, the Lie bracket $[f, g]$ is defined by

$$[f, g](x) = \frac{\partial g(x)}{\partial x} f(x) - \frac{\partial g(x)}{\partial x} f(x)$$

and we can introduce a notation which is more convenient for high-order Lie brackets as

$$ad_f^0 g(x) = g(x),$$
$$ad_f^1 g(x) = [f, g](x),$$
$$ad_f^k g(x) = [f, ad_f^{k-1} g](x).$$

For the convenient of presentation, let us denote

$$dh = \frac{\partial h}{\partial x}$$

which is a row vector. With this notation, we can write

$$L_f h = \ <dh, f>.$$

Based on the definition of Lie Bracket, it can be obtained by a direct evaluation that, for a smooth function $h : \mathbb{R}^n \to \mathbb{R}$,

$$L_{[f,g]} h = L_f L_g h - L_g L_f h = L_f <dh, g> - <dL_f h, g>,$$

that is

$$<dh, [f, g]> = L_f <dh, g> - <dL_f h, g>.$$

Similarly it can be obtained that, for any non-negative integers k and l,

$$<dL_f^k h, ad_f^{l+1} g> = L_f <dL_f^k h, ad_f^l g> - <dL_f^{k+1} h, ad_f^l g> \qquad (6.9)$$

By now, we have enough tools and notations to show that $\frac{\partial \xi}{\partial x}$ has full rank. From the definition of the relative degree,

$$< dL_f^k h, g > = 0, \quad \text{for } k = 0, \ldots, \rho - 2$$

and by using (6.9) iteratively, we can show that

$$< dL_f^k h, ad_f^l g > = 0, \quad \text{for } k + l \leq \rho - 2, \tag{6.10}$$

and

$$< dL_f^k h, ad_f^l g > = (-1)^l < dL_f^{\rho-1} h, g >, \quad \text{for } k + l = \rho - 1. \tag{6.11}$$

From (6.10) and (6.11), we have

$$\begin{bmatrix} dh(x) \\ dL_f h(x) \\ \vdots \\ dL_f^{\rho-1} h \end{bmatrix} \begin{bmatrix} g(x) & ad_f g(x) & \cdots & ad_f^{\rho-1} g(x) \end{bmatrix}$$

$$= \begin{bmatrix} 0 & \cdots & 0 & (-1)^{\rho-1} r(x) \\ 0 & \cdots & (-1)^{\rho-2} r(x) & * \\ \vdots & \vdots & \vdots & \vdots \\ r(x) & \cdots & * & * \end{bmatrix} \tag{6.12}$$

where $r(x) = < dL_f^{\rho-1} h, g >$. Therefore, we conclude that

$$\frac{\partial \xi}{\partial x} = \begin{bmatrix} dh(x) \\ dL_f h(x) \\ \vdots \\ dL_f^{\rho-1} h \end{bmatrix}$$

has full rank. We summarise the result about input–output feedback linearisation in the following theorem.

Theorem 6.1. *If the system in (6.3) has a well-defined relative degree ρ in $\mathcal{D}$, the input–output dynamics of the system can be linearised by the feedback control law*

$$u = \frac{1}{L_g L_f^{\rho-1} h(x)} (-L_f^\rho h(x) + v) \tag{6.13}$$

and the linearised input–output dynamics are described by

$$\dot{\xi}_1 = \xi_2$$

$$\vdots$$

$$\dot{\xi}_{\rho-1} = \xi_\rho \qquad (6.14)$$

$$\dot{\xi}_\rho = v$$

with a partial state transformation

$$\xi_i = L_f^{i-1} h(x) \quad for \ i = 1, \ldots, \rho. \qquad (6.15)$$

When $\rho = n$, the whole nonlinear system dynamics are fully linearised.

The results shown in (6.10) and (6.11) can also be used to conclude the following result which is needed in the next section.

Lemma 6.2. *For any functions $f, g : D \subset \mathbb{R}^n \to \mathbb{R}^n$, and $h : D \subset \mathbb{R}^n \to \mathbb{R}$, all differentiable to certain orders, the following two statements are equivalent for $r > 0$:*

- $L_g h(x) = L_g L_f h(x) = \cdots = L_g L_f^r h(x) = 0,$
- $L_g h(x) = L_{[f,g]} h(x) = \cdots = L_{ad_f^r g} h(x) = 0.$

Remark 6.2. The input–output dynamics can be linearised based on Theorem 6.1. In the case of $\rho < n$, the system for $\rho < n$ can be transformed under certain conditions to the normal form

$$\dot{z} = f_0(z, \xi),$$

$$\dot{\xi}_1 = \xi_2,$$

$$\vdots$$

$$\dot{\xi}_{\rho-1} = \xi_\rho,$$

$$\dot{\xi}_\rho = L_f^\rho h + u L_g L_f^{\rho-1} h,$$

$$y = \xi_1$$

where $z \in \mathbb{R}^{n-\rho}$ is the part of the state variables which are not in the input–output dynamics of the system, and $f_0 : \mathbb{R}^n \to \mathbb{R}^{n-\rho}$ is a smooth function. It is clear that when $\rho < n$, the input–output linearisation does not linearise the dynamics $\dot{z} = f_0(z, \xi)$. Also note that the dynamics $\dot{z} = f_0(z, 0)$ are referred to as the zero dynamics of the system. ◁

To conclude this section, we use an example to demonstrate the input–output linearisation for the case $\rho < n$.

Example 6.4. Consider the nonlinear system

$$\dot{x}_1 = x_1^3 + x_2$$
$$\dot{x}_1 = x_1^2 + x_3 + u$$
$$\dot{x}_2 = x_1^2 + u$$
$$y = x_1.$$

For this system we have

$$f(x) = \begin{bmatrix} x_1^3 + x_2 \\ x_1^2 + x_3 \\ x_1^2 \end{bmatrix}, \quad g(x) = \begin{bmatrix} 0 \\ 1 \\ 1 \end{bmatrix}, \quad h(x) = x_1.$$

It is easy to check that

$$L_f h = x_1^3 + x_2,$$
$$L_g h = 0,$$
$$L_g L_f h = 1,$$
$$L_f^2 h = 3x_1^2(x_1^3 + x_2) + x_1^2 + x_3$$

and therefore the relative degree $\rho = 2$. For input–output linearisation, we set

$$u = \frac{1}{L_g L_f h}(-L_f^2 h + v)$$
$$= -3x_1^2(x_1^3 + x_2) - x_1^2 - x_3 + v.$$

Introduce the partial state transformation

$$\xi_1 = h = x_1,$$
$$\xi_2 = L_f h = x_1^3 + x_2,$$

and it is easy to see that dh and $dL_f h$ are linearly independent. The linearised input–output dynamics are described by

$$\dot{\xi}_1 = \xi_2,$$
$$\dot{\xi}_2 = v.$$

If we like to transform the system to the normal form shown in Remark 6.2, we need to introduce another state, in addition to ξ_1 and ξ_2. For this, we have the additional state

$$z = x_3 - x_2.$$

The inverse transformation is given by

$$x_1 = \xi_1,$$
$$x_2 = \xi_2 - \xi_1^3,$$
$$x_3 = z + \xi_2 - \xi_1^3.$$

With the coordinates z, ξ_1 and ξ_2, we have the system in the normal form

$$\dot{z} = -z - \xi_2 + \xi_1^3$$
$$\dot{\xi}_1 = \xi_2$$
$$\dot{\xi}_2 = z + \xi_2 + \xi_1^2 - \xi_1^3 + 3\xi_1^2\xi_2 + u.$$

It is clear that the input–output linearisation does not linearise the dynamics of z. Also note that the zero dynamics for this system are described by

$$\dot{z} = -z.$$

◁

6.2 Full-state feedback linearisation

Consider a nonlinear system

$$\dot{x} = f(x) + g(x)u, \tag{6.16}$$

where $x \in \mathcal{D} \subset \mathbb{R}^n$ is the state of the system; $u \in \mathbb{R}$ is the input; and $f, g : \mathcal{D} \subset \mathbb{R}^n \to \mathbb{R}^n$ are smooth functions. The full-state linearisation problem is to find a feedback control design

$$u = \alpha(x) + \beta(x)v \tag{6.17}$$

with $\beta(x) \neq 0$ for $x \in \mathcal{D}$ such that the entire system dynamics are linearised by a state transformation with v as the new control input.

It is clear from the results shown in the input–output linearisation that the complete linearisation can only be achieved when the relative degree of the system equals the order of the system. If we can find an output function $h(x)$ for the system (6.16), the input–output linearisation result shown in the previous section can be applied to solve the full-state feedback linearisation problem. Therefore, we need to find an output function $h(x)$ such that

$$L_g L_f^k h(x) = 0 \quad \text{for } k = 0, \dots, n - 2 \tag{6.18}$$

$$L_g L_f^{n-1} h(x) \neq 0 \tag{6.19}$$

Based on Lemma 6.2, the condition specified in (6.18) is equivalent to the condition

$$L_{ad_f^k g} h(x) = 0 \quad \text{for } k = 0, \dots, n - 2 \tag{6.20}$$

and furthermore, the condition in (6.19) is equivalent to

$$L_{ad_f^{n-1} g} h(x) \neq 0. \tag{6.21}$$

The output function $h(x)$ that satisfies the condition shown in (6.20) is a solution of the partial differential equation

$$[g, \, ad_f g, \, \dots, \, ad_f^{n-2} g] \frac{\partial h}{\partial x} = 0. \tag{6.22}$$

To discuss the solution of this partial differential equation, we need a few notations and results. We refer to a collection of vector fields as a distribution. For example if $f_1(x), \ldots, f_k(x)$ are vector fields, with k a positive integer,

$$\Delta = \text{span}\{f_1(x), \ldots, f_k(x)\}$$

is a distribution. The dimension of distribution is defined as

$$\dim(\Delta(x)) = \text{rank}[f_1(x), \ldots, f_n(x)].$$

A distribution Δ is said to be involutive, if for any two vector fields $f_1, f_2 \in \Delta$, we have $[f_1, f_2] \in \Delta$. Note that not all the distributions are involutive, as shown in the following example.

Example 6.5. Consider the distribution

$$\Delta = \text{span}\{f_1(x), f_2(x)\}$$

where

$$f_1(x) = \begin{bmatrix} 2x_2 \\ 1 \\ 0 \end{bmatrix}, \quad f_2(x) = \begin{bmatrix} 1 \\ 0 \\ x_2 \end{bmatrix}.$$

A direct evaluation gives

$$\begin{aligned}
[f_1, f_2] &= \frac{\partial f_2}{\partial x} f_1 - \frac{\partial f_1}{\partial x} f_2 \\
&= \begin{bmatrix} 0 & 0 & 0 \\ 0 & 0 & 0 \\ 0 & 1 & 0 \end{bmatrix} \begin{bmatrix} 2x_2 \\ 1 \\ 0 \end{bmatrix} - \begin{bmatrix} 0 & 2 & 0 \\ 0 & 0 & 0 \\ 0 & 0 & 0 \end{bmatrix} \begin{bmatrix} 1 \\ 0 \\ x_2 \end{bmatrix} \\
&= \begin{bmatrix} 0 \\ 0 \\ 1 \end{bmatrix}.
\end{aligned}$$

It can be shown that $[f_1, f_2] \notin \Delta$, and therefore Δ is not involutive. Indeed, the rank of the matrix $[f_1, f_2, [f_1, f_2]]$ is 3, which means that $[f_1, f_2]$ is linearly independent of f_1 and f_2, and it cannot be a vector field in Δ. The rank of $[f_1, f_2, [f_1, f_2]]$ can be verified by its non-zero determinant as

$$|[f_1, f_2, [f_1, f_2]]| = \begin{vmatrix} 2x_2 & 1 & 0 \\ 1 & 0 & 0 \\ 0 & x_2 & 1 \end{vmatrix} = -1.$$

◁

We say that a distribution Δ is integrable if there exists a non-trivial $h(x)$ such that for any vector field $f \in \Delta$, $< dh, f > = 0$. The relationship between an involutive distribution and its integrability is stated in the following theorem.

Theorem 6.3 (Frobenius theorem). *A distribution is integrable if and only if it is involutive.*

Now we are ready to state the main result for full-state feedback linearisation.

Theorem 6.4. *The system (6.16) is full-state feedback linearisable if and only if* $\forall x \in \mathcal{D}$

- *the matrix* $G = [g(x), ad_f g(x), \ldots, ad_f^{n-1} g(x)]$ *has full rank*
- *the distribution* $\mathcal{G}_{n-1} = span\{g(x), ad_f g(x), \ldots, ad_f^{n-2} g(x)\}$ *is involutive*

Proof. For the sufficiency, we only need to show that there exists a function $h(x)$ such that the relative degree of the system by viewing $h(x)$ as the output is n, and the rest follows from Theorem 6.1. From the second condition that $\mathcal{G}_{n-1}$ is involutive, and Frobenius theorem, there exists a function $h(x)$ such that

$$\frac{\partial h}{\partial x}[g, \ ad_f g, \ \ldots, ad_f^{n-2} g] = 0$$

which is equivalent to, from Lemma 6.2,

$$L_g L_f^k h(x) = 0, \quad \text{for} \quad k = 0, \ldots, n - 2.$$

From the condition that G is full rank, we can establish

$$L_g L_f^{n-1} h(x) \neq 0.$$

In fact, if $L_g L_f^{n-1} h(x) = 0$, then from Lemma 6.2, we have

$$\frac{\partial h}{\partial x} G = 0$$

which is a contradiction as G has full rank.

For necessity, we show that if the system is full-state feedback linearisable, then the two conditions hold. From the problem formulation, we know that the full-state linearisability is equivalent to the existence of an output function $h(x)$ with relative degree n. From the definition of the relative degree and Lemma 6.2, we have

$$L_{ad_f^k g} h(x) = 0, \quad \text{for} \quad k = 0, \ldots, n - 2,$$

which implies

$$\frac{\partial h}{\partial x}[g, \ ad_f g, \ \ldots, ad_f^{n-2} g] = 0.$$

From Frobenius theorem, we conclude that $\mathcal{G}_{n-1}$ is involutive. Furthermore, from the fact that the system with $h(x)$ as the output has a relative degree n, we can show, in the same way as the discussion that leading to Lemma 6.2, that

$$
\begin{bmatrix} dh(x) \\ dL_f h(x) \\ \vdots \\ dL_f^{n-1}h \end{bmatrix} \begin{bmatrix} g(x) & ad_f g(x) & \cdots & ad_f^{n-1}g(x) \end{bmatrix}
$$
$$
= \begin{bmatrix} 0 & \cdots & 0 & (-1)^{n-1}r(x) \\ 0 & \cdots & (-1)^{n-2}r(x) & * \\ \vdots & \vdots & \vdots & \vdots \\ r(x) & \cdots & * & * \end{bmatrix}
$$

where $r(x) = L_g L_f^{n-1}h(x)$. This implies that G has rank n. This concludes the proof. □

Remark 6.3. Let us see the conditions in Theorem 6.4 for linear systems with

$$ f(x) = Ax, \quad g(x) = b. $$

where A is a constant matrix and b is a constant vector. A direct evaluation gives

$$ [f,g] = -Ab, $$

and

$$ ad_f^k g = (-1)^k A^k b $$

for $k > 0$. Therefore, we have

$$ G = [b, -Ab, \ldots, (-1)^{n-1}A^{n-1}b]. $$

It can be seen that the full rank condition of G is equivalent to the full controllability of the linear system. ◁

   In the next example, we consider the dynamics of the system that was considered in Example 6.4 for input–output linearisation for the input $h(x) = x_1$. We will show that the full-state linearisation can be achieved by finding a suitable output function $h(x)$.

Example 6.6. Consider the nonlinear system

$$\dot{x}_1 = x_1^3 + x_2$$
$$\dot{x}_1 = x_1^2 + x_3 + u$$
$$\dot{x}_2 = x_1^2 + u,$$

for full-state feedback linearisation. With

$$f(x) = \begin{bmatrix} x_1^3 + x_2 \\ x_1^2 + x_3 \\ x_1^2 \end{bmatrix}, \quad g(x) = \begin{bmatrix} 0 \\ 1 \\ 1 \end{bmatrix},$$

we have

$$[f, g] = \begin{bmatrix} -1 \\ -1 \\ 0 \end{bmatrix},$$

and

$$ad_f^2 g = \begin{bmatrix} 3x_1^2 + 1 \\ 2x_1 \\ 2x_1 \end{bmatrix}.$$

Hence, we have

$$\mathcal{G}_2 = \text{span} \left\{ \begin{bmatrix} 0 \\ 1 \\ 1 \end{bmatrix}, \begin{bmatrix} -1 \\ -1 \\ 0 \end{bmatrix} \right\}$$

and

$$G = \begin{bmatrix} 0 & -1 & 3x_1^2 + 1 \\ 1 & -1 & 2x_1 \\ 1 & 0 & 2x_1 \end{bmatrix}.$$

The distribution $\mathcal{G}_2$ is involutive, as it is spanned by constant vectors. The matrix G has full rank, as shown by its determinant

$$|G| = 3x_1^2 + 1 \neq 0.$$

Hence, the conditions in Theorem 6.4 are all satisfied, and the system is full-state linearisable. Indeed, we can find

$$h(x) = x_1 - x_2 + x_3$$

by solving

$$\frac{\partial h}{\partial x} \begin{bmatrix} 0 & -1 \\ 1 & -1 \\ 1 & 0 \end{bmatrix} = 0.$$

For this $h(x)$, it is easy to check that

$$L_g h = 0,$$
$$L_f h = x_2 + x_1^3 - x_3,$$
$$L_g L_f h = 0,$$
$$L_f^2 h = 3x_1^2(x_1^3 + x_2) + x_3,$$
$$L_g L_f^2 h = 3x_1^2 + 1,$$
$$L_f^3 h = (15x_1^4 + 6x_1 x_2)(x_1^3 + x_2) + 3x_1^2(x_3 + x_1^2) + x_1^2.$$

It can be seen that the relative degree, indeed, equals 3 as

$$L_g L_f^2 h = 3x_1^2 + 1 \neq 0.$$

For the full-state linearisation, we have the state transformation

$$\xi_1 = x_1 - x_2 + x_3,$$
$$\xi_2 = x_2 + x_1^3 - x_3,$$
$$\xi_2 = 3x_1^2(x_1^3 + x_2) + x_3,$$

and the feedback law

$$u = \frac{1}{3x_1^2 + 1} \left(v - (15x_1^4 + 6x_1 x_2)(x_1^3 + x_2) - 3x_1^2(x_3 + x_1^2) - x_1^2 \right).$$

◁

Chapter 7

Adaptive control of linear systems

The principle of feedback control is to maintain a consistent performance when there are uncertainties in the system or changes in the setpoints through a feedback controller using the measurements of the system performance, mainly the outputs. Many controllers are with fixed controller parameters, such as the controllers designed by normal state feedback control, and H_∞ control methods. The basic aim of adaptive control also is to maintain a consistent performance of a system in the presence of uncertainty or unknown variation in plant parameters, but with changes in the controller parameters, adapting to the changes in the performance of the control system. Hence, there is an adaptation in the controller setting subject to the performance of the closed-loop system. How the controller parameters change is decided by the adaptive laws, which are often designed based on the stability analysis of the adaptive control system.

A number of design methods have been developed for adaptive control. Model Reference Adaptive Control (MRAC) consists of a reference model which produces the desired output, and the difference between the plant output and the reference output is then used to adjust the control parameters and the control input directly. MRAC is often in continuous-time domain, and for deterministic plants. Self-Tuning Control (STC) estimates system parameters and then computes the control input from the estimated parameters. STC is often in discrete-time and for stochastic plants. Furthermore, STC often has a separate identification procedure for estimation of the system parameters, and is referred to as indirect adaptive control, while MRAC adapts to the changes in the controller parameters, and is referred to as direct adaptive control. In general, the stability analysis of direct adaptive control is less involved than that of indirect adaptive control, and can often be carried out using Lyapunov functions. In this chapter, we focus on the basic design method of MRAC.

Compared with the conventional control design, adaptive control is more involved, with the need to design the adaptation law. MRAC design usually involves the following three steps:

- Choose a control law containing variable parameters.
- Design an adaptation law for adjusting those parameters.
- Analyse the stability properties of the resulting control system.

7.1   MRAC of first-order systems

The basic design idea can be clearly demonstrated by first-order systems. Consider a first-order system

$$\dot{y} + a_p y = b_p u, \tag{7.1}$$

where y and $u \in \mathbb{R}$ are the system output and input respectively, and a_p and b_p are unknown constant parameters with $\text{sgn}(b_p)$ known. The output y is to follow the output of the reference model

$$\dot{y}_m + a_m y_m = b_m r. \tag{7.2}$$

The reference model is stable, i.e., $a_m > 0$. The signal r is the reference input. The design objective is to make the tracking error $e = y - y_m$ converge to 0.

Let us first design a Model Reference Control (MRC), that is, the control design assuming all the parameters are known, to ensure that the output y follows y_m. Rearrange the system model as

$$\dot{y} + a_m y = b_p \left(u - \frac{a_p - a_m}{b_p} y \right)$$

and therefore we obtain

$$\dot{e} + a_m e = b_p \left(u - \frac{a_p - a_m}{b_p} y - \frac{b_m}{b_p} r \right)$$

$$:= b_p \left(u - a_u y - a_r r \right),$$

where

$$a_y = \frac{a_p - a_m}{b_p},$$

$$a_r = \frac{b_m}{b_p}.$$

If all the parameters are known, the control law is designed as

$$u = a_r r + a_y y \tag{7.3}$$

and the resultant closed-loop system is given by

$$\dot{e} + a_m e = 0.$$

The tracking error converges to zero exponentially.

One important design principle in adaptive control is the so-called *the certainty equivalence principle*, which suggests that the unknown parameters in the control design are replaced by their estimates. Hence, when the parameters are unknown, let $\hat{a}_r$ and $\hat{a}_y$ denote their estimates of a_r and a_y, and the control law, based on the certainty equivalence principle, is given by

$$u = \hat{a}_r r + \hat{a}_y y. \tag{7.4}$$

Note that the parameters a_r and a_y are the parameters of the controllers, and they are related to the original system parameters a_p and b_p, but not the original system parameters themselves.

The certainty equivalence principle only suggests a way to design the adaptive control input, not how to update the parameter estimates. Stability issues must be considered when deciding the adaptive laws, i.e., the way how estimated parameters are updated. For first-order systems, the adaptive laws can be decided from Lyapunov function analysis.

With the proposed adaptive control input (7.4), the closed-loop system dynamics are described by

$$\dot{e} + a_m e = b_p(-\tilde{a}_y y - \tilde{a}_r r), \tag{7.5}$$

where $\tilde{a}_r = a_r - \hat{a}_r$ and $\tilde{a}_y = a_y - \hat{a}_y$. Consider the Lyapunov function candidate

$$V = \frac{1}{2}e^2 + \frac{|b_p|}{2\gamma_r}\tilde{a}_r^2 + \frac{|b_p|}{2\gamma_y}\tilde{a}_y^2, \tag{7.6}$$

where γ_r and γ_y are constant positive real design parameters. Its derivative along the trajectory (7.5) is given by

$$\dot{V} = -a_m e^2 + \tilde{a}_r\left(|b_p|\frac{\dot{\tilde{a}}_r}{\gamma_r} - eb_p r\right) + \tilde{a}_y\left(|b_p|\frac{\dot{\tilde{a}}_y}{\gamma_y} - eb_p y\right).$$

If we can set

$$|b_p|\frac{\dot{\tilde{a}}_r}{\gamma_r} - eb_p r = 0, \tag{7.7}$$

$$|b_p|\frac{\dot{\tilde{a}}_y}{\gamma_y} - eb_p y = 0, \tag{7.8}$$

we have

$$\dot{V} = -a_m e^2. \tag{7.9}$$

Noting that $\dot{\tilde{a}}_r = -\dot{\hat{a}}_r$ and $\dot{\tilde{a}}_y = -\dot{\hat{a}}_y$, the conditions in (7.7) and (7.8) can be satisfied by setting the adaptive laws as

$$\dot{\hat{a}}_r = -\mathrm{sgn}(b_p)\gamma_r er, \tag{7.10}$$

$$\dot{\hat{a}}_y = -\mathrm{sgn}(b_p)\gamma_y ey. \tag{7.11}$$

The positive real design parameters γ_r and γ_y are often referred to as adaptive gains, as they can affect the speed of parameter adaptation.

From (7.9) and Theorem 4.2, we conclude that the system is Lyapunov stable with all the variables e, $\tilde{a}_r$ and $\tilde{a}_y$ bounded, and hence the boundedness of $\hat{a}_r$ and $\hat{a}_y$.

However, based on the stability theorems introduced in Chapter 4, we cannot conclude anything about the tracking error e other than its boundedness. In order to do it, we need to introduce an important lemma for stability analysis of adaptive control systems.

Lemma 7.1 (Barbalat's lemma). *If a function $f(t) : \mathbb{R} \longrightarrow \mathbb{R}$ is uniformly continuous for $t \in [0, \infty)$, and $\int_0^\infty f(t)dt$ exists, then $\lim_{t \to \infty} f(t) = 0$.*

From (7.9), we can show that

$$\int_0^\infty e^2(t)dt = \frac{V(0) - V(\infty)}{a_m} < \infty. \tag{7.12}$$

Therefore, we have established that $e \in L_2 \cap L_\infty$ and $\dot{e} \in L_\infty$. Since $\dot{e}$ and e are bounded, e^2 is uniformly continuous. Therefore, we can conclude from Barbalat's lemma that $\lim_{t \to \infty} e^2(t) = 0$, and hence $\lim_{t \to \infty} e(t) = 0$.

We summarise the stability result in the following lemma.

Lemma 7.2. *For the first-order system (7.1) and the reference model (7.2), the adaptive control input (7.4) together with the adaptive laws (7.10) and (7.11) ensures the boundedness of all the variables in the closed-loop system, and the convergence to zero of the tracking error.*

Remark 7.1. The stability analysis ensures the convergence to zero of the tracking error, but nothing can be told about the convergence of the estimated parameters. The estimated parameters are assured to be bounded from the stability analysis. In general, the convergence of the tracking error to zero and the boundedness of the adaptive parameters are stability results that we can establish for MRAC. The convergence of the estimated parameters may be achieved by imposing certain conditions of the reference signal to ensure the system is excited enough. This is similar to the concept of persistent excitation for system identification.   ◁

Example 7.1. Consider a first-order system

$$G_p = \frac{b}{s + a},$$

where $b = 1$ and a is an unknown constant parameter. We will design an adaptive controller such that the output of the system follows the output of the reference model

$$G_m = \frac{1}{s + 2}.$$

We can directly use the result presented in Lemma 7.2, i.e., we use the adaptive laws (7.10) and (7.11) and the control input (7.4). Since b is known, we only have one unknown parameter, and it is possible to design a simpler control based on the same design principle.

From the system model, we have

$$\dot{y} + ay = u,$$

which can be changed to

$$\dot{y} + 2y = u - (a - 2)y.$$

Subtracting the reference model

$$\dot{y}_m + 2y_m = r,$$

we obtain that

$$\dot{e} + 2e = u - a_y y - r.$$

where $a_y = a - 2$. We then design the adaptive law and control input as

$$\dot{\hat{a}}_y = -\gamma_y e y,$$
$$u = \hat{a}_y y + r.$$

The stability analysis follows the same discussion that leads to Lemma 7.2. Simulation study has been carried out with $a = -1$, $\gamma = 10$ and $r = 1$. The simulation results are shown in Figure 7.1. The figure shows that the estimated parameter converges to the true value $a_y = -3$. The convergence of the estimated parameters is not guaranteed by Lemma 7.2. Indeed, some strong conditions on the input or reference signal are needed to generate enough excitation for the parameter estimation to achieve the convergence of the estimated parameters in general. ◁

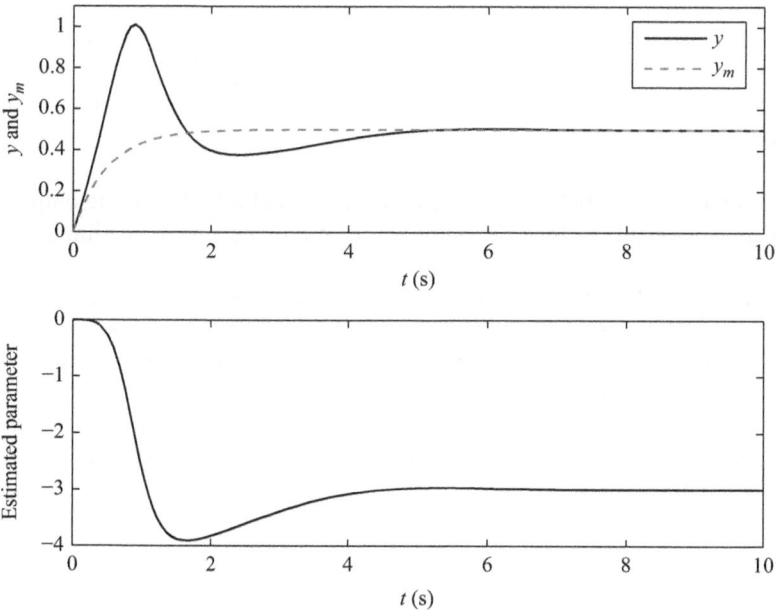

Figure 7.1 Simulation results of Example 7.1

7.2   Model reference control

It is clear from MRAC design for first-order systems that an MRC input is designed first which contains unknown parameters, and the adaptive control input is then obtained based on the certainty equivalence principle. Hence, MRC design is the first step for MRAC. Furthermore, MRC itself deserves a brief introduction, as it is different from the classical control design methods shown in standard undergraduate texts. In this section, we will start with MRC for systems with relative degree 1, and then move on to MRC of systems with high-order relative degrees.

Consider an nth-order system with the transfer function

$$y(s) = k_p \frac{Z_p(s)}{R_p(s)} u(s), \tag{7.13}$$

where $y(s)$ and $u(s)$ denote the system output and input in frequency domain; k_p is the high frequency gain; and Z_p and R_p are monic polynomials with orders of $n - \rho$ and n respectively with ρ as the relative degree. The reference model is chosen to have the same relative degree of the system, and is described by

$$y_m(s) = k_m \frac{Z_m(s)}{R_m(s)} r(s), \tag{7.14}$$

where $y_m(s)$ is the reference output for $y(s)$ to follow; $r(s)$ is a reference input; and $k_m > 0$ and Z_m and R_m are monic Hurwitz polynomials.

Remark 7.2. A monic polynomial is a polynomial whose leading coefficient, the coefficient of the highest power, is 1. A polynomial is said to be Hurwitz if all its roots are with negative real parts, i.e., its roots locate in the open left half of the complex plane. The high-frequency gain is the leading coefficient of the numerator of a transfer function.   ◁

The objective of MRC is to design a control input u such that the output of the system asymptotically follows the output of the reference model, i.e., $\lim_{t \to \infty} (y(t) - y_m(t)) = 0$.

Note that in this chapter, we abuse the notations of y, u and r by using same notations for the functions in time domain and their Laplace transformed functions in the frequency domain. It should be clear from the notations that $y(s)$ is the Laplace transform of $y(t)$ and similarly for u and r.

To design MRC for systems with $\rho = 1$, we follow a similar manipulation to the first-order system by manipulating the transfer functions. We start with

$$y(s)R_p(s) = k_p Z_p(s)u(s)$$

and then

$$y(s)R_m(s) = k_p Z_p(s)u(s) - (R_p(s) - R_m(s))y(s).$$

Note that $R_p(s) - R_m(s)$ is a polynomial with order $n - 1$, and $\frac{R_m(s) - R_p(s)}{Z_m(s)}$ is a proper transfer function, as $R_p(s)$ and $R_m(s)$ are monic polynomials. Hence, we can write

$$y(s)R_m(s) = k_p Z_m(s) \left(\frac{Z_p(s)}{Z_m(s)} u(s) + \frac{R_m(s) - R_p(s)}{Z_m(s)} y(s) \right).$$

If we parameterise the transfer functions as

$$\frac{Z_p(s)}{Z_m(s)} = 1 - \frac{\theta_1^T \alpha(s)}{Z_m(s)},$$

$$\frac{R_m(s) - R_p(s)}{Z_m(s)} y(s) = -\frac{\theta_2^T \alpha(s)}{Z_m(s)} y(s) - \theta_3,$$

where $\theta_1 \in \mathbb{R}^{n-1}$, $\theta_2 \in \mathbb{R}^{n-1}$ and $\theta_3 \in \mathbb{R}$ are constants and

$$\alpha(s) = [s^{n-2}, \ldots, 1]^T,$$

we obtain that

$$y(s) = k_p \frac{Z_m(s)}{R_m(s)} \left(u(s) - \frac{\theta_1^T \alpha(s)}{Z_m(s)} u(s) - \frac{\theta_2^T \alpha(s)}{Z_m(s)} y(s) - \theta_3 y(s) \right). \tag{7.15}$$

Hence, we have the dynamics of tracking error given by

$$e_1(s) = k_p \frac{Z_m(s)}{R_m(s)} \left(u(s) - \frac{\theta_1^T \alpha(s)}{Z_m(s)} u(s) - \frac{\theta_2^T \alpha(s)}{Z_m(s)} y(s) - \theta_3 y(s) - \theta_4 r \right), \tag{7.16}$$

where $e_1 = y - y_m$ and $\theta_4 = \frac{k_m}{k_p}$.

The control input for MRC is given by

$$u(s) = \frac{\theta_1^T \alpha(s)}{Z_m(s)} u(s) + \frac{\theta_2^T \alpha(s)}{Z_m(s)} y + \theta_3 y + \theta_4 r(s)$$

$$:= \theta^T \omega, \tag{7.17}$$

where

$$\theta^T = [\theta_1^T, \theta_4^T, \theta_3, \theta_4],$$
$$\omega = [\omega_1^T, \omega_2^T, y, r]^T,$$

with

$$\omega_1 = \frac{\alpha(s)}{Z_m(s)} u,$$

$$\omega_2 = \frac{\alpha(s)}{Z_m(s)} y.$$

Remark 7.3. The control design shown in (7.17) is a dynamic feedback controller. Each element in the transfer matrix $\dfrac{\alpha(s)}{Z_m(s)}$ is strictly proper, i.e., with relative degree greater than or equal to 1. The total number of parameters in θ equals $2n$.

◁

Lemma 7.3. *For the system (7.13) with relative degree 1, the control input (7.17) solves MRC problem with the reference model (7.14) and $\lim_{t\to\infty} (y(t) - y_m(t)) = 0$.*

Proof. With the control input (7.17), the closed-loop dynamics are given by

$$e_1(s) = k_p \frac{Z_m(s)}{R_m(s)} \epsilon(s),$$

where $\epsilon(s)$ denotes exponentially convergent signals due to non-zero initial values. The reference model is stable, and then the track error $e_1(t)$ converges to zero exponentially.

□

Example 7.2. Design MRC for the system

$$y(s) = \frac{s+1}{s^2 - 2s + 1} u(s)$$

with the reference model

$$y_m(s) = \frac{s+3}{s^2 + 2s + 3} r(s).$$

We follow the procedures shown early to obtain the MRC control. From the transfer function of the system, we have

$$y(s)(s^2 + 2s + 3) = (s+1)u(s) + (4s+2)y(s),$$

which leads to

$$\begin{aligned}
y(s) &= \frac{s+3}{s^2 + 2s + 3}\left(\frac{s+1}{s+3}u(s) + \frac{4s+2}{s+3}y(s)\right) \\
&= \frac{s+3}{s^2 + 2s + 3}\left(u(s) - \frac{2}{s+3}u(s) - \frac{10}{s+3}y(s) + 4y(s)\right).
\end{aligned}$$

Subtracting it by the reference model, we have

$$e_1(s) = \frac{s+3}{s^2 + 2s + 3}\left(u(s) - \frac{2}{s+3}u(s) - \frac{10}{s+3}y(s) + 4y(s) - r(s)\right),$$

which leads to the MRC control input

$$\begin{aligned}
u(s) &= \frac{2}{s+3}u(s) + \frac{10}{s+3}y(s) - 4y(s) + r(s) \\
&= [2\ 10\ -4\ 1][\omega_1(s)\ \omega_2(s)\ y(s)\ r(s)]^T,
\end{aligned}$$

where

$$\omega_1(s) = \frac{1}{s+3}u(s),$$

$$\omega_2(s) = \frac{1}{s+3}y(s).$$

Note that the control input in the time domain is given by

$$u(s) = [2 \ 10 \ -4 \ 1][\omega_1(t) \ \omega_2(t) \ y(t) \ r(t)]^T,$$

where

$$\dot{\omega}_1 = -3\omega_1 + u,$$

$$\dot{\omega}_2 = -3\omega_2 + y.$$

◁

For a system with $\rho > 1$, the input in the same format as (7.17) can be obtained. The only difference is that Z_m is of order $n - \rho < n - 1$. In this case, we let $P(s)$ be a monic and Hurwitz polynomial with order $\rho - 1$ so that $Z_m(s)P(s)$ is of order $n - 1$. We adopt a slightly different approach from the case of $\rho = 1$.

Consider the identity

$$y(s) = \frac{Z_m(s)}{R_m(s)}\left(\frac{R_m(s)P(s)}{Z_m(s)P(s)}y(s)\right)$$

$$= \frac{Z_m(s)}{R_m(s)}\left(\frac{Q(s)R_p(s) + \Delta(s)}{Z_m(s)P(s)}y(s)\right). \tag{7.18}$$

Note that the second equation in (7.18) follows from the identity

$$R_m(s)P(s) = Q(s)R_p(s) + \Delta(s),$$

where $Q(s)$ is a monic polynomial with order $n - \rho - 1$, and $\Delta(s)$ is a polynomial with order $n - 1$. In fact $Q(s)$ can be obtained by dividing $R_m(s)P(s)$ by $R_p(s)$ using long division, and $\Delta(s)$ is the remainder of the polynomial division. From the transfer function of the system, we have

$$R_p(s)y(s) = k_p Z_p(s)u(s).$$

Substituting it into (7.18), we have

$$y(s) = k_p \frac{Z_m(s)}{R_m(s)}\left(\frac{Q(s)Z_p(s)}{Z_m(s)P(s)}u + \frac{k_p^{-1}\Delta(s)}{Z_m(s)P(s)}y(s)\right).$$

Similar to the case for $\rho = 1$, if we parameterise the transfer functions as

$$\frac{Q(s)Z_p(s)}{Z_m(s)P(s)} = 1 - \frac{\theta_1^T\alpha(s)}{Z_m(s)P(s)},$$

$$\frac{k_p^{-1}\Delta}{Z_m(s)P(s)} = -\frac{\theta_2^T\alpha(s)}{Z_m(s)P(s)} - \theta_3$$

where $\theta_1 \in \mathbb{R}^{n-1}$ and $\theta_2 \in \mathbb{R}^{n-1}$ and $\theta_3 \in \mathbb{R}$ are constants and

$$\alpha(s) = [s^{n-2}, \ldots, 1]^T,$$

we obtain that

$$y(s) = k_p \frac{Z_m(s)}{R_m(s)} \left(u(s) - \frac{\theta_1^T \alpha(s)}{Z_m(s)P(s)} u(s) - \frac{\theta_2^T \alpha(s)}{Z_m(s)P(s)} y(s) - \theta_3 y(s) \right).$$

Hence, we have the dynamics of tracking error given by

$$e_1(s) = k_p \frac{Z_m(s)}{R_m(s)} \left(u(s) - \frac{\theta_1^T \alpha(s)}{Z_m(s)P(s)} u(s) - \frac{\theta_2^T \alpha(s)}{Z_m(s)P(s)} y(s) - \theta_3 y(s) - \theta_4 r \right),$$

where $e_1 = y - y_m$ and $\theta_4 = \frac{k_m}{k_p}$. The control input is designed as

$$u = \frac{\theta_1^T \alpha(s)}{Z_m(s)P(s)} u + \frac{\theta_2^T \alpha(s)}{Z_m(s)P(s)} y + \theta_3 y + \theta_4 r$$

$$:= \theta^T \omega \tag{7.19}$$

with the same format as (7.17) except

$$\omega_1 = \frac{\alpha(s)}{Z_m(s)P(s)} u,$$

$$\omega_2 = \frac{\alpha(s)}{Z_m(s)P(s)} y.$$

Remark 7.4. The final control input is in the same format as shown for the case $\rho = 1$. The filters for w_1 and w_2 are in the same order as in the case for $\rho = 1$, as the order of $Z_m(s)P(s)$ is still $n - 1$. ◁

Lemma 7.4. *For the system (7.13) with relative degree $\rho > 1$, the control input (7.19) solves MRC problem with the reference model (7.14) and $\lim_{t\to\infty} (y(t) - y_m(t)) = 0$.*

The proof is the same as the proof for Lemma 7.3.

Example 7.3. Design MRC for the system

$$y(s) = \frac{1}{s^2 - 2s + 1} u$$

with the reference model

$$y_m(s) = \frac{1}{s^2 + 2s + 3} r.$$

The relative degree of the system is 2. We set $P = s + 1$. Note that

$$(s^2 + 2s + 3)(s + 1) = (s + 5)(s^2 - 2s + 1) + (14s - 2).$$

From the reference model, we have

$$
\begin{aligned}
y(s) &= \frac{1}{s^2 + 2s + 3} \left(\frac{(s^2 + 2s + 3)(s + 1)}{s + 1} y(s) \right) \\
&= \frac{1}{s^2 + 2s + 3} \left(\frac{(s + 5)(s^2 - 2s + 1)y(s) + (14s - 2)y(s)}{s + 1} \right) \\
&= \frac{1}{s^2 + 2s + 3} \left(\frac{(s + 5)u(s) + (14s - 2)y(s)}{s + 1} \right) \\
&= \frac{1}{s^2 + 2s + 3} \left(u(s) + \frac{4}{s + 1}u(s) - 16\frac{1}{s + 1}y(s) + 14y(s) \right).
\end{aligned}
$$

The dynamics of the tacking error are given by

$$
e_1(s) = \frac{1}{s^2 + 2s + 3} \left(u(s) + \frac{4}{s + 1}u(s) - 16\frac{1}{s + 1}y(s) + 14y(s) - r(s) \right).
$$

We can then design the control input as

$$
u = [-1, \ -16, \ 14, \ 1][\omega_1, \ \omega_2, \ y, \ r]^T
$$

with

$$
\omega_1 = \frac{1}{s + 1}u,
$$

$$
\omega_2 = \frac{1}{s + 1}y.
$$

◁

7.3 MRAC of linear systems with relative degree 1

Adaptive control deals with uncertainties in terms of unknown constant parameters. It may be used to tackle some changes or variations in model parameters in adaptive control application, but the stability analysis will be carried under the assumption the parameters are constants. There are other common assumptions for adaptive control which are listed below:

- the known system order n
- the known relative degree ρ
- the minimum phase of the plant
- the known sign of the high frequency gain $\text{sgn}(k_p)$

In this section, we present MRAC design for linear systems with relative degree 1.

Consider an nth-order system with the transfer function

$$
y(s) = k_p \frac{Z_p(s)}{R_p(s)} u(s), \tag{7.20}
$$

where $y(s)$ and $u(s)$ denote the system output and input in frequency domain; k_p is the high frequency gain; and Z_p and R_p are monic polynomials with orders of $n - 1$ and n respectively. This system is assumed to be minimum phase, i.e., $Z_p(s)$ is a Hurwitz polynomial, and the sign of the high-frequency gain, $\mathrm{sgn}(k_p)$, is known. The coefficients of the polynomials and the value of k_p are constants and unknown. The reference model is chosen to have the relative degree 1 and strictly positive real, and is described by

$$y_m(s) = k_m \frac{Z_m(s)}{R_m(s)} r(s), \tag{7.21}$$

where $y_m(s)$ is the reference output for $y(s)$ to follow, $r(s)$ is a reference input, and $Z_m(s)$ and $R_m(s)$ are monic polynomials and $k_m > 0$. Since the reference model is strictly positive real, Z_m and R_m are Hurwitz polynomials.

MRC shown in the previous section gives the control design in (7.17). Based on the certainty equivalence principle, we design the adaptive control input as

$$u(s) = \hat{\theta}^T \omega, \tag{7.22}$$

where $\hat{\theta}$ is an estimate of the unknown vector $\theta \in \mathbb{R}^{2n}$, and ω is given by

$$\omega = [\omega_1^T, \omega_2^T, y, r]^T$$

with

$$\omega_1 = \frac{\alpha(s)}{Z_m(s)} u,$$

$$\omega_2 = \frac{\alpha(s)}{Z_m(s)} y.$$

With the designed adaptive control input, it can be obtained, from the tracking error dynamics shown in (7.16), that

$$e_1(s) = k_p \frac{Z_m(s)}{R_m(s)} (\hat{\theta}^T \omega - \theta^T \omega)$$

$$= k_m \frac{Z_m(s)}{R_m(s)} \left(-\frac{k_p}{k_m} \tilde{\theta}^T \omega \right) \tag{7.23}$$

where $\tilde{\theta} = \theta - \hat{\theta}$.

To analyse the stability using a Lyapunov function, we put the error dynamics in the state space form as

$$\dot{e} = A_m e + b_m \left(-\frac{k_p}{k_m} \tilde{\theta}^T \omega \right)$$

$$e_1 = c_m^T e \tag{7.24}$$

where (A_m, b_m, c_m) is a minimum state space realisation of $k_m \frac{Z_m(s)}{R_m(s)}$, i.e.,

$$c_m^T (sI - A_m)^{-1} b_m = k_m \frac{Z_m(s)}{R_m(s)}.$$

Since (A_m, b_m, c_m) is a strictly positive real system, from Kalman–Yakubovich lemma (Lemma 5.4), there exist positive definite matrices P and Q such that

$$A_m^T P_m + P_m A_m = -Q_m, \tag{7.25}$$

$$P_m b_m = c_m. \tag{7.26}$$

Define a Lyapunov function candidate as

$$V = \frac{1}{2} e^T P_m e + \frac{1}{2} \left| \frac{k_p}{k_m} \right| \tilde{\theta}^T \Gamma^{-1} \tilde{\theta},$$

where $\Gamma \in \mathbb{R}^{2n}$ is a positive definite matrix. Its derivative is given by

$$\dot{V} = \frac{1}{2} e^T (A_m^T P_m + P_m A_m) e + e^T P_m b_m \left(-\frac{k_p}{k_m} \tilde{\theta}^T \omega \right) + \left| \frac{k_p}{k_m} \right| \tilde{\theta}^T \Gamma^{-1} \dot{\tilde{\theta}}$$

Using the results from (7.25) and (7.26), we have

$$\dot{V} = -\frac{1}{2} e^T Q_m e + e_1 \left(-\frac{k_p}{k_m} \tilde{\theta}^T \omega \right) + \left| \frac{k_p}{k_m} \right| \tilde{\theta}^T \Gamma^{-1} \dot{\tilde{\theta}}$$

$$= -\frac{1}{2} e^T Q_m e + \left| \frac{k_p}{k_m} \right| \tilde{\theta}^T \left(\Gamma^{-1} \dot{\tilde{\theta}} - \mathrm{sgn}(k_p) e_1 \omega \right).$$

Hence, the adaptive law is designed as

$$\dot{\hat{\theta}} = -\mathrm{sgn}(k_p) \Gamma e_1 \omega, \tag{7.27}$$

which results in

$$\dot{V} = -\frac{1}{2} e^T Q_m e.$$

We can now conclude the boundedness of e and $\hat{\theta}$. Furthermore it can be shown that $e \in L_2$ and $\dot{e}_1 \in L_\infty$. Therefore, from Barbalat's lemma we have $\lim_{t \to \infty} e_1(t) = 0$. The boundedness of other system state variables can be established from the minimum-phase property of the system.

We summarise the stability analysis for MRAC of linear systems with relative degree 1 in the following theorem.

Theorem 7.5. *For the first-order system (7.20) and the reference model (7.21), the adaptive control input (7.22) together with the adaptive law (7.27) ensures the boundedness of all the variables in the closed-loop system, and the convergence to zero of the tracking error.*

Remark 7.5. The stability result shown in Theorem 7.5 only guarantees the convergence of the tracking error to zero, not the convergence of the estimated parameters. In the stability analysis, we use Kalman–Yakubovich lemma for the definition of Lyapunov function and the stability proof. That is why we choose the reference model to be strictly positive real. From the control design point of view, we do not need to know the actual values of P_m and Q_m, as long as they exist, which is guaranteed by the selection of a strictly positive real model. Also it is clear from the stability analysis, that the unknown parameters must be constant. Otherwise, we would not have $\dot{\hat{\theta}} = -\dot{\tilde{\theta}}$.   ◁

7.4   MRAC of linear systems with high relatives

In this section, we will introduce adaptive control design for linear systems with their relative degrees higher than 1. Similar to the case for relative degree 1, the certainty equivalence principle can be applied to the control design, but the designs of the adaptive laws and the stability analysis are much more involved, due to the higher relative degrees. One difficulty is that there is not a clear choice of Lyapunov function candidate as in the case of $\rho = 1$.

Consider an nth-order system with the transfer function

$$y(s) = k_p \frac{Z_p(s)}{R_p(s)} u(s), \tag{7.28}$$

where $y(s)$ and $u(s)$ denote the system output and input in frequency domain, k_p is the high frequency gain, Z_p and R_p are monic polynomials with orders of $n - \rho$ and n respectively, with $\rho > 1$ being the relative degree of the system. This system is assumed to be minimum phase, i.e., $Z_p(s)$ is Hurwitz polynomial, and the sign of the high-frequency gain, $\text{sgn}(k_p)$, is known. The coefficients of the polynomials and the value of k_p are constants and unknown. The reference model is chosen as

$$y_m(s) = k_m \frac{Z_m(s)}{R_m(s)} r(s) \tag{7.29}$$

where $y_m(s)$ is the reference output for $y(s)$ to follow; $r(s)$ is a reference input; and $Z_m(s)$ and $R_m(s)$ are monic polynomials with orders $n - \rho$ and n respectively and $k_m > 0$. The reference model (7.29) is required to satisfy an additional condition that there exists a monic and Hurwitz polynomial $P(s)$ of order $n - \rho - 1$ such that

$$y_m(s) = k_m \frac{Z_m(s)P(s)}{R_m(s)} r(s) \tag{7.30}$$

is strictly positive real. This condition also implies that Z_m and R_m are Hurwitz polynomials.

MRC shown in the previous section gives the control design in (7.19). We design the adaptive control input, again using the certainty equivalence principle, as

$$u = \hat{\theta}^T \omega, \tag{7.31}$$

where $\hat{\theta}$ is an estimate of the unknown vector $\theta \in \mathbb{R}^{2n}$, and ω is given by

$$\omega = [\omega_1^T, \omega_2^T, y, r]^T$$

with

$$\omega_1 = \frac{\alpha(s)}{Z_m(s)P(s)}u,$$

$$\omega_2 = \frac{\alpha(s)}{Z_m(s)P(s)}y.$$

The design of adaptive law is more involved, and we need to examine the dynamics of the tacking error, which are given by

$$e_1 = k_p \frac{Z_m}{R_m}(u - \theta^T\phi)$$

$$= k_m \frac{Z_m P(s)}{R_m}\left(k(u_f - \theta^T\phi)\right), \tag{7.32}$$

where

$$k = \frac{k_p}{k_m}, \quad u_f = \frac{1}{P(s)}u \quad \text{and} \quad \phi = \frac{1}{P(s)}\omega.$$

An auxiliary error is constructed as

$$\epsilon = e_1 - k_m \frac{Z_m P(s)}{R_m}\left(\hat{k}(u_f - \hat{\theta}^T\phi)\right) - k_m \frac{Z_m P(s)}{R_m}\left(\epsilon n_s^2\right), \tag{7.33}$$

where $\hat{k}$ is an estimate of k, $n_s^2 = \phi^T\phi + u_f^2$. The adaptive laws are designed as

$$\dot{\hat{\theta}} = -\text{sgn}(b_p)\Gamma\epsilon\phi, \tag{7.34}$$

$$\dot{\hat{k}} = \gamma\epsilon(u_f - \hat{\theta}^T\phi). \tag{7.35}$$

With these adaptive laws, a stability result can be obtained for the boundedness of parameter estimates and the convergence of the tracking error. For the completeness, we state the theorem below without giving the proof.

Theorem 7.6. *For the system (7.28) and the reference model (7.29), the adaptive control input (7.31) together with the adaptive laws (7.34) and (7.35) ensures the boundedness of all the variables in the closed-loop system, and the convergence to zero of the tracking error.*

7.5 Robust adaptive control

Adaptive control design and its stability analysis have been carried out under the condition that there is only parametric uncertainty in the system. However, many types of non-parametric uncertainties do exist in practice. These include

- high-frequency unmodelled dynamics, such as actuator dynamics or structural vibrations
- low-frequency unmodelled dynamics, such as Coulomb frictions

- measurement noise
- computation roundoff error and sampling delay

Such non-parametric uncertainties will affect the performance of adaptive control systems when they are applied to practical systems. They may cause instability. The difference between adaptive control of linear systems and other design methods is parameter estimation. Without specific requirement for input signals, such as persistent excitation, we can only establish the boundedness of estimated parameters, and the asymptotic convergence to zero of the output error. Therefore, the closed-loop adaptive system is not even asymptotically stable. For other design methods for linear systems without parameter adaptation, the closed-loop systems are normally exponentially stable. For a linear system with exponential stability, the state inherently remain bounded under any bounded input. This is not the case for a linear system under adaptive control, due to the difference in the stability properties.

Many non-parametric uncertainties can be represented by a bounded disturbance to the nominal system. Even a bounded disturbance can cause serious problem in parameter adaptation. Let us consider a simple example.

Consider the system output is described by

$$y = \theta \omega. \tag{7.36}$$

The adaptive law

$$\dot{\hat{\theta}} = \gamma \epsilon \omega, \tag{7.37}$$

where

$$\epsilon = y - \hat{\theta}\omega$$

will render the convergence of the estimate $\hat{\theta}$ by taking

$$V = \frac{1}{2\gamma}\tilde{\theta}^2$$

as a Lyapunov function candidate and the analysis

$$\dot{V} = -\tilde{\theta}(y - \hat{\theta}\omega)\omega$$
$$= -\tilde{\theta}^2\omega^2. \tag{7.38}$$

The boundedness of $\hat{\theta}$ can then be concluded, no matter what the signal ω is.

Now, if the signal is corrupted by some unknown bounded disturbance $d(t)$, i.e.,

$$y = \theta\omega + d(t).$$

the same adaptive will have a problem. In this case,

$$\dot{V} = -\tilde{\theta}(y - \hat{\theta}\omega)\omega$$
$$= -\tilde{\theta}(\theta\omega + d - \hat{\theta}\omega)\omega$$
$$= -\tilde{\theta}^2\omega^2 - \tilde{\theta}d\omega$$
$$= -\frac{\tilde{\theta}^2\omega^2}{2} - \frac{1}{2}(\tilde{\theta}\omega + d)^2 + \frac{d^2}{2}.$$

From the above analysis, we cannot conclude the boundedness of $\tilde{\theta}$ even though ω is bounded. In fact, if we take $\theta = 2$, $\gamma = 1$ and $\omega = (1 + t)^{-1/2} \in L_\infty$ and let

$$d(t) = (1 + t)^{-1/4} \left(\frac{5}{4} - 2(1 + t)^{-1/4} \right),$$

it can then be obtained that

$$y(t) = \frac{5}{4}(1 + t)^{-1/4}, \rightarrow 0 \text{ as } t \rightarrow \infty,$$

$$\dot{\hat{\theta}} = \frac{5}{4}(1 + t)^{-3/4} - \hat{\theta}(1 + t)^{-1}$$

which has a solution

$$\hat{\theta} = (1 + t)^{1/4} \rightarrow \infty \text{ as } t \rightarrow \infty. \tag{7.39}$$

In this example, we have observed that adaptive law designed for the disturbance-free system fails to remain bounded even though the disturbance is bounded and converges to zero as t tends to infinity.

Remark 7.6. If ω is a constant, then from (7.38) we can show that the estimate exponentially converges to the true value. In this case, there is only one unknown parameter. If θ is a vector, the requirement for the convergence is much stronger. In the above example, ω is bounded, but not in a persistent way. It does demonstrate that even a bounded disturbance can cause the estimated parameter divergent. ◁

Robust adaptive control issue is often addressed by modifying parameter adaptive laws to ensure the boundedness of estimated parameters. It is clear from the example shown above that bounded disturbance can cause estimated parameters unbounded. Various robust adaptive laws have been introduced to keep estimated parameters bounded in the presence of bounded disturbances. We will show two strategies using the following simple model:

$$y = \theta\omega + d(t) \tag{7.40}$$

with d as a bounded disturbance. In the following we keep using $\epsilon = y - \hat{\theta}\omega$ and $V = \frac{\tilde{\theta}^2}{2\gamma}$. Once the basic ideas are introduced, it is not difficult to extend the robust adaptive laws to adaptive control of dynamic systems.

Dead-zone modification is a modification to the parameter adaptive law to stop parameter adaptation when the error is very close to zero. The adaptive law is modified as

$$\dot{\hat{\theta}} = \begin{cases} \gamma\epsilon\omega & |\epsilon| > g \\ 0 & |\epsilon| \le g \end{cases} \tag{7.41}$$

where g is a constant satisfying $g > |d(t)|$ for all t. For $\epsilon > g$, we have

$$\begin{aligned}
\dot{V} &= -\tilde{\theta}\epsilon\omega \\
&= -(\theta\omega - \hat{\theta}\omega)\epsilon \\
&= -(y - d(t) - \hat{\theta}\omega)\epsilon \\
&= -(\epsilon - d(t))\epsilon \\
&< 0.
\end{aligned}$$

Therefore, we have

$$\dot{V} \begin{cases} < 0, & |\epsilon| > g \\ = 0, & |\epsilon| \le g \end{cases}$$

and we can conclude that V is bounded. Intuitively, when the error ϵ is small, the bounded disturbance can be more dominant, and therefore, the correct adaptation direction is corrupted by the disturbance. In such a case, a simple strategy would be just to stop parameter adaptation. The parameter adaptation stops in the range $|\epsilon| \le g$, and for this reason, this modification takes the name 'dead-zone' modification. The size of the dead zone depends on the size of the bounded disturbances. One problem with the dead-zone modification is that the adaptive law is discontinuous, and this may not be desirable in some applications.

σ-*Modification* is another strategy to ensure the boundedness of estimated parameters. The adaptive law is modified by adding an additional term $-\gamma\sigma\hat{\theta}$ to the normal adaptive law as

$$\dot{\hat{\theta}} = \gamma\epsilon\omega - \gamma\sigma\hat{\theta} \tag{7.42}$$

where σ is a positive real constant. In this case, we have

$$\begin{aligned}
\dot{V} &= -(\epsilon - d(t))\epsilon + \sigma\tilde{\theta}\hat{\theta} \\
&= -\epsilon^2 + d(t)\epsilon - \sigma\tilde{\theta}^2 + \sigma\tilde{\theta}\theta \\
&\le -\frac{\epsilon^2}{2} + \frac{d_0^2}{2} - \sigma\frac{\tilde{\theta}^2}{2} + \sigma\frac{\theta^2}{2} \\
&\le -\sigma\gamma V + \frac{d_0^2}{2} + \sigma\frac{\theta^2}{2}, \tag{7.43}
\end{aligned}$$

where $d_0 \ge |d(t)|$, $\forall t \ge 0$. Applying Lemma 4.5 (comparison lemma) to (7.43), we have

$$V(t) \le e^{-\sigma\gamma t} + V(0) \int_0^t e^{-\sigma\gamma(t-\tau)} \left(\frac{d_0^2}{2} + \sigma\frac{\theta^2}{2} \right) d\tau,$$

and therefore we can conclude that $V \in L_\infty$, which implies the boundedness of the estimated parameter. A bound can be obtained for the bounded parameter as

$$V(\infty) \le \frac{1}{\sigma \gamma} \left(\frac{d_0^2}{2} + \sigma \frac{\theta^2}{2} \right).$$ (7.44)

Note that this modification does not need a bound for the bounded disturbances, and also it provides a continuous adaptive law. For these reasons, σ-modification is one of the most widely used modifications for parameter adaptation.

Remark 7.7. We re-arrange the adaptive law (7.42) as

$$\dot{\hat{\theta}} + \gamma \sigma \hat{\theta} = \gamma \epsilon \omega.$$

Since $(\gamma \sigma)$ is a positive constant, the adaptive law can be viewed as a stable first-order dynamic system with $(\epsilon \omega)$ as the input and $\hat{\theta}$ as the output. With a bounded input, obviously $\hat{\theta}$ remains bounded. ◁

The robust adaptive laws introduced here can be applied to various adaptive control schemes. We demonstrate the application of a robust adaptive law to MRAC with $\rho = 1$. We start directly from the error model (7.24) with an additional bounded disturbance

$$\dot{e} = A_m e + b_m(-k\tilde{\theta}^T \omega + d(t))$$
$$e_1 = c_m^T e,$$ (7.45)

where $k = k_p/k_m$ and $d(t)$ are a bounded disturbance with the bound d_0, which represents the non-parametric uncertainty in the system. As discussed earlier, we need a robust adaptive law to deal with the bounded disturbances. If we take σ-modification, then the robust adaptive law is

$$\dot{\hat{\theta}} = -\mathrm{sgn}(k_p)\Gamma e_1 \omega - \sigma \Gamma \hat{\theta}.$$ (7.46)

We will show that this adaptive law will ensure the boundedness of the variables.
Let

$$V = \frac{1}{2} e^T P e + \frac{1}{2} |k| \tilde{\theta}^T \Gamma^{-1} \tilde{\theta}.$$

Similar to the analysis leading to Theorem 7.5, the derivative of V is obtained as

$$\dot{V} = -\frac{1}{2} e^T Q e + e_1(-k\tilde{\theta}^T \omega + d) + |k| \tilde{\theta}^T \Gamma^{-1} \dot{\hat{\theta}}$$
$$\le -\frac{1}{2}\lambda_{min}(Q)\|e\|^2 + e_1 d + |k|\sigma \tilde{\theta}^T \hat{\theta}$$
$$\le -\frac{1}{2}\lambda_{min}(Q)\|e\|^2 + |e_1 d| - |k|\sigma \|\tilde{\theta}\|^2 + |k|\sigma \tilde{\theta}^T \theta.$$

Note that

$$|e_1 d| \leq \frac{1}{4}\lambda_{min}(Q)\|e\|^2 + \frac{d_0^2}{\lambda_{min}(Q)},$$

$$|\tilde{\theta}^T \theta| \leq \frac{1}{2}\|\tilde{\theta}\|^2 + \frac{1}{2}\|\theta\|^2.$$

Hence, we have

$$\dot{V} \leq -\frac{1}{4}\lambda_{min}(Q)\|e\|^2 - \frac{|k|\sigma}{2}\|\tilde{\theta}\|^2 + \frac{d_0^2}{\lambda_{min}(Q)} + \frac{|k|\sigma}{2}\|\theta\|^2$$

$$\leq -\alpha V + \frac{d_0^2}{\lambda_{min}(Q)} + \frac{|k|\sigma}{2}\|\theta\|^2,$$

where α is a positive real and

$$\alpha = \frac{\min\{(1/2)\lambda_{min}(Q), |k|\sigma\}}{\max\{\lambda_{max}(P), |k|/\lambda_{min}(\Gamma)\}}.$$

Therefore, we can conclude the boundedness of V from Lemma 4.5 (comparison lemma), which further implies the boundedness of the tracking error e_1 and the estimate $\hat{\theta}$.

From the above analysis, it is clear that the adaptive law with σ-modification ensures the boundedness of all the variables in the closed-loop adaptive control system. It is worth noting that the output tracking error e_1 will not asymptotically converge to zero, even though the bounded disturbance $d(t)$ becomes zero. That is the price to pay for the robust adaptive scheme.

Chapter 8
Nonlinear observer design

Observers are needed to estimate unmeasured state variables of dynamic systems. They are often used for output feedback control design when only the outputs are available for the control design. Observers can also be used for other estimation purposes such as fault detection and diagnostics. There are many results on nonlinear observer design in literature, and in this chapter, we can introduce only a number of results. Observer design for linear systems is briefly reviewed before the introduction of observers with linear error dynamics. We then introduce another observer design method based on Lyapunov's auxiliary theorem, before the observer design for systems with Lipschitz nonlinearities. At the end of this chapter, adaptive observers are briefly described.

8.1 Observer design for linear systems

We briefly review the results for linear systems. Consider

$$\dot{x} = Ax$$
$$y = Cx,$$

(8.1)

where $x \in \mathbb{R}^n$ is the state; $y \in \mathbb{R}^m$ is the system output with $m < n$; and $A \in \mathbb{R}^{n \times n}$ and $C \in \mathbb{R}^{m \times n}$ are constant matrices. From linear system theory, we know that this system, or the pair (A, C), is observable if the matrix

$$\mathcal{P}_o = \begin{bmatrix} C \\ CA \\ \vdots \\ CA^{n-1} \end{bmatrix}$$

has rank n. The observability condition is equivalent to that the matrix

$$\begin{bmatrix} \lambda I - A \\ C \end{bmatrix}$$

has rank n for any value of $\lambda \in \mathbb{C}$.

When the system is observable, an observer can be designed as

$$\dot{\hat{x}} = A\hat{x} + L(y - C\hat{x}),$$

(8.2)

where $\hat{x} \in \mathbb{R}^n$ is the estimate of the state x, and $L \in \mathbb{R}^{n \times m}$ is the observer gain such that $(A - LC)$ is Hurwitz.

For the observer (8.2), it is easy to see the estimate $\hat{x}$ converges to x asymptotically. Let $\tilde{x} = x - \hat{x}$, and we can obtain

$$\dot{\tilde{x}} = (A - LC)\tilde{x}.$$

Remark 8.1. In the observer design, we do not consider control input terms in the system in (8.1), as they do not affect the observer design for linear systems. In fact, if Bu term is added to the right-hand side of the system (8.1), for the observer design, we can simply add it to the right-hand side of the observer in (8.2), and the observer error will still converge to zero exponentially.   ◁

Remark 8.2. The observability condition for the observer design of (8.1) can be relaxed to the detectability of the system, or the condition that (A, C) is detectable, for the existence of an observer gain L such that $(A - LC)$ is Hurwitz. Detectability is weaker than observability, and it basically requires the unstable modes of the system observable. The pair (A, C) is detectable if the matrix

$$\begin{bmatrix} \lambda I - A \\ C \end{bmatrix}$$

has rank n for any λ in the closed right half of the complex plan. Some other design methods shown in this chapter also need only the condition of detectability, although, for simplicity, we state the requirement for the observability.   ◁

There is another approach to full-state observer design for linear systems (8.1). Consider a dynamic system

$$\dot{z} = Fz + Gy, \tag{8.3}$$

where $z \in \mathbb{R}^n$ is the state; $F \in \mathbb{R}^{n \times n}$ is Hurwitz; and $G \in \mathbb{R}^{n \times m}$. If there exists an invertible matrix $T \in \mathbb{R}^{n \times n}$ such that Z converges to Tx, then (8.3) is an observer with the state estimate given by $\hat{x} = T^{-1}z$. Let

$$e = Tx - z.$$

A direction evaluation gives

$$\begin{aligned} \dot{e} &= TAx - (Fz + GCx) \\ &= F(Tx - z) + (TA - FT - GC)x \\ &= Fe + (TA - FT - GC)x. \end{aligned}$$

If we have

$$TA - FT - GC = 0,$$

the system (8.3) is an observer with an exponentially convergent estimation error. We have the following lemma to summarise this observer design.

Lemma 8.1. *The dynamic system (8.3) is an observer for the system (8.1) if and only if F is Hurwitz and there exists an invertible matrix T such that*

$$TA - FT = GC. \tag{8.4}$$

Proof. The sufficiency has been shown in the above analysis. For necessity, we only need to observe that if any of the conditions is not satisfied, we cannot guarantee the convergence of e to zero for a general linear system (8.1). Indeed, if (8.4) is not satisfied, then e will be a state variable with a non-zero input, and we can set up a case such that e does not converge to zero. So does for the condition that F is Hurwitz. □

How to find matrices F and G such that the condition (8.4) is satisfied? We list the result in the following lemma without the proof.

Lemma 8.2. *Suppose that F and A have exclusively different eigenvalues. The necessary condition for the existence of a non-singular solution T to the matrix equation (8.4) is that the pair (A, C) is observable and the pair (F, G) is controllable. This condition is also sufficient when the system (8.1) is single output, i.e., $m = 1$.*

This lemma suggests that we can choose a controllable pair (F, G) and make sure that the eigenvalues of F are different from those of A. An observer can be designed if there is a solution of T from (8.4). For single output system, the solution is guaranteed.

8.2 Linear observer error dynamics with output injection

Now consider

$$\dot{x} = Ax + \phi(y, u)$$
$$y = Cx, \tag{8.5}$$

where $x \in \mathbb{R}^n$ is the state, $y \in \mathbb{R}^m$ is the system output with $m < n$, $u \in \mathbb{R}^s$ is the control input, or other known variables, $A \in \mathbb{R}^{n \times n}$ and $C \in \mathbb{R}^{m \times n}$ are constant matrices and $\phi : \mathbb{R}^m \times \mathbb{R}^s \to \mathbb{R}^n$ is a continuous function. This system is a nonlinear system. However, comparing with the system (8.1), the only difference is the additional term $\phi(y, u)$. The system (8.5) can be viewed as the linear system (8.1) perturbed by the nonlinear term $\phi(y, u)$.

If the pair (A, C) is observable, we can design an observer as

$$\dot{\hat{x}} = A\hat{x} + L(y - C\hat{x}) + \phi(y, u), \tag{8.6}$$

where $\hat{x} \in \mathbb{R}^n$ is the estimate of the state x and $L \in \mathbb{R}^{n \times m}$ is the observer gain such that $(A - LC)$ is Hurwitz. The only difference between this observer and the one in (8.2) is the nonlinear term $\phi(y, u)$. It can be seen that the observer error still satisfies

$$\dot{\tilde{x}} = (A - LC)\tilde{x}.$$

Note that even the system (8.5) is nonlinear, the observer error dynamics are linear. The system in the format of (8.5) is referred to as the system with linear observer errors. More specifically, if we drop the control input u in the function ϕ, the system is referred to as the output injection form for observer design.

Let us summarise the result in the following proposition.

Proposition 8.3. *For the nonlinear system (8.5), a full-state observer can be designed as in (8.6) if (A, C) is observable. Furthermore, the observer error dynamics are linear and exponentially stable.*

For a nonlinear system in a more general form, there may exist a state transformation to put the system in the format of (8.5), and then the proposed observer can be applied.

We will introduce the conditions for the existence of a nonlinear state transformation to put the system in the format shown in (8.6). Here, we only consider single output case, and for the simplicity, we do not consider the system with a control input.

The system under consideration is described by

$$\dot{x} = f(x)$$
$$y = h(x),$$
$$(8.7)$$

where $x \in \mathbb{R}^n$ is the state vector, $y \in \mathbb{R}$ is the output, $f : \mathbb{R}^n \to \mathbb{R}^n$ and $h : \mathbb{R}^n \to \mathbb{R}$ are continuous nonlinear functions with $f(0) = 0$, and $h(0) = 0$. We will show that under what conditions there exists a state transformation

$$z = \Phi(x),$$
$$(8.8)$$

where $\Phi : \mathbb{R}^n \to \mathbb{R}^n$, such that the transformed system is in format

$$\dot{z} = Az + \phi(Cz)$$
$$y = Cz,$$
$$(8.9)$$

where $A \in \mathbb{R}^{n \times n}$, $C \in \mathbb{R}^{1 \times n}$, (A, C) is observable and $\phi : \mathbb{R} \to \mathbb{R}^n$. Without loss of generality, we take the transformed system (8.9) as

$$\begin{aligned}
\dot{z}_1 &= z_2 + \phi_1(y) \\
\dot{z}_2 &= z_3 + \phi_2(y) \\
&\quad\cdots \\
\dot{z}_{n-1} &= z_n + \phi_{n-1}(y) \\
\dot{z}_n &= \phi_n(y) \\
y &= z_1.
\end{aligned}$$
$$(8.10)$$

This system is in the output injection form for nonlinear observer design.

Remark 8.3. Assume that the system (8.9) is different from (8.10) with (A, C) as a general observable pair, instead of having the special format implied by (8.10).

In this case, we use a different variable $\bar{z}$ to denote the state for (8.10), and it can be written as

$$\dot{\bar{z}} = \bar{A}\bar{z} + \bar{\phi}(y)$$
$$y = \bar{C}\bar{z}. \tag{8.11}$$

It is clear that there exists a linear transformation

$$\bar{z} = Tz,$$

which transforms the system (8.9) to the system (8.11), because (A, C) is observable. The transformation from (8.7) to (8.11) is given by

$$\bar{z} = T\phi(x) := \bar{\Phi}(x).$$

Therefore, if there exists a nonlinear transformation from (8.7) to (8.9), there must exist a nonlinear transformation from (8.7) to (8.11), and vise versa. That is why we can consider the transformed system in the format of (8.10) without loss of generality. ◁

If the state transformation transforms the system (8.7) to (8.9), we must have

$$\left[\frac{\partial \Phi(x)}{\partial x} f(x) \right]_{x=\Psi(z)} = Az + \phi(Cz) \tag{8.12}$$
$$h(\Psi(z)) = Cz,$$

where $\Psi = \Phi^{-1}$. For notational convenience, let us denote

$$\bar{f}(z) = Az + \phi(Cz)$$
$$\bar{h}(z) = Cz.$$

From the structure shown in (8.10), we can obtain

$$\bar{h}(z) = z_1,$$
$$L_{\bar{f}}\bar{h}(z) = z_2 + \phi_1(z_1),$$
$$L_{\bar{f}}^2\bar{h}(z) = z_3 + \frac{\partial \phi_1}{\partial z_1}(z_2 + \phi_1(z_1))$$
$$:= z_3 + \bar{\phi}_2(z_1, z_2),$$

$$\cdots$$

$$L_{\bar{f}}^{n-1}\bar{h}(z) = z_n + \sum_{k=1}^{n-2} \frac{\partial \bar{\phi}_{n-2}}{\partial z_k}(z_{k+1} + \phi_k(z_1))$$
$$:= z_n + \bar{\phi}_{n-1}(z_1, \ldots, z_{n-1}).$$

From the above expression, it is clear that

$$
\begin{bmatrix}
\dfrac{\partial \bar h}{\partial z} \\[2mm]
\dfrac{\partial L_{\bar f} \bar h(z)}{\partial z} \\[2mm]
\vdots \\[2mm]
\dfrac{\partial L_{\bar f}^{n-1} \bar h(z)}{\partial z}
\end{bmatrix}
=
\begin{bmatrix}
1 & 0 & \cdots & 0 \\
* & 1 & \cdots & 0 \\
\vdots & \vdots & \ddots & \vdots \\
* & * & \cdots & 1
\end{bmatrix}
$$

This implies that

$$
d\bar h,\ dL_{\bar f}\bar h,\ \ldots,\ dL_{\bar f}^{n-1}\bar h
$$

are linearly independent. This property is invariant under state transformation, and therefore, we need the condition under the coordinate x, that is

$$
dh,\ dL_f h,\ \ldots,\ dL_f^{n-1} h
$$

are linearly independent. Indeed, we have

$$
L_f^{n-1} h(x) = (L_{\bar f}^{n-1} \bar h(z))_{z=\Psi(x)}, \quad \text{for } k = 0, 1, \ldots, n-1
$$

and therefore

$$
\begin{bmatrix}
\dfrac{\partial h(x)}{\partial x} \\[2mm]
\dfrac{\partial L_f h(x)}{\partial x} \\[2mm]
\vdots \\[2mm]
\dfrac{\partial L_f^{n-1} h(x)}{\partial x}
\end{bmatrix}
=
\begin{bmatrix}
\dfrac{\partial \bar h(z)}{\partial z} \\[2mm]
\dfrac{\partial L_{\bar f} \bar h(z)}{\partial z} \\[2mm]
\vdots \\[2mm]
\dfrac{\partial L_{\bar f}^{n-1} \bar h(z)}{\partial z}
\end{bmatrix}_{z=\Psi(x)}
\dfrac{\partial \Psi(x)}{\partial x}.
$$

The linear independence of $dh,\ dL_f h,\ \ldots,\ dL_f^{n-1} h$ is a consequence of the observability of (A, C) in (8.9). In some literature, this linear independence condition is defined as the observability condition for nonlinear system (8.7). Unlike linear systems, this condition is not enough to design a nonlinear observer. We state necessary and sufficient conditions for the transformation to the output injection form in the following theorem.

Theorem 8.4. *The nonlinear system (8.7) can be transformed to the output injection form in (8.10) if and only if*

- *the differentials $dh,\ dL_f h,\ \ldots,\ dL_f^{n-1} h$ are linearly independent*
- *there exists a map $\Psi : \mathbb{R}^n \to \mathbb{R}^n$ such that*

$$\frac{\partial \Psi(z)}{\partial z} = \left[ad_{-f}^{n-1}r, \ldots, ad_{-f}r, r \right]_{x=\Psi(z)}, \tag{8.13}$$

where r is a vector field solved from

$$\begin{bmatrix} dh \\ dL_f h \\ \vdots \\ dL_f^{n-1}h \end{bmatrix} r = \begin{bmatrix} 0 \\ \vdots \\ 0 \\ 1 \end{bmatrix}. \tag{8.14}$$

Proof. *Sufficiency.* From the first condition and (8.14), we can show, in a similar way as for (6.12), that

$$\begin{bmatrix} dh(x) \\ dL_f h(x) \\ \vdots \\ dL_f^{\rho-1}h \end{bmatrix} \left[ad_{-f}^{n-1}r(x) \ \ldots \ ad_{-f}r(x) \ r(x) \right]$$

$$= \begin{bmatrix} 1 & 0 & \ldots & 0 \\ * & 1 & \ldots & 0 \\ \vdots & \vdots & \ddots & \vdots \\ * & * & \ldots & 1 \end{bmatrix}. \tag{8.15}$$

Therefore, $\left[ad_{-f}^{n-1}r(x) \ \ldots \ ad_{-f}r(x) \ r(x) \right]$ has full rank. This implies that there exists an inverse mapping for Ψ. Let us denote it as $\Phi = \Psi^{-1}$, and hence we have

$$\frac{\partial \Phi(x)}{\partial x} \left[ad_{-f}^{n-1}r(x) \ \ldots \ ad_{-f}r(x) \ r(x) \right] = I. \tag{8.16}$$

Let us define the state transformation as $z = \Phi(x)$ and denote the functions after this transformation as

$$\bar{f}(z) = \left[\frac{\partial \Phi(x)}{\partial x} f(x) \right]_{x=\Psi(z)},$$

$$\bar{h}(z) = h(\Psi(z)).$$

We need to show that the functions $\bar{f}$ and $\bar{h}$ are in the format of the output injection form as in (8.10). From (8.16), we have

$$\frac{\partial \Phi(x)}{\partial x} ad_{-f}^{n-k}r(x) = e_k, \quad \text{for } k = 1, \ldots, n,$$

where e_k denotes the kth column of the identity matrix. Hence, we have, for $k = 1, \ldots, n-1$,

$$\left[\frac{\partial \Phi(x)}{\partial x} ad_{-f}^{n-k} r(x)\right]_{x=\Psi(z)} = \left[\frac{\partial \Phi(x)}{\partial x}[-f(x), ad_{-f}^{n-(k+1)} r(x)]\right]_{x=\Psi(z)}$$

$$= \left[-\frac{\partial \Phi(x)}{\partial x}f(x), \frac{\partial \Phi(x)}{\partial x} ad_{-f}^{n-(k+1)} r(x)\right]_{x=\Psi(z)}$$

$$= [-\bar{f}(z), e_{k+1}]$$

$$= \frac{\partial \bar{f}(z)}{\partial z_{k+1}}.$$

This implies that

$$\frac{\partial \bar{f}(z)}{\partial z_{k+1}} = e_k, \quad \text{for } k = 1, \dots, n-1,$$

i.e.,

$$\frac{\partial \bar{f}(z)}{\partial z} = \begin{bmatrix} * & 1 & 0 & \cdots & 0 \\ * & 0 & 1 & \cdots & 0 \\ \vdots & \vdots & \vdots & \ddots & \vdots \\ * & 0 & 0 & \cdots & 1 \\ * & 0 & 0 & \cdots & 0 \end{bmatrix}.$$

Therefore, we have shown that $\bar{f}$ is in the output injection form.
From the second condition, we have

$$\frac{\partial \Psi(z)}{\partial z_k} = [ad_{-f(x)}^{n-k} r(x)]_{x=\Psi(z)} \quad \text{for } k = 1, \dots, n.$$

Hence, we obtain that, for $k = 1, \dots, n$,

$$\frac{\partial \bar{h}(z)}{\partial z_k} = \left[\frac{\partial h(x)}{\partial x}\right]_{x=\Psi(z)} \frac{\partial \Psi(z)}{\partial z_k}$$

$$= \left[\frac{\partial h(x)}{\partial x}\right]_{x=\Psi(z)} [ad_{-f(x)}^{n-k} r(x)]_{x=\Psi(z)}$$

$$= [L_{ad_{-f(x)}^{n-k} r(x)} h(x)]_{x=\Psi(z)}.$$

Furthermore from (8.15), we have

$$L_{ad_{-f(x)}^{n-1} r(x)} h(x) = 1,$$

$$L_{ad_{-f(x)}^{n-k} r(x)} h(x) = 0, \quad \text{for } k = 2, \dots, n.$$

Therefore, we have

$$\frac{\partial \bar{h}(z)}{\partial z} = [1, 0, \dots, 0].$$

This concludes the proof for sufficiency.

Necessity. The discussion prior to this theorem shows that the first condition is necessary. Assume that there exists a state transformation $z = \Phi(x)$ to put the system in the output injection form, and once again, we denote

$$\bar{f}(z) = \left[\frac{\partial \Phi(x)}{\partial x} f(x) \right]_{x=\Psi(z)}$$

$$\bar{h}(z) = h(\Psi(z)),$$

where $\Psi = \Phi^{-1}$. We need to show that when the functions $\bar{f}$ and $\bar{h}$ are in the format of the output injection form, the second condition must hold. Let

$$g(x) = \left[\frac{\partial \Psi(z)}{\partial z_n} \right]_{z=\Phi(x)}.$$

From the state transformation, we have

$$f(x) = \left[\frac{\partial \Psi(z)}{\partial z} \bar{f}(z) \right]_{z=\Phi(x)}.$$

Therefore, we can obtain

$$[-f(x), g(x)] = \left[-\left[\frac{\partial \Psi(z)}{\partial z} \bar{f}(z) \right]_{z=\Phi(x)}, \left[\frac{\partial \Psi(z)}{\partial z_n} \right]_{z=\Phi(x)} \right]$$

$$= \left[\frac{\partial \Psi(z)}{\partial z} \right]_{z=\Phi(x)} [-\bar{f}(z), e_n]_{z=\Phi(x)}$$

$$= \left[\frac{\partial \Psi(z)}{\partial z} \right]_{z=\Phi(x)} \frac{\partial \bar{f}(z)}{\partial z_n}$$

$$= \left[\frac{\partial \Psi(z)}{\partial z} \right]_{z=\Phi(x)} e_{n-1}$$

$$= \left[\frac{\partial \Psi(z)}{\partial z_{n-1}} \right]_{z=\Phi(x)}.$$

Similarly, we can show that

$$ad_{-f}^{n-k} g = \left[\frac{\partial \Psi(z)}{\partial z_{n-k}} \right]_{z=\Phi(x)}, \quad \text{for } k = n-2, \ldots, 1.$$

Hence, we have established that

$$\frac{\partial \Psi(z)}{\partial z} = \left[ad_{-f}^{n-1} g, \ldots, ad_{-f} g, g \right]_{x=\Psi(z)}. \tag{8.17}$$

The remaining part of the proof is to show that $g(x)$ coincides with $r(x)$ in (8.14). From (8.17), we have

$$\frac{\partial \bar{h}(z)}{\partial z} = \frac{\partial h(x)}{\partial x} \frac{\partial \Psi(z)}{\partial z}$$

$$= \frac{\partial h(x)}{\partial x}\left[ad_{-f}^{n-1}g, \ldots, ad_{-f}g, g\right]_{x=\Psi(z)}$$

$$= \left[L_{ad_{-f}^{n-1}g}h(x), \ldots, L_{ad_{-f}g}h(x), L_g h(x)\right]_{x=\Psi(z)}.$$

Since $\bar{h}(z)$ is in the output feedback form, the above expression implies that

$$L_{ad_{-f}^{n-1}g}h(x) = 1,$$

$$L_{ad_{-f}^{n-k}g}h(x) = 0, \quad \text{for } k = 2, \ldots, n,$$

which further imply that

$$L_g L^{n-1}h(x) = 1,$$

$$L_g L^{n-k}h(x) = 0, \quad \text{for } k = 2, \ldots, n,$$

i.e.,

$$\begin{bmatrix} dh \\ dL_f h \\ \vdots \\ dL_f^{n-1}h \end{bmatrix} g = \begin{bmatrix} 0 \\ \vdots \\ 0 \\ 1 \end{bmatrix}.$$

This completes the proof of necessity. $\qquad\qquad\square$

Remark 8.4. It would be interesting to revisit the transformation for linear single-output systems to the observer canonical form, to reveal the similarities between the linear case and the conditions stated in Theorem 8.4. For a single output system

$$\begin{aligned} \dot{x} &= Ax \\ y &= Cx, \end{aligned} \qquad\qquad (8.18)$$

where $x \in \mathbb{R}^n$ is the state; $y \in \mathbb{R}$ is the system output; and $A \in \mathbb{R}^{n\times n}$ and $C \in \mathbb{R}^{1\times n}$ are constant matrices. When the system is observable, we have $\mathcal{P}_o$ full rank. Solving r from $\mathcal{P}_o r = e_1$, i.e.,

$$\begin{bmatrix} C \\ CA \\ \vdots \\ CA^{n-1} \end{bmatrix} r = \begin{bmatrix} 0 \\ \vdots \\ 0 \\ 1 \end{bmatrix},$$

and the state transformation matrix T is then given by

$$T^{-1} = [A^{n-1}r, \ldots, Ar, r]. \qquad\qquad (8.19)$$

We can show that the transformation $z = Tx$ which transforms the system to the observer canonical form

$$\dot{z} = TAT^{-1}z := \bar{A}z$$

$$y = CT^{-1}z := \bar{C}z$$

where

$$\bar{A} = \begin{bmatrix} -a_1 & 1 & 0 & \ldots & 0 \\ -a_2 & 0 & 1 & \ldots & 0 \\ \vdots & \vdots & \vdots & \ddots & \vdots \\ -a_{n-1} & 0 & 0 & \ldots & 1 \\ -a_n & 0 & 0 & \ldots & 0 \end{bmatrix},$$

$$\bar{C} = [1 \ 0 \ \ldots \ 0 \ 0],$$

with constants a_i, $i = 1, \ldots, n$, being the coefficients of the characteristic polynomial

$$|sI - A| = s^n + a_1 s^{n-1} + \cdots + a_{n-1}s + a_n.$$

Indeed, from (8.19), we have

$$AT^{-1} = [A^n r, \ldots, A^2 r, Ar] \tag{8.20}$$

Then from Cayley–Hamilton theorem, we have

$$A^n = -a_1 A^{n-1} - \cdots - a_{n-1}A - a_n I.$$

Substituting this into the previous equation, we obtain that

$$AT^{-1} = [A^{n-1}r, \ldots, Ar, r] \begin{bmatrix} -a_1 & 1 & 0 & \ldots & 0 \\ -a_2 & 0 & 1 & \ldots & 0 \\ \vdots & \vdots & \vdots & \ddots & \vdots \\ -a_{n-1} & 0 & 0 & \ldots & 1 \\ -a_n & 0 & 0 & \ldots & 0 \end{bmatrix} = T^{-1}\bar{A},$$

which gives

$$TAT^{-1} = \bar{A}.$$

Again from (8.19), we have

$$CT^{-1} = [CA^{n-1}r, \ldots, CAr, Cr] = [1, \ 0, \ldots, 0] = \bar{C}.$$

If we identify $f(x) = Ax$ and $h(x) = Cx$, the first condition of Theorem 8.4 is that $\mathcal{P}_o$ has full rank, and the condition in (8.13) is identical as (8.19). ◁

When the conditions for Theorem 8.4 are satisfied, the transformation $z = \Phi(x)$ exists. In such a case, an observer can be designed for the nonlinear system (8.7) as

$$\dot{\hat{z}} = A\hat{z} + L(y - C\hat{z}) + \phi(y)$$

$$\hat{x} = \Psi(\hat{z}), \tag{8.21}$$

where $\hat{x} \in \mathbb{R}^n$ is the estimate of the state x of (8.7); $\hat{z} \in \mathbb{R}^n$ is an estimate of z in (8.10); and $L \in \mathbb{R}^n$ is the observer gain such that $(A - LC)$ is Hurwitz.

Corollary 8.5. *For nonlinear system (8.7), if the conditions for Theorem 8.4 are satisfied, the observer in (8.21) provides an asymptotic state estimate of the system.*

Proof. From Theorem 8.4, we conclude that (A, C) is observable, and there exists an L such that $(A - LC)$ is Hurwitz. From (8.21) and (8.10), we can easily obtain that

$$\dot{\tilde{z}} = (A - LC)\tilde{z}$$

where $\tilde{z} = z - \hat{z}$. We conclude that $\hat{x}$ asymptotically converges to x as $\lim_{t \to \infty} \tilde{z}(t) = 0$. $\qquad \square$

8.3   Linear observer error dynamics via direct state transformation

Observers presented in the last section achieve the linear observer error dynamics by transforming the nonlinear system to the output injection form, for which an observer with linear observer error dynamics can be designed to estimate the transformed state, and the inverse transformation is used to transform the estimate for the original state. In this section, we take a different approach by introducing a state transformation which directly leads to a nonlinear observer design with linear observer error dynamics, and show that this observer can be directly implemented in the original state space.

We consider the same system (8.7) as in the previous section, and describe it here under a different equation number for the convenience of presentation

$$\begin{aligned} \dot{x} &= f(x) \\ y &= h(x), \end{aligned} \tag{8.22}$$

where $x \in \mathbb{R}^n$ is the state vector, $y \in \mathbb{R}$ is the output, $f : \mathbb{R}^n \to \mathbb{R}^n$ and $h : \mathbb{R}^n \to \mathbb{R}$ are continuous nonlinear functions with $f(0) = 0$, and $h(0) = 0$. We now consider if there exists a state transformation

$$z = \Phi(x), \tag{8.23}$$

where $\Phi : \mathbb{R}^n \to \mathbb{R}^n$, such that the transformed system is in format

$$\dot{z} = Fz + Gy, \tag{8.24}$$

for a chosen controllable pair (F, G) with $F \in \mathbb{R}^{n \times n}$ Hurwitz, and $G \in \mathbb{R}^n$. Comparing (8.24) with (8.9), the matrices F and G in (8.24) are chosen for the observer design, while in (8.9), A and C are any observable pair which depends on the original system. Therefore, the transformation in (8.24) is more specific. There is an extra benefit gained from this restriction in the observer design as shown later.

From (8.22) and (8.24), the nonlinear transformation must satisfy the following partial differential equation

$$\frac{\partial \Phi(x)}{\partial x}f(x) = F\Phi(x) + Gh(x).$$
(8.25)

Our discussion will be based on a neighbourhood around the origin.

Definition 8.1. Let $\lambda_i(A)$ for $i = 1, \ldots, n$ are the eigenvalues of a matrix $A \in \mathbb{R}^n$. For another matrix $F \in \mathbb{R}^n$, an eigenvalue of F is resonant with the eigenvalues of A if there exists an integer $q = \sum_{i=1}^{n} q_i > 0$ with q_i being non-negative integers such that for some j with $1 \leq j \leq n$

$$\lambda_j(F) = \sum_{i=1}^{n} q_i \lambda_i(A).$$

The following theorem states a result concerning with the existence of a nonlinear state transformation around the origin for (8.22).

Theorem 8.6. *For the nonlinear system (8.22), there exists a state transformation, i.e., a locally invertible solution to the partial differential equation (8.25) if*

- *the linearised model of (8.22) around the origin is observable*
- *the eigenvalues of F are not resonant with the eigenvalues of $\frac{\partial f}{\partial x}(0)$*
- *the convex hall of $\left\{\lambda_1\left(\frac{\partial f}{\partial x}(0)\right), \ldots, \lambda_n\left(\frac{\partial f}{\partial x}(0)\right)\right\}$ does not contain the origin*

Proof. From the non-resonant condition and the exclusion of the origin of the convex hall of the eigenvalues of $\frac{\partial f}{\partial x}(0)$, we can establish the existence of a solution to the partial differential equation (8.25) by invoking Lyapunov Auxiliary Theorem. That the function Φ is invertible around the origin is guaranteed by the observability of the linearised model and the the controllability of (F, G). Indeed, with the observability of $\left(\frac{\partial f}{\partial x}(0), \frac{\partial h}{\partial x}(0)\right)$ and the controllability of (F, G), we can apply Lemma 8.2 to establish that $\frac{\partial \Phi}{\partial x}(0)$ is invertible. $\square$

With the existence of the nonlinear transformation to put the system in the form of (8.24), an observer can be designed as

$$\dot{\hat{z}} = F\hat{z} + Gy$$
$$\hat{x} = \Psi^{-1}(\hat{z}),$$
(8.26)

where Ψ is the inverse transformation of Φ. It is easy to see that the observer error dynamics are linear as

$$\dot{\tilde{z}} = F\tilde{z}.$$

Note that once the transformation is obtained, the observer is directly given without designing observer gain, unlike the observer design based on the output injection form.

The observer can also be implemented directly in the original state as

$$\dot{\hat{x}} = f(\hat{x}) + \left(\frac{\partial \Phi}{\partial \hat{x}}(\hat{x})\right)^{-1} G(y - h(\hat{x})), \tag{8.27}$$

which is in the same structure as the standard Luenberger observer for linear systems by viewing $\left(\frac{\partial \Phi}{\partial \hat{x}}(\hat{x})\right)^{-1} G$ as the observer gain. In the following theorem, we show that this observer also provides an asymptotic estimate of the system state.

Theorem 8.7. *For the nonlinear system (8.22), if the state transformation in (8.25) exists, the observer (8.27) provides an asymptotic estimate. Furthermore, the dynamics of the transformed observer error* $(\Phi(x) - \Phi(\hat{x}))$ *are linear.*

Proof. Let $e = \Phi(x) - \Phi(\hat{x})$. Direct evaluation gives

$$\dot{e} = \frac{\partial \Phi(x)}{\partial x} f(x) - \frac{\partial \Phi(\hat{x})}{\partial \hat{x}} \left(f(\hat{x}) + \left(\frac{\partial \Phi}{\partial \hat{x}}(\hat{x})\right)^{-1} G(y - h(\hat{x}))\right)$$

$$= \frac{\partial \Phi(x)}{\partial x} f(x) - \frac{\partial \Phi(\hat{x})}{\partial \hat{x}} f(\hat{x}) - G(y - h(\hat{x}))$$

$$= F\Phi(x) + Gy - (F\Phi(\hat{x}) + Gh(\hat{x})) - G(y - h(\hat{x}))$$

$$= F(\Phi(x) - \Phi(\hat{x}))$$

$$= Fe.$$

Therefore, the dynamics of the transformed observer error are linear, and the transformed observer error converges to zero exponentially, which implies the asymptotic convergence of $\hat{x}$ to x. □

Remark 8.5. For linear systems, we have briefly introduced two ways to design observers. For the observer shown in (8.2), we have introduced the observer shown in (8.6) to deal with nonlinear systems. The nonlinear observer in (8.27) can be viewed as a nonlinear version of (8.3). For both cases, the observer error dynamics are linear. ◁

8.4 Observer design for Lipschitz nonlinear systems

In this section, we will deal with observer design for nonlinear systems with Lipschitz nonlinearity. We introduced this definition for a time-varying function in Chapter 2 for the existence of a unique solution for a nonlinear system. For a time-invariant function, the definition is similar, and we list below for the convenience of presentation.

Definition 8.2. A function $\phi : \mathbb{R}^n \times \mathbb{R}^s \to \mathbb{R}^n$ is Lipschitz with a Lipschitz constant γ if for any vectors $x, \hat{x} \in \mathbb{R}^n$ and $u \in \mathbb{R}^s$

$$\|\phi(x, u) - \phi(\hat{x}, u)\| \le \gamma \|x - \hat{x}\|, \tag{8.28}$$

with $\gamma > 0$.

Once again we consider linear systems perturbed by nonlinear terms as

$$\begin{aligned}\dot{x} &= Ax + \phi(x, u) \\ y &= Cx,\end{aligned} \tag{8.29}$$

where $x \in \mathbb{R}^n$ is the state; $y \in \mathbb{R}^m$ is the system output with $m < n$; $u \in \mathbb{R}^s$ is the control input, or other known variables, $A \in \mathbb{R}^{n \times n}$ and $C \in \mathbb{R}^{m \times n}$ are constant matrices with (A, C) observable; and $\phi : \mathbb{R}^n \times \mathbb{R}^s \to \mathbb{R}^n$ is a continuous function with Lipschitz constant γ with respect to the state variable x. Comparing with the system (8.5), the only difference is the nonlinear term $\phi(x, u)$. Here, it is a function of state variable, not only the output, as in (8.5), and therefore the observer design by output injection does not work.

We can still design an observer based on the linear part of the system, and replace the unknown state in the nonlinear function by its estimate, that is,

$$\dot{\hat{x}} = A\hat{x} + L(y - C\hat{x}) + \phi(\hat{x}, u), \tag{8.30}$$

where $\hat{x} \in \mathbb{R}^n$ is the estimate of the state x and $L \in \mathbb{R}^{n \times m}$ is the observer gain. However, the condition that $(A - LC)$ is Hurwitz is not enough to guarantee the convergence of the observer error to zero. Indeed, a stronger condition is needed, as shown in the following theorem.

Theorem 8.8. *The observer (8.30) provides an exponentially convergent state estimate if for the observer gain L, there exists a positive definite matrix $P \in \mathbb{R}^{n \times n}$ such that*

$$(A - LC)^T P + P(A - LC) + \gamma^2 PP + I + \epsilon I = 0, \tag{8.31}$$

where γ is the Lipschitz constant of ϕ and ϵ is any positive real constant.

Proof. Let $\tilde{x} = x - \hat{x}$. From (8.29) and (8.31), we have

$$\dot{\tilde{x}} = (A - LC)\tilde{x} + \phi(x, u) - \phi(\hat{x}, u).$$

Let

$$V = \tilde{x}^T P \tilde{x}.$$

Its derivative along the observer error dynamics is obtained as

$$\dot{V} = \tilde{x}^T ((A - LC)^T P + P(A - LC))\tilde{x} + 2\tilde{x}^T P(\phi(x, u) - \phi(\hat{x}, u)).$$

For the term involved with the nonlinear function ϕ, we have

$$2\tilde{x}^T P(\phi(x,u) - \phi(\hat{x},u))$$

$$\leq \gamma^2 \tilde{x}^T PP\tilde{x} + \frac{1}{\gamma^2}(\phi(x,u) - \phi(\hat{x},u))^T (\phi(x,u) - \phi(\hat{x},u))$$

$$= \gamma^2 \tilde{x}^T PP\tilde{x} + \frac{1}{\gamma^2}\|\phi(x,u) - \phi(\hat{x},u)\|^2$$

$$\leq \gamma^2 \tilde{x}^T PP\tilde{x} + \|\tilde{x}\|^2$$

$$= \tilde{x}^T (\gamma^2 PP + I)\tilde{x}.$$

Applying the condition shown in (8.31), we have the derivative of V satisfying

$$\dot{V} \leq \tilde{x}^T ((A - LC)^T P + P(A - LC) + \gamma^2 PP + I)\tilde{x}$$

$$= -\epsilon \tilde{x}^T \tilde{x}.$$

This implies that the observer error converges to zero exponentially. □

In the condition (8.31), ϵ is an arbitrary positive real number. In this case, we can use inequality to replace the equality, that is

$$(A - LC)^T P + P(A - LC) + \gamma^2 PP + I < 0. \tag{8.32}$$

Using the same Lyapunov function candidate as in the proof of Theorem 8.8, we can show that

$$\dot{V} < 0$$

which implies that the observer error converges to zero asymptotically. We summarise this result below.

Corollary 8.9. *The observer (8.30) provides an asymptotically convergent state estimate if for the observer gain L, there exists a positive definite matrix $P \in \mathbb{R}^{n \times n}$ that satisfies the inequality (8.32).*

Remark 8.6. By using the inequality (8.32) instead of the equality (8.31), we only establish the asymptotic convergence to zero of the observer error, not the exponential convergence that is established in Theorem 8.8 using (8.31). Furthermore, establishing the asymptotic convergence of the observer error from $\dot{V} < 0$ requires the stability theorems based on invariant sets, which are not covered in this book. ◁

The condition shown in (8.32) can be relaxed if one-side Lipschitz constant is used instead of the Lipschitz constant.

Definition 8.3. A function $\phi : \mathbb{R}^n \times \mathbb{R}^s \to \mathbb{R}^n$ is one-sided Lipschitz with respect to P and one-sided Lipschitz constant v if for any vectors $x, \hat{x} \in \mathbb{R}^n$ and $u \in \mathbb{R}^s$

$$(x - \hat{x})^T P(\phi(x,u) - \phi(\hat{x},u)) \leq v\|x - \hat{x}\|^2 \tag{8.33}$$

where $P \in \mathbb{R}^{n \times n}$ is a positive real matrix, and v is a real number.

Note that the one-sided Lipschitz constant ν can be negative. It is easy to see from the definition of the one-sided Lipschitz condition that the term $(x - \hat{x})^T P(\phi(x, u) - \phi(\hat{x}, u))$ is exactly the cross-term in the proof Theorem 8.8 which causes the term $\gamma^2 PP + I$ in (8.32). Hence, with the Lipschitz constant ν with respect to P, the condition shown in (8.32) can be replaced by

$$(A - LC)^T P + P(A - LC) + 2\nu I < 0. \tag{8.34}$$

This condition can be further manipulated to obtain the result shown in the following theorem.

Theorem 8.10. *The observer (8.30) provides an asymptotically convergent state estimate if the following conditions hold:*

$$L = \sigma P^{-1} C^T, \tag{8.35}$$

$$A^T P + PA + 2\nu I - 2\sigma C^T C < 0, \tag{8.36}$$

where $P \in \mathbb{R}^{n \times n}$ is a positive real matrix, σ is a positive real constant and ν is the one-sided Lipschitz constant of ϕ with respect to x and P.

Proof. From (8.35), we have

$$\left(\sqrt{\sigma} CP^{-1} - \frac{L^T}{\sqrt{\sigma}} \right)^T \left(\sqrt{\sigma} CP^{-1} - \frac{L^T}{\sqrt{\sigma}} \right) = 0,$$

which gives

$$P^{-1} C^T L^T + LCP^{-1} = \sigma P^{-1} C^T CP^{-1} + \frac{LL^T}{\sigma}.$$

Using (8.35) and multiplying the above equation by P on both sides, we obtain the identity

$$C^T L^T P + PLC = 2\sigma C^T C.$$

From this identity and (8.36), we can easily obtain the inequality (8.34). Similar to the proof of Theorem 8.8, we let

$$V = \tilde{x}^T P \tilde{x},$$

and obtain, using the one-sided Lipschitz condition of ϕ,

$$\dot{V} = \tilde{x}^T ((A - LC)^T P + P(A - LC)) \tilde{x} + 2\tilde{x}^T P(\phi(x, u) - \phi(\hat{x}, u))$$
$$\leq \tilde{x}^T ((A - LC)^T P + P(A - LC)) \tilde{x}. + 2\nu \tilde{x}^T \tilde{x}.$$

Applying the inequality (8.34) to the above expression, we have

$$\dot{V} < 0,$$

which implies that the observer error asymptotically converges to zero.   □

To end this section, we consider a class of systems with nonlinear Lipschitz output function

$$\dot{x} = Ax$$
$$y = h(x), \tag{8.37}$$

where $x \in \mathbb{R}^n$ is the state vector, $y \in \mathbb{R}^m$ is the output, $A \in \mathbb{R}^{n \times n}$ is a constant matrix and $h : \mathbb{R}^n \to \mathbb{R}^m$ is a continuous function. We can write the nonlinear function h as $h = Hx + h_1(x)$ with Hx denoting a linear part of the output, and the nonlinear part h_1 with Lipschitz constant γ.

An observer can be designed as

$$\dot{\hat{x}} = A\hat{x} + L(y - h(\hat{x})), \tag{8.38}$$

where the observer gain $L \in \mathbb{R}^{n \times m}$ is a constant matrix.

Theorem 8.11. *The observer (8.38) provides an exponentially convergent state estimate of (8.37) if the observer gain L can be chosen to satisfy the following conditions:*

$$L = \frac{1}{\gamma^2} P^{-1} H^T, \tag{8.39}$$

$$PA + A^T P - \frac{H^T H}{\gamma^2} + (1 + \epsilon)I = 0, \tag{8.40}$$

where $P \in \mathbb{R}^{n \times n}$ is a positive definite matrix and ϵ is a positive real constant.

Proof. Let $\tilde{x} = x - \hat{x}$. From (8.37) and (8.38), we have

$$\dot{\tilde{x}} = (A - LH)\tilde{x} + L(h_1(x) - h_1(\hat{x})).$$

Let

$$V = \tilde{x}^T P \tilde{x},$$

where $\tilde{x} = x - \hat{x}$. It can be obtained that

$$\dot{V} = \tilde{x}^T((A - LH)^T P + P(A - LH))\tilde{x} + 2\tilde{x}^T PL(h_1(x) - h_1(\hat{x}))$$
$$\leq \tilde{x}^T((A - LH)^T P + P(A - LH))\tilde{x} + \tilde{x}^T(I + \gamma^2 PLL^T P)\tilde{x}$$
$$= \tilde{x}^T \left(A^T P + PA - \frac{H^T H}{\gamma^2} + I \right) \tilde{x} + \tilde{x}^T \left(\frac{H}{\gamma} - \gamma L^T P \right)^T \left(\frac{H}{\gamma} - \gamma L^T P \right) \tilde{x}$$
$$= -\epsilon \tilde{x}^T \tilde{x}.$$

Therefore, we can conclude that $\tilde{x}$ converges to zero exponentially. □

Remark 8.7. The nonlinearity in the output function with linear dynamics may occur in some special cases such as modelling a periodic signal as the output of a second-order linear system. This kind of formulation is useful for internal model design to asymptotically reject some general periodic disturbances, as shown in Chapter 10. ◁

8.5 Reduced-order observer design

The observers introduced in the previous sections all have the same order as the original systems. One might have noticed that an observer provides estimate of the entire state variables, which include the system output. Only the variables which are not contained in the output are needed for state estimation from an observer. This was noticed at the very early stage of the development of observer design of linear systems, and leads to the design of observers with less order than the original systems. The observers with less order than the original system are referred to as reduced-order observers. This section devotes to reduced-order observer design for nonlinear systems.

Consider a nonlinear system

$$\dot{x} = f(x, u)$$
$$y = h(x),$$
(8.41)

where $x \in \mathbb{R}^n$ is the state vector, $u \in \mathbb{R}^s$ is the known input, $y \in \mathbb{R}^m$ is the output and $f : \mathbb{R}^n \times \mathbb{R}^s \rightarrow \mathbb{R}^n$ is a nonlinear smooth vector field. To design a reduced-order observer, we need to study the dynamics of other state variables other than the output. We can define a partial-state transformation.

Definition 8.4. A function $g : \mathbb{R}^n \rightarrow \mathbb{R}^{n-m}$ is an output-complement transformation if the function T given by

$$T(x) := \begin{bmatrix} h(x) \\ g(x) \end{bmatrix}$$

is a diffeomorphism, where $h(x)$ is the output function. The transformed states are referred to as output-complement states.

Clearly an output-complement transformation defines a state transformation together with the output function. If the output-complement states are known, then the state variables can be fully determined. In fact, if we can design an observer for the output-complement state, this observer is a reduced-order observer.

Definition 8.5. The dynamic model

$$\dot{z} = p(z, y) + q(y, u)$$
(8.42)

is referred to as reduced-order observer form for the system (8.41) if the $z = g(x)$ is output-complement transformation, and the dynamic system

$$\dot{z} = p(z, y)$$

is differentially stable.

For a system which can be transformed to a reduced-order observer form, we can propose a reduced-order observer design as

$$\dot{\hat{z}} = p(\hat{z}, y) + q(y, u)$$

$$\hat{x} = T^{-1}\left(\begin{bmatrix} y \\ \hat{z} \end{bmatrix}\right). \tag{8.43}$$

Theorem 8.12. *If the system (8.41) can be transformed to the reduced-order observer form (8.42), the state estimate $\hat{x}$ provided by the reduced-order observer in (8.43) asymptotically converges to the state variable of (8.41).*

Proof. First, let us establish the boundedness of z. Since $\dot{z} = p(z, y)$ is differentially stable, from the definition of the differential stability in Chapter 5, there exists a Lyapunov function, $V(z)$, such that the conditions (5.41) are satisfied. Take the same function $V(z)$ here as the Lyapunov function candidate with $\hat{z} = 0$. From (5.41) and (8.42), we have

$$\dot{V} \leq -\gamma_3(\|z\|) + \frac{\partial V}{\partial z} q(y, u)$$

$$\leq -\gamma_3(\|z\|) + \|\frac{\partial V}{\partial z}\| \|q(y, u)\|. \tag{8.44}$$

Let us recall Young's inequality in a simplified form that for any $a \in \mathbb{R}$ and $b \in \mathbb{R}$, and a pair of constants $p > 1$ and $q > 1$ with $\frac{1}{p} + \frac{1}{q} = 1$, we have

$$|ab| \leq \frac{\epsilon^p}{p}|a|^p + \frac{1}{q\epsilon^q}|b|^q$$

for any positive real constant ϵ.

Applying Young's inequality to the second term on the right-hand side of (8.44) gives

$$\left\|\frac{\partial V(z)}{\partial z}\right\| \|q(y, u)\| \leq \frac{c_4^{c_2}}{c_2}\left\|\frac{\partial V(z)}{\partial z}\right\|^{c_2} + \frac{1}{c_3 c_4^{c_3}}\|q(y, u)\|^{c_3}, \tag{8.45}$$

where $c_3 = \frac{c_2}{c_2 - 1}$, and c_4 is an arbitrary positive real constant. We set $c_4 = (\frac{c_1 c_2}{2})^{1/c_2}$, which results in

$$\left\|\frac{\partial V(z)}{\partial x}\right\| \|q(y, u)\| \leq \frac{c_1}{2}\left\|\frac{\partial V(z)}{\partial z}\right\|^{c_2} + \frac{1}{2}c_5\|q(y, u)\|^{c_3}, \tag{8.46}$$

where $c_5 = \frac{1}{c_3}(\frac{1}{2}c_1 c_2)^{-\frac{c_3}{c_2}}$. Substituting (8.46) into (8.44), we have

$$\dot{V} \leq -\frac{1}{2}\gamma_3(\|z\|) + \frac{1}{2}c_5\|q(y, u)\|^{c_3}. \tag{8.47}$$

Since $q(y, u)$ is continuous, there exists a class $\mathcal{K}$ function $\bar{g}$ such that, for all $y \in \mathbb{R}^m$ and $u \in \mathbb{R}^s$, we have

$$\|q(y, u)\| \leq \bar{g}(\|y\| + \|u\|).$$

We then choose $\chi(\cdot) = \gamma_3^{-1}(2c_5(\bar{g}(\cdot))^{c_3})$. For $\|z\| \geq \chi(\|y\| + \|u\|)$, we have

$$\gamma_3(\|z\|) \geq 2c_5(\bar{g}(\|y\| + \|u\|))^{c_3}$$
$$\geq 2c_5\|q(y,u)\|^{c_3},$$

which further implies that

$$\dot{V} \leq -\frac{1}{4}\gamma_3(\|z\|). \tag{8.48}$$

Hence, $V(z)$ is an ISS-Lyapunov function, and therefore z is bounded when y is bounded.

The dynamics of e are given by

$$\dot{e} = p(z(t),y) - p((z(t) - e),y).$$

Taking $V(e)$ as the Lyapunov function candidate, we have

$$\dot{V} = \frac{\partial V}{\partial e}(p(z(t),y) - p((z(t) - e),y))$$
$$\leq -\gamma_3(\|e\|).$$

Therefore, we can conclude that the estimation error asymptotically converges to zero. With $\hat{z}$ as a convergent estimate of z, we can further conclude that $\hat{x}$ is an asymptotic estimate of x. $\qquad\square$

To demonstrate the proposed reduced-order observer design, let us consider an example.

Example 8.1. Consider a second-order system

$$\dot{x}_1 = x_1^2 - 3x_1^2x_2 - x_1^3$$
$$\dot{x}_2 = -x_2 + x_1^2 - 6x_2x_1^2 + 3x_2^2x_1 - x_2^3$$
$$y = x_1.$$

Let us check if the system can be transformed to the reduced-order observer form. For this, we need to find $g(x)$. Take

$$z = g(x) = x_2 - x_1.$$

We have

$$\dot{z} = -x_2 - (x_2 - x_1)^3$$
$$= -(1 + z^2)z + y.$$

Comparing with the format shown in (8.42), we have $p(z,y) = -(1+z^2)z + y$. Note that for the system without input, we always have $\dot{z} = p(z,y)$. Let $V = \frac{1}{2}z^2$. It is easy to see the first and the third conditions in (5.41) are satisfied. For the second condition, we have

$$\frac{\partial V(z-\hat{z})}{\partial z}(p(z) - p(\hat{z}))$$

$$= -(z-\hat{z})(z-\hat{z}+z^3-\hat{z}^3)$$

$$= -(z-\hat{z})^2(1+z^2-z\hat{z}+\hat{z}^2)$$

$$= -(z-\hat{z})^2(1+\frac{1}{2}(z^2+\hat{z}^2+(z-\hat{z})^2))$$

$$\leq -(z-\hat{z})^2.$$

Therefore, the system satisfies the conditions specified in (5.41). We design the reduced-order observer as

$$\dot{\hat{z}} = -(1+\hat{z}^2)\hat{z}+y$$

$$\hat{x}_2 = \hat{z}+y.$$

Simulation study has been carried out, and the simulation results are shown in Figures 8.1 and 8.2. ◁

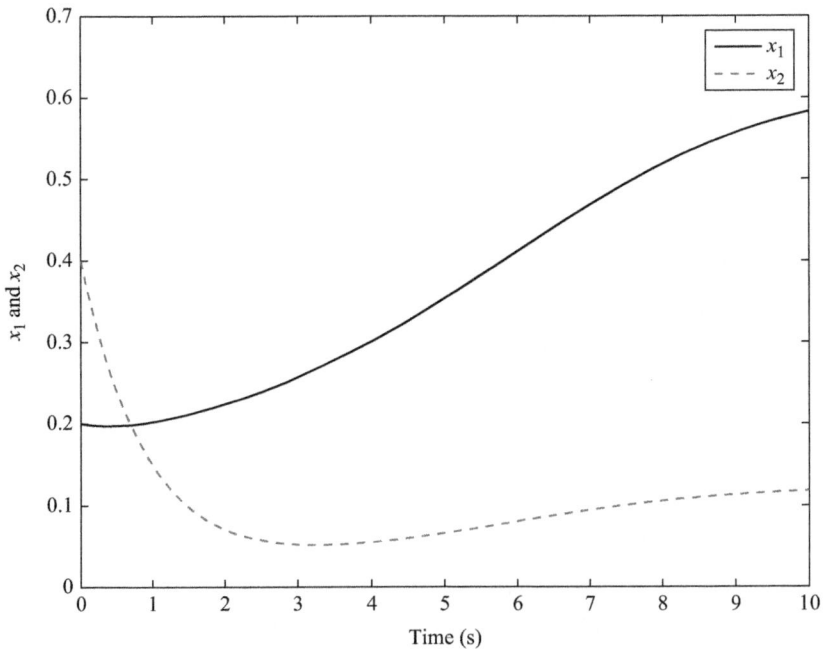

Figure 8.1   State variables

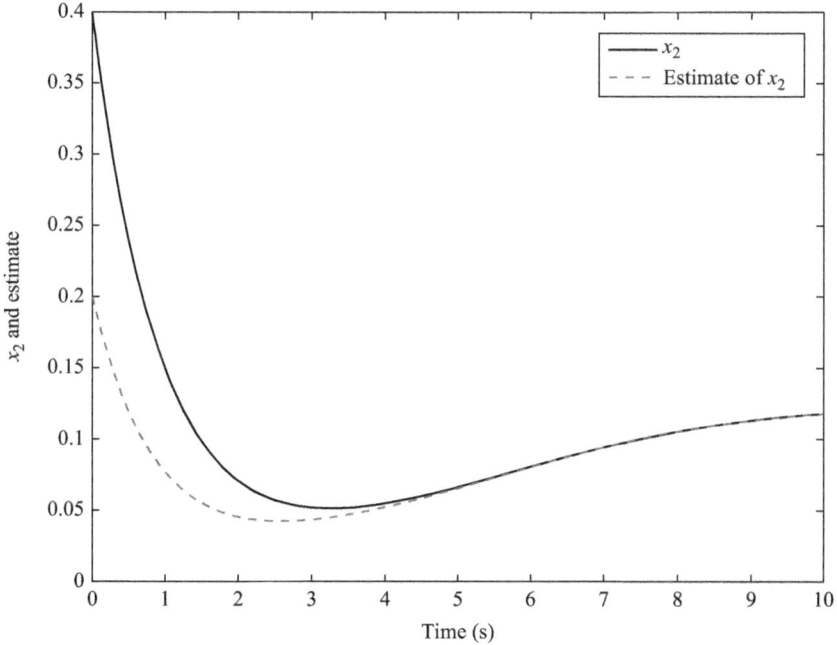

Figure 8.2   Unmeasured state and its estimate

There are lack of systematic design methods for nonlinear observers with global convergence when there are general nonlinear terms of unmeasured state variables in the systems. With the introduction of the reduced-order observer form, we like to further explore the class of nonlinear systems which can be transformed to the format shown in (8.42), and therefore a nonlinear observer can then be designed accordingly.

We consider a multi-output (MO) nonlinear system

$$
\begin{aligned}
\dot{x} &= Ax + \phi(y, u) + E\varphi(x, u) \\
y &= Cx,
\end{aligned}
\tag{8.49}
$$

where $x \in \mathbb{R}^n$ is the state; $y \in \mathbb{R}^m$ is the output; $u \in \mathbb{R}^s$ is the control; ϕ is a known nonlinear smooth vector field; $\varphi : \mathbb{R}^n \times \mathbb{R}^s \to \mathbb{R}^m$ is a smooth nonlinear function; $C \in \mathbb{R}^{m \times n}$, $E \in \mathbb{R}^{n \times m}$ and $A \in \mathbb{R}^{n \times n}$ are constant matrices, with (A, C) observable.

When $\varphi = 0$, the system (8.49) degenerates to the well-known form of the nonlinear systems with the linear observer error dynamics, and nonlinear observer can be easily designed by using nonlinear output injection. With this additional term φ, the nonlinear output injection term can no longer be used to generate a linear observer error dynamics. We will convert the system to the reduced-order observer form considered, and then apply the reduced-order observer design for state observation.

Without loss of generality, we can assume that C has full row rank. There exists a nonsingular state transformation M such that

$$CM^{-1} = [I_m, 0_{m \times (n-m)}].$$

If span$\{E\}$ is a complement subspace of ker$\{C\}$ in $\mathbb{R}^n$, we have $(CM^{-1})(ME) = CE$ invertible. If we partition the matrix ME as

$$ME := \begin{bmatrix} E_1 \\ E_2 \end{bmatrix}$$

with $E_1 \in \mathbb{R}^{m \times m}$, then we have $CE = E_1$. Furthermore, if we partition Mx as

$$Mx := \begin{bmatrix} \chi_1 \\ \chi_2 \end{bmatrix}$$

with $\chi_1 \in \mathbb{R}^m$, we have

$$z = g(x) = \chi_2 - E_2 E_1^{-1} \chi_1. \tag{8.50}$$

Note that we have $\chi_1 = y$. With the partition of

$$MAM^{-1} := \begin{bmatrix} A_{1,1} & A_{1,2} \\ A_{2,1} & A_{2,2} \end{bmatrix}$$

and

$$M\phi := \begin{bmatrix} \phi_1 \\ \phi_2 \end{bmatrix},$$

we can write the dynamics χ_1 and χ_2 as

$$\dot{\chi}_1 = A_{1,1} \chi_1 + A_{1,2} \chi_2 + \phi_1 + E_1 \varphi$$
$$\dot{\chi}_2 = A_{2,1} \chi_1 + A_{2,2} \chi_2 + \phi_2 + E_2 \varphi.$$

Then we can obtain the dynamics of z as

$$\dot{z} = A_{2,1} \chi_1 + A_{2,2} \chi_2 + \phi_2 - E_2 E_1^{-1}(A_{1,1} \chi_1 + A_{1,2} \chi_2 + \phi_1)$$

$$= (A_{2,2} - E_2 E_1^{-1} A_{1,2}) \chi_2 + (A_{2,1} - E_2 E_1^{-1} A_{1,1}) \chi_1 + \phi_2 - E_2 E_1^{-1} \phi_1$$

$$= (A_{2,2} - E_2 E_1^{-1} A_{1,2}) z + q(y, u), \tag{8.51}$$

where

$$q(y, u) = (A_{2,2} - E_2 E_1^{-1} A_{1,2}) E_2 E_1^{-1} y + (A_{2,1} - E_2 E_1^{-1} A_{1,1}) y$$
$$+ \phi_2(y, u) - E_2 E_1^{-1} \phi_1(y, u).$$

Note that the nonlinear function $\varphi(x, u)$ does not appear in the dynamics of z in (8.51) due to the particular choice of z in (8.50).

Remark 8.8. After the state transformation, (8.51) is in the same format as (8.42). Therefore, we can design a reduced-order observer if $\dot{z} = (A_{2,2} - E_2 E_1^{-1} A_{1,2}) z$ is differentially stable. Notice that it is a linear system. Hence, it is differentially stable if it is asymptotically stable, which means the eigenvalues of $(A_{2,2} - E_2 E_1^{-1} A_{1,2})$ are with negative real parts. ◁

Following the format shown in (8.43), a reduced-order observer can then be designed as

$$\dot{\hat{z}} = (A_{2,2} - E_2 E_1^{-1} A_{1,2})\hat{z} + q(y, u) \tag{8.52}$$

and the estimate of x is given by

$$\hat{x} = M^{-1} \begin{bmatrix} y \\ z + E_2 E_1^{-1} y \end{bmatrix}. \tag{8.53}$$

Theorem 8.13. *For a system (8.49), if*

- *C has full row rank, and span$\{E\}$ is a complement subspace of ker$\{C\}$ in $\mathbb{R}^n$*
- *all the invariant zeros of (A, E, C) are with negative real parts*

it can be transformed to the reduced-order observer form. The estimates $\hat{z}$ given in (8.52) and $\hat{x}$ (8.53) converge to the respective state variables z and x exponentially.

Proof. We only need to show that $\dot{z} = p(z, y)$ is differentially stable, and then we can apply Theorem 8.12 to conclude the asymptotic convergence of the reduced-order observer error. The exponential convergence comes as a consequence of the linearity in the reduced-order observer error dynamics. For (8.52), we have

$$p(z, y) = (A_{2,2} - E_2 E_1^{-1} A_{1,2})z,$$

which is linear in z. Therefore, the proof can be completed by proving that the matrix $(A_{2,2} - E_2 E_1^{-1} A_{1,2})$ is Hurwitz. It can be shown that the eigenvalues of $(A_{2,2} - E_2 E_1^{-1} A_{1,2})$ are the invariant zeros of (A, E, C). Indeed, we have

$$\begin{bmatrix} M & 0 \\ 0 & I_m \end{bmatrix} \begin{bmatrix} sI - A & E \\ C & 0 \end{bmatrix} \begin{bmatrix} M^{-1} & 0 \\ 0 & I_m \end{bmatrix}$$

$$= \begin{bmatrix} sI - MAM^{-1} & ME \\ CM^{-1} & 0 \end{bmatrix}$$

$$= \begin{bmatrix} sI_m - A_{1,1} & -A_{1,2} & E_1 \\ -A_{2,1} & sI_{n-m} - A_{2,2} & E_2 \\ I_m & 0 & 0 \end{bmatrix}. \tag{8.54}$$

Let us multiply the above matrix in the left by the following matrix to perform a row operation:

$$\begin{bmatrix} I_m & 0 & 0 \\ -E_2 E_1^{-1} & I_{n-m} & 0 \\ 0 & 0 & I_m \end{bmatrix},$$

we result in the following matrix:

$$\begin{bmatrix} sI_m - A_{1,1} & -A_{1,2} & E_1 \\ \Delta & sI_{n-m} - (A_{2,2} - E_2 E_1^{-1} A_{1,2}) & 0 \\ I_m & 0 & 0 \end{bmatrix}$$

with

$$\Delta = -E_2 E_1^{-1}(sI_m - A_{1,1}) - A_{2,1}.$$

Since E_1 is invertible, any values of s which make the matrix

$$\begin{bmatrix} sI - A & E \\ C & 0 \end{bmatrix}$$

rank deficient must be the eigenvalues of $(A_{2,2} - E_2 E_1^{-1} A_{1,2})$. From the second condition that all the invariant zeros of (A, E, C) are with negative real part, we can conclude that $(A_{2,2} - E_2 E_1^{-1} A_{1,2})$ is Hurwitz.   □

Example 8.2. Consider a third-order system

$$\dot{x}_1 = -x_1 + x_2 - y_1 + u + x_2 x_3 - x_1 x_3$$
$$\dot{x}_2 = -x_1 + x_2 + x_3 - 2y_1 + u + y_1 y_2 + x_2 x_3$$
$$\dot{x}_3 = -y_1^2 + x_1 x_3 + x_2 x_3$$
$$y_1 = x_1$$
$$y_2 = -x_1 + x_2.$$

We can identify

$$\phi = \begin{bmatrix} -y_1 + u \\ -2y_1 + y_1 y_2 + u \\ -y_1^2 \end{bmatrix}, \quad \varphi = \begin{bmatrix} x_2 x_3 - x_1 x_3 \\ x_1 x_3 \end{bmatrix}, \tag{8.55}$$

and we have

$$A = \begin{bmatrix} -1 & 1 & 0 \\ -1 & 1 & 1 \\ 0 & 0 & 0 \end{bmatrix}, \quad E = \begin{bmatrix} 1 & 0 \\ 1 & 1 \\ 1 & 2 \end{bmatrix}, \quad C = \begin{bmatrix} 1 & 0 & 0 \\ -1 & 1 & 0 \end{bmatrix}. \tag{8.56}$$

It can be easily checked that (A, C) is observable and the invariant zero of (A, E, C) is at -2. Therefore, the conditions in Theorem 8.13 are satisfied. It is also easy to see that $x_2 = y_2 + y_1$. Therefore, the only unmeasured state variable is x_3. Indeed, following the procedures introduced earlier, we have

$$M = \begin{bmatrix} 1 & 0 & 0 \\ -1 & 1 & 0 \\ 0 & 0 & 1 \end{bmatrix}, \quad E_1 = \begin{bmatrix} 1 & 0 \\ 0 & 1 \end{bmatrix}, \quad E_2[1\ 2],$$

$$\chi_1 = \begin{bmatrix} x_1 \\ x_2 - x_1 \end{bmatrix}, \quad \chi_2 = x_3,$$

and

$$z = x_3 - y_1 - 2y_2 = x_3 + x_1 - 2x_2.$$

With

$$A_{1,1} = \begin{bmatrix} 0 & 1 \\ 0 & 0 \end{bmatrix}, \quad A_{1,2} = \begin{bmatrix} 0 \\ 1 \end{bmatrix}, \quad A_{2,1} \begin{bmatrix} 0 & 0 \end{bmatrix}, \quad A_{2,2} = 0,$$

$$\phi_1 = \begin{bmatrix} -y_1 + u \\ -y_1 + y_1 y_2 \end{bmatrix}, \quad \phi_2 = -y_1^2,$$

we have the dynamics of z as

$$\dot{z} = -2z + q(y, u),$$

where

$$q(y, u) = y_1 - 5y_2 - y_1^2 - 2y_1 y_2 - u.$$

A reduced-order observer can then be designed as

$$\dot{\hat{z}} = -2\hat{z} + q(y, u)$$
$$\hat{x}_3 = \hat{z} + y_1 + 2y_2.$$

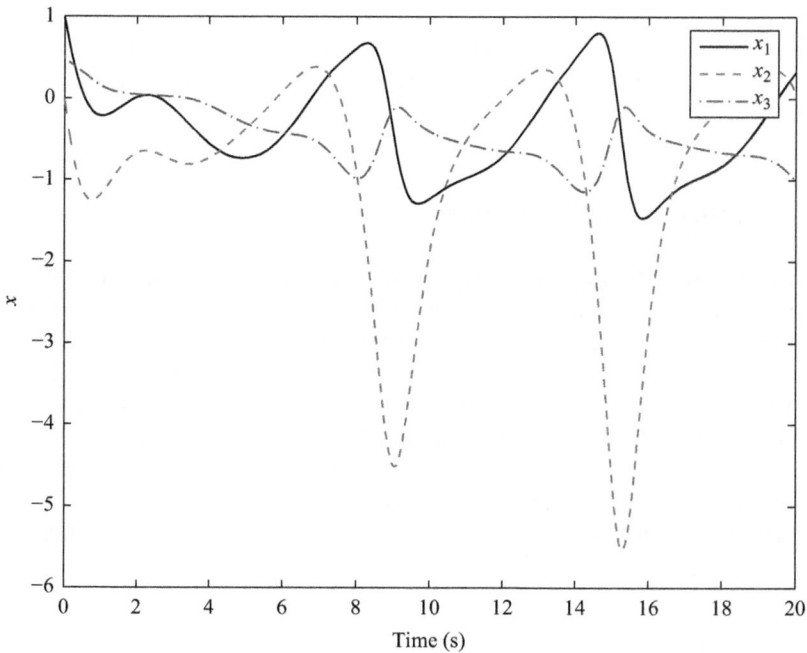

Figure 8.3   State variables

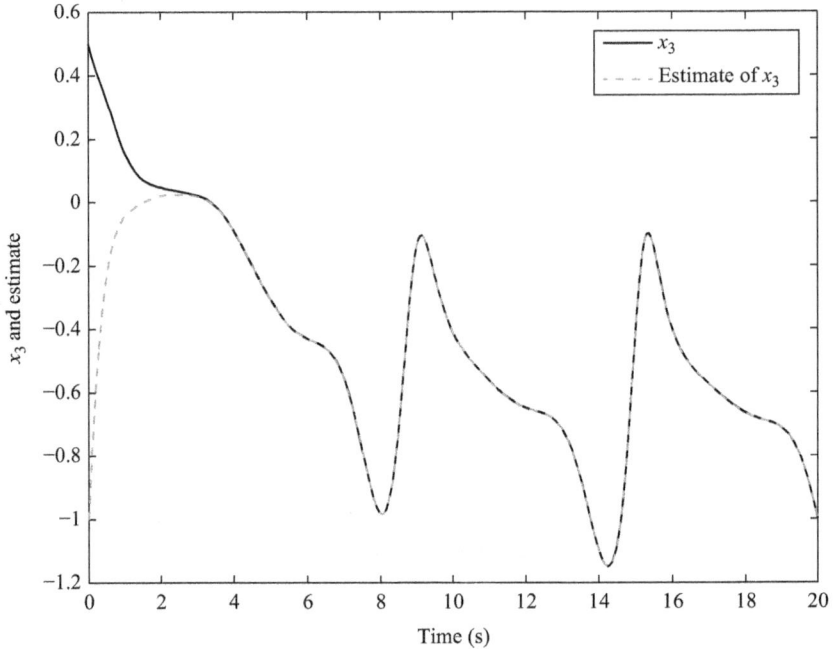

Figure 8.4   Unmeasured state and its estimate

Simulation study has been carried with $x(0) = [1, 0, 0.5]^T$ and $u = \sin t$. The plots of the state variables are shown in Figure 8.3 and the estimated state is shown in Figure 8.4 .   ◁

8.6   Adaptive observer design

When there are unknown parameters in dynamic systems, observers can still be designed under certain conditions to provide estimates of unknown states by using adaptive parameters. These observers are referred to as adaptive observers. One would expect more stringent conditions imposed on nonlinear systems for which adaptive observers can be designed. In this section, we will consider two classes of nonlinear systems. The first one is based on the nonlinear systems in the output injection form or output feedback form with unknown parameters and the other class of the systems is nonlinear Lipschitz systems with unknown parameters.

Consider nonlinear systems which can be transformed to a single-output system

$$
\begin{aligned}
\dot{x} &= Ax + \phi_0(y, u) + b\phi^T(y, u)\theta \\
y &= Cx,
\end{aligned}
$$

(8.57)

where $x \in \mathbb{R}^n$ is the state, $y \in \mathbb{R}$ is the system output, $u \in \mathbb{R}^s$ is the control input, or other known variables, $A \in \mathbb{R}^{n \times n}$, $b \in \mathbb{R}^{n \times 1}$ and $C \in \mathbb{R}^{1 \times n}$ are constant matrices, with (A, C) observable, $\theta \in \mathbb{R}^r$ is an unknown constant vector, $\phi_0 : \mathbb{R}^m \times \mathbb{R}^s \to \mathbb{R}^n$ and $\phi : \mathbb{R}^m \times \mathbb{R}^s \to \mathbb{R}^r$ are continuous functions. If the parameter vector θ is known, the system is in the output-injection form.

One would expect that additional conditions are required for the design of an adaptive observer. If the linear system characterised by (A, b, C) is of relative degree 1 and minimum phase, we can design an observer gain $L \in \mathbb{R}^n$ such that the linear system characterised by $(A - LC, b, C)$ is a positive real system. In such a case, there exist positive real matrices P and Q such that

$$(A - LC)^T P + P(A - LC) = -Q$$
$$Pb = C^T. \tag{8.58}$$

Let us consider the observer designed as

$$\dot{\hat{x}} = A\hat{x} + \phi_0(y, u) + b\phi^T(y, u)\hat{\theta} + L(y - C\hat{x}), \tag{8.59}$$

where $\hat{x}$ is the estimate of x and $\hat{\theta}$ is the estimate of θ. We need to design an adaptive law for $\hat{\theta}$. Let $\tilde{x} = x - \hat{x}$. The observer error dynamics are obtained as

$$\dot{\tilde{x}} = (A - LC)\tilde{x} + b\phi^T(y, u)\tilde{\theta}, \tag{8.60}$$

where $\tilde{\theta} = \theta - \hat{\theta}$.

Consider a Lyapunov function candidate

$$V = \tilde{x}^T P \tilde{x} + \tilde{\theta}^T \Gamma^{-1} \tilde{\theta},$$

where $\Gamma \in \mathbb{R}^{n \times n}$ is a positive definite matrix. From (8.58) and (8.60), we have

$$\dot{V} = \tilde{x}^T ((A - LC)^T P + P(A - LC))\tilde{x} + 2\tilde{x}^T P b \phi(y, u)^T \tilde{\theta} + 2\dot{\tilde{\theta}}^T \Gamma^{-1} \tilde{\theta}$$
$$= -\tilde{x}^T Q \tilde{x} + 2\tilde{x}^T C^T \phi(y, u)^T \tilde{\theta} - 2\dot{\hat{\theta}}^T \Gamma^{-1} \tilde{\theta}$$
$$= -\tilde{x}^T Q \tilde{x} - 2(\dot{\hat{\theta}} - (y - C\hat{x})\Gamma\phi(y, u))^T \Gamma^{-1} \tilde{\theta}$$

If we set the adaptive law as

$$\dot{\hat{\theta}} = (y - C\hat{x})\Gamma\phi(y, u),$$

we obtain

$$\dot{V} = -\tilde{x}^T Q \tilde{x}.$$

Then similar to stability analysis of adaptive control systems, we can conclude that $\lim_{t\to\infty} \tilde{x}(t) = 0$, and $\hat{\theta}$ is bounded. The above analysis leads to the following theorem.

Theorem 8.14. *For the observable single-output system (8.57), if the linear system characterised by (A, b, C) is minimum phase and has relative degree 1, there exists an* *observer gain $L \in \mathbb{R}^n$ that satisfies the conditions in (8.58) and an adaptive observer designed as*

$$\dot{\hat{x}} = A\hat{x} + \phi_0(y, u) + b\phi^T(y, u)\hat{\theta} + L(y - C\hat{x})$$
$$\dot{\hat{\theta}} = (y - C\hat{x})\Gamma\phi(y, u),$$

(8.61)

where $\Gamma \in \mathbb{R}^{n \times n}$ is a positive definite matrix and provides an asymptotically convergent state estimate with adaptive parameter vector remaining bounded.

Remark 8.9. In this remark, we will show a particular choice of the observer gain for the adaptive control observer for a system that satisfies the conditions in Theorem 8.14. Without loss of generality, we assume

$$A = \begin{bmatrix} 0 & 1 & 0 & \cdots & 0 \\ 0 & 0 & 1 & \cdots & 0 \\ \vdots & \vdots & \vdots & \ddots & \vdots \\ 0 & 0 & 0 & \cdots & 1 \\ 0 & 0 & 0 & \cdots & 0 \end{bmatrix}, \quad b = \begin{bmatrix} b_1 \\ b_2 \\ \vdots \\ b_{n-1} \\ b_n \end{bmatrix},$$

$$C = [1 \quad 0 \quad \cdots \quad 0 \quad 0],$$

where $b_1 \neq 0$, since the relative degree is 1. Indeed, for an observer system $\{A, b, C\}$ with relative degree 1, there exists a state transformation, as shown in Remark 8.4, to transform the system to the observable canonical form. Furthermore, if we move the first column of A in the canonical form, and combine it with $\phi_0(y, u)$, we have A as in the format shown above. Since the system is minimum phase, b is Hurwitz, i.e.,

$$B(s) := b_1 s^{n-1} + b_2 s^{n-2} + \cdots + b_{n-1}s + b_n = 0$$

has all the solutions in the left half of the complex plane. In this case, we can design the observer gain to cancel all the zeros of the system. If we denote $L = [l_1, l_2, \ldots, l_n]$, we can choose L to satisfy

$$s^n + l_1 s^{n-1} + l_2 s^{n-2} + \cdots + l_{n-1}s + l_n = B(s)(s + \lambda),$$

where λ is a positive real constant. The above polynomial equation implies

$$L = (\lambda I + A)b.$$

This observer gain ensures

$$C (sI - (A - LC))^{-1} b = \frac{1}{s + \lambda},$$

which is a strict positive real transfer function.   ◁

Now we will consider adaptive observer design for a class of Lipschitz nonlinear systems. Consider

$$\dot{x} = Ax + \phi_0(x, u) + b\phi^T(x, u)\theta$$
$$y = Cx,$$

(8.62)

where $x \in \mathbb{R}^n$ is the state, $y \in \mathbb{R}$ is the system output, $u \in \mathbb{R}^s$ is the control input, or other known variables, $A \in \mathbb{R}^{n \times n}$, $b \in \mathbb{R}^{n \times 1}$ and $C \in \mathbb{R}^{1 \times n}$ are constant matrices, with (A, C) observable, $\theta \in \mathbb{R}^r$ is an unknown constant vector, $\phi_0 : \mathbb{R}^n \times \mathbb{R}^s \to \mathbb{R}^n$ and $\phi : \mathbb{R}^m \times \mathbb{R}^s \to \mathbb{R}^r$ are Lipschitz nonlinear functions with Lipschitz constants γ_1 and γ_2 with respect to x.

This system is only different in nonlinear functions from the system (8.57) considered earlier for adaptive observer design, where the nonlinear functions are restricted to the functions of the system output only, but the functions do not have to be globally Lipschitz for a globally convergent observer.

We propose an adaptive observer for (8.62) as

$$\dot{\hat{x}} = A\hat{x} + \phi_0(\hat{x}, u) + b\phi^T(\hat{x}, u)\hat{\theta} + L(y - C\hat{x})$$
$$\dot{\hat{\theta}} = (y - C\hat{x})\Gamma\phi(\hat{x}, u),$$

$$(8.63)$$

where $L \in \mathbb{R}^n$ is an observer gain and $\Gamma \in \mathbb{R}^{n \times n}$ is a positive definite matrix.

When state variables other than the system output are involved in the nonlinear functions, we often impose the conditions that the nonlinear function are Lipschitz, and the conditions are more stringent for designing an observer than the nonlinear systems with only nonlinear functions of the system output. We have considered adaptive observer design for a class of nonlinear systems (8.57) with only the nonlinear functions of the system output. For the adaptive observer (8.63) to work for the system (8.62), we will expect stronger conditions for adaptive observer design. For this, we state the following result.

Theorem 8.15. *The adaptive observer proposed in (8.63) provides an asymptotically convergent state estimate for the system (8.62) if there exists a positive definite matrix P such that*

$$(A - LC)^T P + P(A - LC) + (\gamma_1 + \gamma_2\gamma_3\|b\|)(PP + I) + \epsilon I \leq 0$$
$$Pb = C^T,$$

$$(8.64)$$

where $\gamma_3 \geq \|\theta\|$, and ϵ is a positive real constant.

Proof. Let $\tilde{x} = x - \hat{x}$. The observer error dynamics are obtained as

$$\dot{\tilde{x}} = (A - LC)\tilde{x} + \phi_0(x, u) - \phi_0(\hat{x}, u) + b(\phi^T(y, u)\theta - \phi^T(\hat{x}, u)^T\hat{\theta})$$
$$= (A - LC)\tilde{x} + \phi_0(x, u) - \phi_0(\hat{x}, u) + b(\phi(x, u) - \phi(\hat{x}, u))^T\theta$$
$$\quad + b\phi^T(\hat{x}, u)\tilde{\theta},$$

$$(8.65)$$

where $\tilde{\theta} = \theta - \hat{\theta}$. Consider a Lyapunov function candidate

$$V = \tilde{x}^T P\tilde{x} + \tilde{\theta}^T\Gamma^{-1}\tilde{\theta}.$$

Its derivative along the dynamics in (8.65) is obtained as

$$\dot{V} = \tilde{x}^T((A - LC)^T P + P(A - LC))\tilde{x} + 2\tilde{x}^T P(\phi_0(x, u) - \phi_0(\hat{x}, u))$$
$$\quad + 2\tilde{x}^T Pb(\phi(x, u) - \phi(\hat{x}, u))^T\theta + 2\tilde{x}^T Pb\phi^T(\hat{x}, u)\tilde{\theta} + 2\dot{\tilde{\theta}}^T\Gamma^{-1}\tilde{\theta}.$$

From the Lipschitz constants of ϕ_0 and ϕ, we have

$$2\tilde{x}^T P(\phi_0(x, u) - \phi_0(\hat{x}, u)) \leq \gamma_1 \tilde{x}^T (PP + I)\tilde{x},$$
$$2\tilde{x}^T Pb(\phi(x, u) - \phi(\hat{x}, u))^T \theta \leq \gamma_2 \gamma_3 \|b\| \tilde{x}^T (PP + I)\tilde{x}.$$

Then from the adaptive law and (8.64), we have

$$\dot{V} \leq \tilde{x}^T ((A - LC)^T P + P(A - LC) + (\gamma_1 + \gamma_2 \gamma_3 \|b\|)(PP + I))\tilde{x}^T$$
$$+ 2\tilde{x}^T C^T \phi^T (\hat{x}, u)\tilde{\theta} - 2\dot{\hat{\theta}}^T \Gamma^{-1}\tilde{\theta}$$
$$\leq -\epsilon \tilde{x}^T \tilde{x}.$$

This implies that the variables $\tilde{x}$ and $\tilde{\theta}$ are bounded, and furthermore $\tilde{x} \in L_2 \cap L_\infty$. Since all the variables are bounded, $\dot{\tilde{x}}$ is bounded, from $\tilde{x} \in L_2 \cap L_\infty$ and the boundedness of $\dot{\tilde{x}}$, we conclude $\lim_{t \to \infty} \tilde{x}(t) = 0$ by invoking Babalat's Lemma.   □

Chapter 9
Backstepping design

For a nonlinear system, the stability around an equilibrium point can be established if one can find a Lyapunov function. Nonlinear control design can be carried out by exploring the possibility of making a Lyapunov function candidate as a Lyapunov function through control design. In Chapter 7, parameter adaptive laws are designed in this way by setting the parameter adaptive laws to make the derivative of a Lyapunov function candidate negative semi-definite. Backstepping is a nonlinear control design method based on Lyapunov functions. It enables a designed control to be extended to an augmented system, provided that the system is augmented in some specific way. One scheme is so-called adding an integrator in the sense that if a control input is designed for a nonlinear system, then one can design a control input for the augmented system of which an integrator is added between the original system input and the input to be designed. This design strategy can be applied iteratively. There are a few systematic control design methods for nonlinear systems, and backstepping is one of them. In this chapter, we start with the fundamental form of adding an integrator, and then introduce the method for iterative backstepping with state feedback. We also introduce backstepping using output feedback, and adaptive backstepping for certain nonlinear systems with unknown parameters.

9.1 Integrator backstepping

Consider

$$\begin{aligned} \dot{x} &= f(x) + g(x)\xi \\ \dot{\xi} &= u, \end{aligned} \tag{9.1}$$

where $x \in \mathbb{R}^n$ and $\xi \in \mathbb{R}$ are the state variables; $u \in \mathbb{R}$ is the control input; and $f : \mathbb{R}^n \to \mathbb{R}^n$ with $f(0) = 0$ and $g : \mathbb{R}^n \to \mathbb{R}^n$ are continuous functions. Viewing the first equation of (9.1) as the original system with x as the state and ξ as the input, the integration of u gives ξ, which means that an integrator is added to the original system.

We consider the control design problem under the assumption that a known control input exists for the original system. Furthermore, we assume that the Lyapunov function is also known, associated with the known control for x-subsystem. Suppose that control input for the x-subsystem is $\alpha(x)$ with α differentiable and $\alpha(0) = 0$, and the associated Lyapunov function is $V(x)$. We assume that

$$\frac{\partial V}{\partial x}(f(x) + g(x)\alpha(x)) \leq -W(x),\tag{9.2}$$

where $W(x)$ is positive definite.

Condition (9.2) implies that the system $\dot{x} = f(x) + g(x)\alpha(x)$ is asymptotically stable. Consider

$$\dot{x} = f(x) + g(x)\alpha(x) + g(x)(\xi - \alpha(x)).$$

Intuitively, if we can design a control input $u = u(x, \xi)$ to force ξ to converge to $\alpha(x)$, we have a good chance to ensure the stability of the entire system. Let us define

$$z = \xi - \alpha(x).$$

It is easy to obtain the dynamics under the coordinates (x, z) as

$$\dot{x} = f(x) + g(x)\alpha(x) + g(x)z$$

$$\dot{z} = u - \dot{\alpha} = u - \frac{\partial \alpha}{\partial x}(f(x) + g(x)\xi).$$

Consider a Lyapunov function candidate

$$V_c(x, z) = V(x) + \frac{1}{2}z^2.\tag{9.3}$$

Its derivative is given by

$$\begin{aligned}
\dot{V}_c &= \frac{\partial V}{\partial x}(f(x) + g(x)\alpha(x)) + \frac{\partial V}{\partial x}g(x)z \\
&\quad + z\left(u - \frac{\partial \alpha}{\partial x}(f(x) + g(x)\xi)\right) \\
&= -W(x) + z\left[u + \frac{\partial V}{\partial x}g(x) - \frac{\partial \alpha}{\partial x}(f(x) + g(x)\xi)\right].
\end{aligned}$$

Let

$$u = -cz - \frac{\partial V}{\partial x}g(x) + \frac{\partial \alpha}{\partial x}(f(x) + g(x)\xi)\tag{9.4}$$

with $c > 0$ which results in

$$\dot{V}_c = -W(x) - cz^2.\tag{9.5}$$

It is clear that $-W(x) - cz^2$ is negative definite with respect to variables (x, z). Hence, we can conclude that $V_c(x, z)$ is a Lyapunov function, and that $(0, 0)$ in the coordinates (x, z) is a globally asymptotic equilibrium. From $\alpha(0) = 0$, we can conclude that $(0, 0)$ in the coordinates (x, ξ) is also a globally asymptotic equilibrium, which means that the system (9.1) is globally asymptotically stable under the control input (9.4). We summarise the above result in the following lemma.

Lemma 9.1. *For a system described in (9.1), if there exist differentiable function $\alpha(x)$ and a positive-definite function $V(x)$ such that (9.2) holds, the control design given in (9.4) ensures the global asymptotic stability of the closed-loop system.*

Remark 9.1. Considering the structure of (9.1), if ξ is viewed as the control input for the x-subsystem, $\xi = \alpha(x)$ is the desired control, ignoring the ξ-system. This is why ξ can be referred to as a virtual control input for the x-subsystem. The control input u for the overall system is designed with the consideration of the dynamics back to the control design for the x-subsystem, and it may suggest the name of this particular design method as *backstepping*. ◁

Example 9.1. Consider

$$\dot{x}_1 = x_1^2 + x_2$$
$$\dot{x}_2 = u.$$

We design a control input using backstepping. Comparing with (9.1), we can identify

$$x \Rightarrow x_1, \quad \xi \Rightarrow x_2, \quad f(x) \Rightarrow x_1^2, \quad g(x) \Rightarrow 1.$$

First, we need to design $\alpha(x_1)$ to stabilise

$$\dot{x}_1 = x_1^2 + \alpha(x_1).$$

An obvious choice is

$$\alpha(x_1) = -c_1 x_1 - x_1^2$$

with $c_1 > 0$ a constant. This design leads to

$$\dot{x}_1 = x_1^2 + \alpha(x_1) = -c_1 x_1.$$

Hence, we take

$$V(x_1) = \frac{1}{2} x_1^2$$

with

$$\dot{V}(x_1) = -c_1 x_1^2.$$

Therefore, the condition specified in Lemma 9.1 is satisfied with $\alpha(x_1) = -c_1 x_1 - x_1^2$, $V(x_1) = \frac{1}{2} x_1^2$ and $W(x_1) = -c_1 x_1^2$. The control input u can then be obtained from (9.4) by substituting proper functions in the equation. Alternatively, we can obtain the control input by directly following the backstepping method. Indeed, let $z = x_2 - \alpha(x_1)$. The dynamics of the system in coordinate (x_1, z) are obtained as

$$\dot{x}_1 = -c_1 x_1 + z$$
$$\dot{z} = u - \frac{\partial \alpha(x_1)}{\partial x_1}(x_1^2 + x_2),$$

where

$$\frac{\partial \alpha(x_1)}{\partial x_1} = -c_1 - 2x_1.$$

Let

$$V_c(x_1, z) = \frac{1}{2} x_1^2 + \frac{1}{2} z^2.$$

Its derivative along the system dynamics is obtained as

$$\dot{V}_c(x_1, z) = -c_1 x_1^2 + x_1 z + z\left(u - \frac{\partial \alpha(x_1)}{\partial x_1}(x_1^2 + x_2)\right).$$

Designing the control u as

$$u = -x_1 - c_2 z + \frac{\partial \alpha(x_1)}{\partial x_1}(x_1^2 + x_2)$$

results in

$$\dot{V}_c(x_1, z) = -c_1 x_1^2 - c_2 z^2.$$

Hence, the system is asymptotically stable with (x_1, z). As $\alpha(0) = 0$, we conclude $\lim_{t \to \infty} x_2(t) = \lim_{t \to \infty} (z(t) + \alpha(x_1(t))) = 0$, which implies that the system is asymptotically stable in the equilibrium $(0, 0)$ in (x_1, x_2). Note that the closed-loop system in (x_1, z) is written as

$$\begin{bmatrix} \dot{x}_1 \\ \dot{z} \end{bmatrix} = \begin{bmatrix} -c_1 & 1 \\ -1 & -c_2 \end{bmatrix} \begin{bmatrix} x_1 \\ z \end{bmatrix}.$$

◁

   In Example 9.1, backstepping design has been used to design a control input for a nonlinear system with unmatched nonlinearities. When a nonlinear function appears in the same line as the control input, it is referred as a matched nonlinear function, and it can be cancelled by adding the same term in u with an opposite sign. From the system considered in Example 9.1, the nonlinear function x_1^2 does not appear in the same line as the control input u, and therefore it is unmatched. However, it is in the same line as x_2, which is viewed as a virtual control. As a consequence, $\alpha(x_1)$, which is often referred to as a stabilising function, can be designed to cancel the nonlinearity which matches with the virtual control, and backstepping method enables the control in the next line to be designed to stabilise the entire system. This process can be repeated by identifying a virtual control, designing a stabilising function and using backstepping to design control input for more complicated nonlinear systems.

9.2   Iterative backstepping

Consider a nonlinear system

$$\dot{x}_1 = x_2 + \phi_1(x_1)$$

$$\dot{x}_2 = x_3 + \phi_2(x_1, x_2)$$

$$\ldots \tag{9.6}$$

$$\dot{x}_{n-1} = x_n + \phi_{n-1}(x_1, x_2, \ldots, x_{n-1})$$

$$\dot{x}_n = u + \phi_n(x_1, x_2, \ldots, x_n)$$

where $x_i \in \mathbb{R}$ for $i = 1, \ldots, n$ are state variables; $\phi_i : \overbrace{\mathbb{R} \times \cdots \times \mathbb{R}}^{i} \to \mathbb{R}$ for $i = 1, \ldots, n$ are differentiable functions up to the order $n - i$ with $\phi_i(0, \ldots, 0) = 0$; and $u \in \mathbb{R}$ is the control input.

When x_{i+1} in the ith equation of (9.6) is identified as the virtual control, the non-linear function ϕ_i is then matched with respect to the virtual control, and backstepping method can be applied to move down the control design to $(i + 1)$th equation with x_{i+2} as the next virtual control. This process starts from $i = 1$ and can be repeated until $i = n - 1$ when u is reached. We will present this iterative backstepping design for the system (9.6) in n steps. In each step, we could show the Lyapunov function and other details for which Lemma 9.1 can be applied to. Although, for the simplicity of the control design, we leave the stability analysis to the end, stability is considered in designing the stabilising function at each step. The control design will be shown in n steps.

Let

$$z_1 = x_1,$$
$$z_i = x_i - \alpha_{i-1}(x_1, \ldots, x_{i-1}), \quad \text{for } i = 2, \ldots, n,$$

where α_{i-1} for $i = 2, \ldots, n$ are stabilising functions obtained in the iterative backstepping design.

Step 1. For the design of the stabilising function, we arrange the dynamics of z_1 as

$$\dot{z}_1 = (x_2 - \alpha_1) + \alpha_1 + \phi_1(x_1)$$
$$= z_2 + \alpha_1 + \phi_1(x_1).$$

Let

$$\alpha_1 = -c_1 z_1 - \phi_1(x_1). \tag{9.7}$$

The resultant dynamics of z_1 are

$$\dot{z}_1 = -c_1 z_1 + z_2. \tag{9.8}$$

Step 2. The dynamics of z_2 are obtained as

$$\dot{z}_2 = \dot{x}_2 - \dot{\alpha}_1$$
$$= x_3 + \phi_2(x_1, x_2) - \frac{\partial \alpha_1}{\partial x_1}(x_2 + \phi_1(x_1))$$
$$= z_3 + \alpha_2 + \phi_2(x_1, x_2) - \frac{\partial \alpha_1}{\partial x_1}(x_2 + \phi_1(x_1)).$$

Design α_2 as

$$\alpha_2 = -z_1 - c_2 z_2 - \phi_2(x_1, x_2) + \frac{\partial \alpha_1}{\partial x_1}(x_2 + \phi_1(x_1)). \tag{9.9}$$

The resultant dynamics of z_2 are given by

$$\dot{z}_2 = -z_1 - c_2 z_2 + z_3. \tag{9.10}$$

Note that the term $-z_1$ in α_2 is used to tackle a cross-term caused by z_2 in the dynamics of z_1 in the stability analysis. Other terms in α_2 are taken to cancel the nonlinear terms and stabilise the dynamics of z_2.

Step i. For $2 < i < n$, the dynamics of z_i are given by

$$
\begin{aligned}
\dot{z}_i &= \dot{x}_i - \dot{\alpha}_{i-1}(x_1, \ldots, x_{i-1}) \\
&= x_{i+1} + \phi_i(x_1, \ldots, x_i) \\
&\quad - \sum_{j=1}^{i-1} \frac{\partial \alpha_{i-1}}{\partial x_j}(x_{j+1} + \phi_j(x_1, \ldots, x_j)) \\
&= z_{i+1} + \alpha_i + \phi_i(x_1, \ldots, x_i) \\
&\quad - \sum_{j=1}^{i-1} \frac{\partial \alpha_{i-1}}{\partial x_j}(x_{j+1} + \phi_j(x_1, \ldots, x_j)).
\end{aligned}
$$

Design α_i as

$$
\begin{aligned}
\alpha_i &= -z_{i-1} - c_i z_i - \phi_i(x_1, \ldots, x_i) \\
&\quad + \sum_{j=1}^{i-1} \frac{\partial \alpha_{i-1}}{\partial x_j}(x_{j+1} + \phi_j(x_1, \ldots, x_j)).
\end{aligned}
\tag{9.11}
$$

The resultant dynamics of z_i are given by

$$
\dot{z}_i = -z_{i-1} - c_i z_i + z_{i+1}.
\tag{9.12}
$$

Note that similar to the design of α_2, the term $-z_{i-1}$ is used to tackle a cross-term caused by z_i in the dynamics of z_{i-1} in the stability analysis, and the other terms in α_i are used to stabilise the dynamics of z_i.

Step n. At the final step, we have

$$
\begin{aligned}
\dot{z}_n &= \dot{x}_n - \dot{\alpha}_{n-1}(x_1, \ldots, x_{n-1}) \\
&= u + \phi_n(x_1, \ldots, x_n) \\
&\quad - \sum_{j=1}^{n-1} \frac{\partial \alpha_{n-1}}{\partial x_j}(x_{j+1} + \phi_j(x_1, \ldots, x_j)).
\end{aligned}
$$

Design the control input as

$$
\begin{aligned}
u &= -z_{n-1} - c_n z_n - \phi_n(x_1, \ldots, x_n) \\
&\quad + \sum_{j=1}^{n-1} \frac{\partial \alpha_{n-1}}{\partial x_j}(x_{j+1} + \phi_j(x_1, \ldots, x_j)).
\end{aligned}
\tag{9.13}
$$

The resultant dynamics of z_n are given by

$$
\dot{z}_n = -z_{n-1} - c_n z_n.
\tag{9.14}
$$

Note that we can write $u = \alpha_n$ by setting $i = n$ in the expression of α_i in (9.11).

Let us establish the stability of the closed-loop system under the proposed control. The closed-loop dynamics of the system in coordinate $z = [z_1, \ldots, z_n]^T$ can be written as

$$\dot{z} = \begin{bmatrix} -c_1 & 1 & 0 & \cdots & 0 \\ -1 & -c_2 & 1 & \ddots & 0 \\ 0 & -1 & -c_3 & \ddots & 0 \\ \vdots & \ddots & \ddots & \ddots & 1 \\ 0 & 0 & 0 & \ddots & -c_n \end{bmatrix} z := A_z z.$$

Notice that all the off-diagonal A_z are skew symmetric. Let

$$V = \frac{1}{2} z^T z. \tag{9.15}$$

Its derivative along the dynamics of z is obtained as

$$\dot{V} = -\sum_{i=1}^{n} c_i z_i^2 \leq -2 \min_{i=1}^{n} c_i V. \tag{9.16}$$

Therefore, we can conclude that the system is exponentially stable in z-coordinate. From the property that $\phi_i(0, \ldots, 0) = 0$ for $i = 1, \ldots, n$, we can establish that $\alpha_i(0, \ldots, 0) = 0$ for $i = 1, \ldots, n-1$ and $u(0, \ldots, 0) = 0$, which implies that $\lim_{t \to \infty} x_i(t) = 0$ for $i = 1, \ldots, n$. Hence, we have established the following result.

Theorem 9.2. *For a system in the form of (9.6), the control input (9.13) renders the closed-loop system asymptotically stable.*

9.3 Observer backstepping

We have presented integrator backstepping and iterative backstepping based on state feedback. In this section, we present a control design for nonlinear systems using output feedback. An observer is designed, and the observer state, or estimate of the state variables of the original system, is then used for backstepping design.

Consider a nonlinear system which can be transformed to

$$\begin{aligned} \dot{x} &= A_c x + bu + \phi(y) \\ y &= Cx \end{aligned} \tag{9.17}$$

with

$$A_c = \begin{bmatrix} 0 & 1 & 0 & \cdots & 0 \\ 0 & 0 & 1 & \cdots & 0 \\ \vdots & \vdots & \vdots & \ddots & \vdots \\ 0 & 0 & 0 & \cdots & 1 \\ 0 & 0 & 0 & \cdots & 0 \end{bmatrix}, \quad C = \begin{bmatrix} 1 \\ 0 \\ \vdots \\ 0 \end{bmatrix}^T, \quad b = \begin{bmatrix} 0 \\ \vdots \\ 0 \\ b_\rho \\ \vdots \\ b_n \end{bmatrix},$$

where $x \in \mathbb{R}^n$ is the state vector; $u \in \mathbb{R}$ is the control; $\phi : \mathbb{R} \to \mathbb{R}^n$ with $\phi(0) = 0$ is a nonlinear function with element ϕ_i being differentiable up to the $(n - i)$th order; and $b \in \mathbb{R}^n$ is a known constant Hurwitz vector with $b_\rho \neq 0$, which implies that the relative degree of the system is ρ. This form of the system is often referred to as the output feedback form. Since b is Hurwitz, the linear system characterised by (A_c, b, C) is minimum phase. Note that a vector is said Hurwitz if its corresponding polynomial is Hurwitz.

Remark 9.2. For a system in the output feedback form, if the input is zero, the system is in exactly the same form as (8.10). We have shown the geometric conditions in Chapter 8 for systems to be transformed to the output injection form (8.10), and similar geometric conditions can be specified for nonlinear systems to be transformed to the output feedback form. Clearly we can see that for the system (9.17) with any observable pair (A, C), there exists a linear transformation to put the system in the form of (9.17) with the specific (A_c, C). ◁

Since the system (9.17) is in the output injection form, we design an observer as

$$\dot{\hat{x}} = A_c x + bu + \phi(y) + L(y - C\hat{x}), \tag{9.18}$$

where $\hat{x} \in \mathbb{R}^n$ is the state estimate and $L \in \mathbb{R}^n$ is an observer gain designed such that $(A - LC)$ is Hurwitz. Let $\tilde{x} = x - \hat{x}$, and it is easy to see

$$\dot{\tilde{x}} = (A_c - LC)\tilde{x}. \tag{9.19}$$

The backstepping design can be carried out with the state estimate $\hat{x}$ in ρ steps. From the structure of the system (9.17) we have $y = x_1$. The backstepping design will start with the dynamics of y. In the following design, we assume $\rho > 1$. In the case of $\rho = 1$, control input can be designed directly without using backstepping.

To apply the observer backstepping through $\hat{x}$ in (9.18), we define

$$z_1 = y,$$
$$z_i = \hat{x}_i - \alpha_{i-1}, \quad i = 2, \ldots, \rho, \tag{9.20}$$
$$z_{\rho+1} = \hat{x}_{\rho+1} + b_\rho u - \alpha_\rho,$$

where $\alpha_i, i = 1, \ldots, \rho$, are stabilising functions decided in the control design.

Consider the dynamics of z_1

$$\dot{z}_1 = x_2 + \phi_1(y). \tag{9.21}$$

We use $\hat{x}_2$ to replace the unmeasurable x_2 in (9.21), resulting at

$$\dot{z}_1 = \hat{x}_2 + \tilde{x}_2 + \phi_1(y)$$
$$= z_2 + \alpha_1 + \tilde{x}_2 + \phi_1(y). \tag{9.22}$$

We design α_1 as

$$\alpha_1 = -c_1 z_1 - k_1 z_1 - \phi_1(y), \tag{9.23}$$

where c_i and k_i for $i = 1, \ldots, \rho$ are positive real design parameters. Comparing the backstepping design using the output feedback with the one using state feedback, we have one additional term, $-k_1 z_1$, which is used to tackle the observer error $\tilde{x}_2$ in the closed-loop system dynamics. Then from (9.22) and (9.23), we have

$$\dot{z}_1 = z_2 - c_1 z_1 - k_1 z_1 + \tilde{x}_2. \tag{9.24}$$

Note that α_1 is a function of y, i.e., $\alpha_1 = \alpha_1(y)$.

For the dynamics of z_2, we have

$$\dot{z}_2 = \dot{\hat{x}}_2 - \dot{\alpha}_1$$
$$= \hat{x}_3 + \phi_2(y) + l_2(y - \hat{x}_1) - \frac{\partial \alpha_1}{\partial y}(x_2 + \phi_1(y))$$
$$= z_3 + \alpha_2 + \phi_2(y) + l_2(y - \hat{x}_1) - \frac{\partial \alpha_1}{\partial y}(\hat{x}_2 + \tilde{x}_2 + \phi_1(y)).$$

where l_2 is the second element of the observer gain L, and in the subsequent design, l_i is the ith element of L. We design α_2 as

$$\alpha_2 = -z_1 - c_2 z_2 - k_2 \left(\frac{\partial \alpha_1}{\partial y} \right)^2 z_2 - \phi_2(y) - l_2(y - \hat{x}_1)$$
$$+ \frac{\partial \alpha_1(y)}{\partial y}(\hat{x}_2 + \phi_1(y)). \tag{9.25}$$

Note that $\alpha_2 = \alpha_2(y, \hat{x}_1, \hat{x}_2)$. The resultant dynamics of z_2 are given by

$$\dot{z}_2 = -z_1 - c_2 z_2 - k_2 \left(\frac{\partial \alpha_1}{\partial y} \right)^2 z_2 + z_3 - \frac{\partial \alpha_1}{\partial y}\tilde{x}_2.$$

For the dynamics of z_i, $2 < i \leq \rho$, we have

$$\dot{z}_i = \dot{\hat{x}}_i - \dot{\alpha}_{i-1}$$
$$= z_{i+1} + \alpha_i + \phi_i(y) + l_i(y - \hat{x}_1)$$
$$- \frac{\partial \alpha_{i-1}}{\partial y}(\hat{x}_2 + \tilde{x}_2 + \phi_1(y)) - \sum_{j=1}^{i-1} \frac{\partial \alpha_{i-1}}{\partial \hat{x}_j}\dot{\hat{x}}_j.$$

We design α_i, $2 < i \leq \rho$, as

$$\alpha_i = -z_{i-1} - c_i z_i - k_i \left(\frac{\partial \alpha_{i-1}}{\partial y} \right)^2 z_i - \phi_i(y) - l_i(y - \hat{x}_1)$$

$$+ \frac{\partial \alpha_{i-1}}{\partial y} (\hat{x}_2 + \phi_1(y)) + \sum_{j=1}^{i-1} \frac{\partial \alpha_{i-1}}{\partial \hat{x}_j} \dot{\hat{x}}_j. \tag{9.26}$$

Note that $\alpha_i = \alpha_i(y, \hat{x}_1, \dots, \hat{x}_i)$. The resultant dynamics of z_i, $2 < i \leq \rho$, are given by

$$\dot{z}_i = -z_{i-1} - c_i z_i - k_i \left(\frac{\partial \alpha_{i-1}}{\partial y} \right)^2 z_i + z_{i+1} - \frac{\partial \alpha_{i-1}}{\partial y} \tilde{x}_2. \tag{9.27}$$

When $i = \rho$, the control input appears in the dynamics of z_i, and it is included in $z_{\rho+1}$, as shown in the definition (9.20). We design the control input by setting $z_{\rho+1} = 0$, which gives

$$u = \frac{\alpha_\rho(y, \hat{x}_1, \dots, \hat{x}_\rho) - \hat{x}_{\rho+1}}{b_\rho}. \tag{9.28}$$

The stability result of the above control design is given in the following theorem.

Theorem 9.3. *For a system in the form of (9.17), the dynamic output feedback control with the input (9.28) and the observer (9.18) asymptotically stabilise the system.*

Proof. From the observer error dynamics, we know that the error exponentially converges to zero. Since $(A_c - LC)$ is Hurwitz, there exists a positive definite matrix $P \in \mathbb{R}^{n \times n}$ such that

$$(A_c - LC)^T P + P(A_c - LC) = -I.$$

This implies that for

$$V_e = \tilde{x}^T P \tilde{x},$$

we have

$$\dot{V}_e = -\|\tilde{x}\|^2. \tag{9.29}$$

Let

$$V_z = \sum_{i=1}^{\rho} z_i^2.$$

From the dynamics of z_1, z_2 and z_i, we can obtain that

$$\dot{V}_z = \sum_{i=1}^{\rho} \left(-c_i z_i^2 - k_i \left(\frac{\partial \alpha_{i-1}}{\partial y} \right)^2 z_i^2 - \frac{\partial \alpha_{i-1}}{\partial y} z_i \tilde{x}_2 \right),$$

where we define $\alpha_0 = y$ for notational convenience. Note that if we ignore the two terms concerning with k_i and $\tilde{x}_2$ in the dynamics of z_i, the evaluation of the derivative of V_z will be exactly the same as the stability analysis that leads to Theorem 9.2. For the cross-term concerning with $\tilde{x}_2$, we have, from Young's inequality,

$$\left| \frac{\partial \alpha_{i-1}}{\partial y} z_i \tilde{x}_2 \right| \leq k_i \left(\frac{\partial \alpha_{i-1}}{\partial y} \right)^2 z_i^2 + \frac{1}{4k_i} \tilde{x}_2^2.$$

Hence, we obtain that

$$\dot{V}_z \leq \sum_{i=1}^{\rho} \left(-c_i z_i^2 + \frac{1}{4k_i} \tilde{x}_2^2 \right). \tag{9.30}$$

Let

$$V = V_z + \left(1 + \frac{1}{4d} \right) V_e,$$

where $d = \min_{i=1}^{\rho} k_i$. From (9.29) and (9.30), we have

$$\dot{V} \leq \sum_{i=1}^{\rho} \left(-c_i z_i^2 + \frac{1}{4k_i} \tilde{x}_2^2 \right) - \left(1 + \frac{1}{4d} \right) \|\tilde{x}\|^2$$

$$\leq - \sum_{i=1}^{\rho} c_i z_i^2 - \|\tilde{x}\|^2.$$

Therefore, we can conclude that z_i for $i = 1, \ldots, \rho$ and $\tilde{x}$ exponentially converge to zero. Noticing that $y = z_1$, and $\alpha_1(0) = 0$, we can conclude that $\lim_{t \to \infty} \hat{x}_2(t) = 0$, which further implies that $\lim_{t \to \infty} x_2(t) = 0$. Following the same process, we can show that $\lim_{t \to \infty} x_i(t) = 0$ for $i = 1, \ldots, \rho$.

We still need to establish the stability property for x_i, with $i = \rho + 1, \ldots, n$. Let

$$\xi = \begin{bmatrix} x_{\rho+1} \\ \vdots \\ x_n \end{bmatrix} - \begin{bmatrix} b_{\rho+1} \\ \vdots \\ b_n \end{bmatrix} \frac{x_\rho}{b_\rho},$$

and from the system dynamics (9.17), it can be shown that

$$\dot{\xi} = B\xi + \begin{bmatrix} \phi_{\rho+1}(y) \\ \vdots \\ \phi_n(y) \end{bmatrix} - \begin{bmatrix} b_{\rho+1} \\ \vdots \\ b_n \end{bmatrix} \frac{\phi_\rho(y)}{b_\rho} + B \begin{bmatrix} b_{\rho+1} \\ \vdots \\ b_n \end{bmatrix} \frac{x_\rho}{b_\rho}, \tag{9.31}$$

where

$$B = \begin{bmatrix} -b_{\rho+1}/b_\rho & 1 & \cdots & 0 \\ -b_{\rho+2}/b_\rho & 0 & \ddots & 0 \\ \vdots & \vdots & \vdots & \vdots \\ -b_{n-1}/b_\rho & 0 & \cdots & 1 \\ -b_n/b_\rho & 0 & \cdots & 0 \end{bmatrix}$$

is a companion matrix associated with vector b. Since b is Hurwitz, the matrix B is Hurwitz. All the terms other than $B\xi$ in the right-hand side of (9.31) converge to zero, which implies that $\lim_{t\to\infty} \xi(t) = 0$. Therefore, we can conclude $\lim_{t\to\infty} x_i(t) = 0$ for $i = \rho + 1, \ldots, n$. This concludes the proof.   □

9.4   Backstepping with filtered transformation

We have presented backstepping design for a class of dynamic systems using output feedback in the previous section. The control design starts from the system output and the subsequent steps in the backstepping design are carried out with estimates of state variables provided by an observer. The backstepping design, often referred to as observer backstepping, completes in ρ steps, with ρ being the relative degree of the system, and other state estimates of x_i for $i > \rho$ are not used in the design. Those estimates are redundant, and make the dynamic order of the controller higher. There is an alternative design method to observer backstepping for nonlinear systems in the output feedback form, of which the resultant order of the controller is exactly $\rho - 1$. In this section, we present backstepping design with filtered transformation for the system (9.17), of which the main equations are shown here again for the convenience of presentation,

$$\dot{x} = A_c x + bu + \phi(y)$$
$$y = Cx. \tag{9.32}$$

Define an input filter

$$\dot{\xi}_1 = -\lambda_1 \xi_1 + \xi_2$$
$$\vdots \tag{9.33}$$
$$\dot{\xi}_{\rho-1} = -\lambda_{\rho-1}\xi_{\rho-1} + u,$$

where $\lambda_i > 0$ for $i = 1, \ldots, \rho - 1$ are the design parameters. Define the filtered transformation

$$\bar{\zeta} = x - \sum_{i=1}^{\rho-1} \bar{d}_i \xi_i, \tag{9.34}$$

where $\bar{d}_i \in \mathbb{R}^n$ for $i = 1, \ldots, \rho - 1$ and they are generated recursively by

$$\bar{d}_{\rho-1} = b,$$
$$\bar{d}_i = (A_c + \lambda_{i+1} I)\bar{d}_{i+1} \quad \text{for } i = \rho - 2, \ldots, 1.$$

We also denote

$$d = (A_c + \lambda_1 I)\bar{d}_1.$$

From the filtered transformation, we have

$$
\begin{aligned}
\dot{\bar{\zeta}} &= A_c x + b u + \phi(y) - \sum_{i=1}^{\rho-2} \bar{d}_i(-\lambda_i \xi_i + \xi_{i+1}) - \bar{d}_{\rho-1}(-\lambda_i \xi_{\rho-1} + u) \\
&= A_c \bar{\zeta} + \sum_{i=1}^{\rho-1} A \bar{d}_i \xi_i + \phi(y) + \sum_{i=1}^{\rho-1} \bar{d}_i \lambda_i \xi_i - \sum_{i=1}^{\rho-2} \bar{d}_i \xi_{i+1} \\
&= A_c \bar{\zeta} + \sum_{i=1}^{\rho-1} (A + \lambda_i I) \bar{d}_i \xi_i + \phi(y) - \sum_{i=1}^{\rho-2} \bar{d}_i \xi_{i+1} \\
&= A_c \bar{\zeta} + \phi(y) + d \xi_1.
\end{aligned}
$$

The output under the coordinate $\bar{\zeta}$ is given by

$$
\begin{aligned}
y &= C \bar{\zeta} + \sum_{i=1}^{\rho-1} C \bar{d}_i \xi_i \\
&= C \bar{\zeta}
\end{aligned}
$$

because $C \bar{d}_i = 0$ for $i = 1, \ldots, \rho - 1$, from the fact that $\bar{d}_{i,j} = 0$ for $i = 1, \ldots,$ $\rho - 1$, $1 \le j \le i$. Hence, under the filtered transformation, the system (9.32) is then transformed to

$$
\begin{aligned}
\dot{\bar{\zeta}} &= A_c \bar{\zeta} + \phi(y) + d \xi_1, \\
y &= C \bar{\zeta}.
\end{aligned}
\tag{9.35}
$$

Let us find a bit more information of d. From the definition

$$
\bar{d}_{\rho-2} = (A_c + \lambda_{\rho-1} I) b
$$

we have

$$
\sum_{i=\rho-1}^{n} \bar{d}_{\rho-2,i} s^{n-i} = (s + \lambda_{\rho-1}) \sum_{\rho}^{n} b_i s^{n-i}.
$$

Repeating the process iteratively, we can obtain

$$
\sum_{i=1}^{n} d_i s^{n-i} = \prod_{i=1}^{\rho-1} (s + \lambda_i) \sum_{\rho}^{n} b_i s^{n-i}
\tag{9.36}
$$

which implies that $d_1 = b_\rho$ and that d is Hurwitz if b Hurwitz. In the special form of A_c and C used here, b and d decide the zeros of the linear systems characterised by (A_c, b, C) and (A_c, d, C) respectively as the solutions to the following polynomial equations:

$$\sum_{\rho}^{n} b_i s^{n-i} = 0,$$

$$\sum_{i=1}^{n} d_i s^{n-i} = 0.$$

Hence, the invariant zeros of (A_c, d, C) are the invariant zeros of (A_c, b, C) plus λ_i for $i = 1, \ldots, \rho - 1$. For the transformed system, ξ_1 can be viewed as the new input. In this case, the relative degree with ξ_1 as the input is 1. The filtered transformation lifts the relative degree from ρ to 1.

As the filtered transformation may have its use independent of backstepping design shown here, we summarise the property of the filtered transformation in the following lemma.

Lemma 9.4. *For a system in the form of (9.32) with relative degree ρ, the filtered transformation defined in (9.34) transforms the system to (9.35) of relative degree 1, with the same high frequency gain. Furthermore, the zeros of (9.35) consist of the zeros of the original system (9.32) and λ_i for $i = 1, \ldots, \rho - 1$.*

We introduce another state transform to extract the internal dynamics of (9.35) with $\zeta \in \mathbb{R}^{n-1}$ given by

$$\zeta = \bar{\zeta}_{2:n} - \frac{d_{2:n}}{d_1} y, \tag{9.37}$$

where $\zeta \in \mathbb{R}^{n-1}$ forms the state variable of the transformed system together with y, the notation $(\cdot)_{2:n}$ refers to the vector or matrix formed by the 2nd row to the nth row. With the coordinates (ζ, y), (9.35) is rewritten as

$$\begin{aligned} \dot{\zeta} &= D\zeta + \psi(y) \\ \dot{y} &= \zeta_1 + \psi_y(y) + b_\rho \xi_1, \end{aligned} \tag{9.38}$$

where D is the companion matrix of d given by

$$D = \begin{bmatrix} -d_2/d_1 & 1 & \cdots & 0 \\ -d_3/d_1 & 0 & \ddots & 0 \\ \vdots & \vdots & \vdots & \vdots \\ -d_{n-1}/d_1 & 0 & \cdots & 1 \\ -d_n/d_1 & 0 & \cdots & 0 \end{bmatrix},$$

and

$$\psi(y) = D\frac{d_{2:n}}{d_1} y + \phi_{2:n}(y) - \frac{d_{2:n}}{d_1} \phi_1(y),$$

$$\psi_y(y) = \frac{d_2}{d_1} y + \phi_1(y).$$

If we view ξ_1 as the input, the system (9.35) is of relative degree 1 with the stable zero dynamics. For such a system, there exists an output feedback law to globally and exponentially stabilise the system.

For this, we have the following lemma, stating in a more stand alone manner.

Lemma 9.5. *For a nonlinear system (9.32), if the relative degree is 1, there exist a continuous function $\varphi : \mathbb{R} \to \mathbb{R}$ with $\varphi(0) = 0$ and a positive real constant c such that the control input in the form of*

$$u = -cy - \varphi(y) \tag{9.39}$$

globally and asymptotically stabilises the system.

Proof. Introducing the same transformation as (9.37), i.e.,

$$\zeta = x_{2:n} - \frac{b_{2:n}}{b_1} y,$$

we can obtain exactly the same transformed system as in (9.38) with d being replaced by b and ξ_1 by u. Since D is Hurwitz, there exists a positive definite matrix P such that

$$D^T P + PD = -3I. \tag{9.40}$$

We set

$$\varphi(y) = \psi_y(y) + \frac{\|P\|^2 \|\psi(y)\|^2}{y}. \tag{9.41}$$

Note that $\psi(0) = 0$, and therefore $\|\psi(y)\|^2 / y$ is well defined. The closed-loop system is then obtained as

$$\dot{\zeta} = D\zeta + \psi(y),$$

$$\dot{y} = \zeta_1 - cy - \frac{\|P\|^2 \|\psi(y)\|^2}{y}.$$

Let

$$V = \zeta^T P \zeta + \frac{1}{2} y^2.$$

We have

$$\dot{V} = -cy^2 - 3\|\zeta\|^2 + 2\zeta^T P \psi(y) + y\zeta_1 - \|P\|^2 \|\psi(y)\|^2.$$

With the inequalities of the cross terms

$$|2\zeta^T P \psi(y)| \leq \|\zeta\|^2 + \|P\|^2 \|\psi(y)\|^2,$$

$$|y\zeta_1| \leq \frac{1}{4} y^2 + \|\zeta\|^2,$$

we have

$$\dot{V} \leq -\left(c - \frac{1}{4}\right) y^2 - \|\zeta\|^2.$$

Therefore, the proposed control design with $c > \frac{1}{4}$ and (9.41) exponentially stabilises the system in the coordinate (ζ, y), which implies the exponential stability of the closed-loop system in x coordinate, because the transformation from (ζ, y) to x is linear. □

From Lemma 9.5, we know the desired value of ξ_1. But we cannot directly assign a function to ξ_1 as it is not the actual control input. Here backstepping can be applied to design control input based on the desired function of ξ_1. Together with the filtered transformation, the overall system is given by

$$
\begin{aligned}
\dot{\zeta} &= D\zeta + \psi(y) \\
\dot{y} &= \zeta_1 + \psi_y(y) + b_\rho \xi_1 \\
\dot{\xi}_1 &= -\lambda_1 \xi_1 + \xi_2
\end{aligned}
\tag{9.42}
$$

$$
\cdots
$$

$$
\dot{\xi}_{\rho-1} = -\lambda_{\rho-1} \xi_{\rho-1} + u,
$$

to which the backstepping design is then applied. Indeed, in the backstepping design, ξ_i for $i = 1, \ldots, \rho - 1$ can be viewed as virtual controls.

Let

$$
\begin{aligned}
z_1 &= y, \\
z_i &= \xi_{i-1} - \alpha_{i-1}, \quad \text{for } i = 2, \ldots, \rho \\
z_{\rho+1} &= u - \alpha_\rho,
\end{aligned}
$$

where α_i for $i = 2, \ldots, \rho$ are stabilising functions to be designed. We also use the positive real design parameters c_i and k_i for $i = 1, \ldots, \rho$ and $\gamma > 0$.

Based on the result shown in Lemma 9.5, we have

$$
\alpha_1 = -c_1 z_1 - k_1 z_1 + \psi_y(y) - \frac{\gamma \|P\|^2 \|\psi(y)\|^2}{y}
\tag{9.43}
$$

and

$$
\dot{z}_1 = z_2 - c_1 z_1 - k_1 z_1 - \gamma \|P\|^2 \|\psi(y)\|^2.
\tag{9.44}
$$

For the dynamics of z_2, we have

$$
\begin{aligned}
\dot{z}_2 &= -\lambda_1 \xi_1 + \xi_2 - \frac{\partial \alpha_1}{\partial y} \dot{y} \\
&= z_3 + \alpha_2 - \lambda_1 \xi_1 - \frac{\partial \alpha_1}{\partial y} (\zeta_1 + \psi(y)).
\end{aligned}
$$

The design of α_2 is then given by

$$
\alpha_2 = -z_1 - c_2 z_2 - k_2 \left(\frac{\partial \alpha_1}{\partial y} \right)^2 z_2 + \frac{\partial \alpha_1}{\partial y} \psi(y) + \lambda_1 \xi_1.
\tag{9.45}
$$

The resultant dynamics of z_2 is obtained as

$$\dot{z}_2 = -z_1 - c_2 z_2 - k_2 \left(\frac{\partial \alpha_1}{\partial y} \right)^2 z_2 + z_3 - \frac{\partial \alpha_1}{\partial y} \zeta_1. \tag{9.46}$$

Note that $\alpha_2 = \alpha_2(y, \xi_1)$.

For the subsequent steps for $i = 3, \ldots, \rho$, we have

$$\dot{z}_i = -\lambda_{i-1}\xi_{i-1} + \xi_i - \frac{\partial \alpha_{i-1}}{\partial y}\dot{y} - \sum_{j=1}^{i-2} \frac{\partial \alpha_{i-1}}{\partial \xi_j}\dot{\xi}_j$$

$$= z_{i+1} + \alpha_i - \lambda_{i-1}\xi_{i-1} - \frac{\partial \alpha_{i-1}}{\partial y}(\zeta_1 + \psi(y))$$

$$- \sum_{j=1}^{i-2} \frac{\partial \alpha_{i-1}}{\partial \xi_j}(-\lambda_j \xi_{i-1} + \xi_{j+1}).$$

The design of α_i is given by

$$\alpha_i = -z_{i-1} - c_i z_i - k_i \left(\frac{\partial \alpha_{i-1}}{\partial y} \right)^2 z_i + \frac{\partial \alpha_{i-1}}{\partial y}\psi(y)$$

$$+ \sum_{j=1}^{i-2} \frac{\partial \alpha_{i-1}}{\partial \xi_j}(-\lambda_j \xi_j + \xi_{j+1}) + \lambda_{i-1}\xi_{i-1}. \tag{9.47}$$

The resultant dynamics of z_i is obtained as

$$\dot{z}_i = -z_{i-1} - c_i z_i - k_i \left(\frac{\partial \alpha_{i-1}}{\partial y} \right)^2 z_i + z_{i+1} - \frac{\partial \alpha_{i-1}}{\partial y}\zeta_1. \tag{9.48}$$

Note that $\alpha_i = \alpha_i(y, \xi_1, \ldots, \xi_{i-1})$.

When $i = \rho$, the control input appears in the dynamics of z_i, through the term $z_{\rho+1}$. We design the control input by setting $z_{\rho+1} = 0$, which gives $u = \alpha_\rho$, that is

$$u = -z_{\rho-1} - c_\rho z_\rho - k_\rho \left(\frac{\partial \alpha_{\rho-1}}{\partial y} \right)^2 z_\rho + \frac{\partial \alpha_{\rho-1}}{\partial y}\psi(y)$$

$$+ \sum_{j=1}^{i-2} \frac{\partial \alpha_{\rho-1}}{\partial \xi_j}(-\lambda_j \xi_j + \xi_{j+1}) + \lambda_{\rho-1}\xi_{\rho-1}. \tag{9.49}$$

For the control design parameters, $c_i, i = 1, \ldots, \rho$ and γ can be any positive, and for d_i, the following condition must be satisfied:

$$\sum_{l=1}^{\rho} \frac{1}{4k_i} \leq \gamma.$$

The stability result of the above control design is given in the following theorem.

Theorem 9.6. *For a system in the form of (9.32), the dynamic output feedback control (9.49) obtained through backstepping with the input filtered transformation asymptotically stabilises the system.*

Proof. Let

$$V_z = \sum_{i=1}^{\rho} z_i^2.$$

From the dynamics for z_i shown in (9.44), (9.46) and (9.48) we can obtain that

$$\dot{V}_z = \sum_{i=1}^{\rho} \left(-c_i z_i^2 - k_i \left(\frac{\partial \alpha_{i-1}}{\partial y} \right)^2 z_i^2 - \frac{\partial \alpha_{i-1}}{\partial y} z_i \zeta_1 \right) - \gamma \|P\|^2 \|\psi(y)\|^2,$$

where we define $\alpha_0 = y$ for notational convenience. For the cross-term concerning with ζ_1, we have

$$\left| \frac{\partial \alpha_{i-1}}{\partial y} z_i \zeta_1 \right| \le k_i \left(\frac{\partial \alpha_{i-1}}{\partial y} \right)^2 z_i^2 + \frac{1}{4k_i} \zeta_1^2.$$

Hence, we obtain that

$$\dot{V}_z \le \sum_{i=1}^{\rho} \left(-c_i z_i^2 + \frac{1}{4k_i} \zeta_1^2 \right) - \gamma \|P\|^2 \|\psi(y)\|^2. \tag{9.50}$$

Let

$$V_\zeta = \zeta^T P \zeta$$

and we can obtain, similar to the proof of Lemma 9.5,

$$\dot{V}_\zeta \le -2\|\zeta\|^2 + \|P\|^2 \|\psi(y)\|^2.$$

Let

$$V = V_z + \gamma V_\zeta.$$

and we have

$$\dot{V} \le \sum_{i=1}^{\rho} \left(-c_i z_i^2 + \frac{1}{4k_i} \zeta_1^2 \right) - 2\gamma \|\zeta\|^2 \tag{9.51}$$

$$\le -\sum_{i=1}^{\rho} c_i z_i^2 - \gamma \|\zeta\|^2. \tag{9.52}$$

Therefore, we have shown that the system (9.42) is exponentially stable under the coordinate $(\zeta, z_1, \ldots, z_{\rho-1})$. With $y = z_1$, we can conclude $\lim_{t\to\infty} \bar{\zeta}(t) = 0$. From $y = z_1$, and $\alpha_1(0) = 0$, we can conclude that $\lim_{t\to\infty} \xi_1(t) = 0$. Following the same process, we can show that $\lim_{t\to\infty} \xi(t) = 0$ for $i = 1, \ldots, \rho - 1$. Finally from the filtered transformation (9.34), we can establish $\lim_{t\to\infty} x(t) = 0$. □

9.5 Adaptive backstepping

We have shown backstepping design for two classes of systems with state feedback and output feedback respectively. In this section, we will show the nonlinear adaptive control design for a class of system of which there are unknown parameters.

Consider a first-order nonlinear system described by

$$\dot{y} = u + \phi^T(y)\theta \tag{9.53}$$

where $\phi : \mathbb{R} \to \mathbb{R}^p$ is a smooth nonlinear function, and $\theta \in \mathbb{R}^p$ is an unknown vector of constant parameters. For this system, adaptive control law can be designed as

$$u = -cy - \phi^T(y)\hat{\theta} \tag{9.54}$$

$$\dot{\hat{\theta}} = \Gamma y \phi(y) \tag{9.55}$$

where c is a positive real constant, and $\Gamma \in \mathbb{R}^{p \times p}$ is a positive definite gain matrix.

The closed-loop dynamics is given by

$$\dot{y} = -cy + \phi^T(y)\tilde{\theta}$$

with the usual notation $\tilde{\theta} = \theta - \hat{\theta}$.

For stability analysis, let

$$V = \frac{1}{2}y^2 + \frac{1}{2}\tilde{\theta}^T \Gamma^{-1}\tilde{\theta}$$

and its derivative is obtained as

$$\dot{V} = -cy^2,$$

which ensures the boundedness of y and $\hat{\theta}$. We can show $\lim_{t \to \infty} y(t) = 0$ in the same way by invoking Babalat's Lemma as in the stability analysis of adaptive control systems shown in Chapter 7.

Remark 9.3. The system considered above is nonlinear with unknown parameters. However, the unknown parameters are linearly parameterised, i.e., the terms relating to the unknown parameters, $\phi^T(y)\theta$ are linear with the unknown parameters, instead of some nonlinear functions $\phi(y, \theta)$, which are referred to as nonlinearly parameterised. Obviously, nonlinear parameterised unknown parameters are much more difficult to deal with in adaptive control. In this book, we only consider linearly parameterised unknown parameters. ◁

For the first-order system, the control input is matched with the uncertainty and the nonlinear function. Backstepping can also be used with adaptive control to deal with nonlinear and unknown parameters which are not in line with the input, or, unmatched.

Consider a nonlinear system

$$\dot{x}_1 = x_2 + \phi_1(x_1)^T\theta$$
$$\dot{x}_1 = x_3 + \phi_2(x_1, x_2)^T\theta$$
$$\cdots \tag{9.56}$$
$$\dot{x}_{n-1} = x_n + \phi_{n-1}(x_1, x_1, \ldots, x_{n-1})^T\theta$$
$$\dot{x}_n = u + \phi_n(x_1, x_2, \ldots, x_n)^T\theta,$$

where $x_i \in \mathbb{R}$ for $i = 1, \ldots, n$ are state variables; $\theta \in \mathbb{R}^p$ is an unknown vector of constant parameters; $\phi_i : \overbrace{\mathbb{R} \times \cdots \times \mathbb{R}}^{i} \to \mathbb{R}^p$ for $i = 1, \ldots, n$ are differentiable functions up to the order $n - i$ with $\phi_i(0, \ldots, 0) = 0$; and $u \in \mathbb{R}$ is the control input.

Note that this system is exactly the same as (9.6) if the parameter vector θ is known.

The backstepping design method will be applied iteratively in a similar way to the control design for (9.6), with only difference of including adaptive parameters in the control design.

Let

$$z_1 = x_1,$$
$$z_i = x_i - \alpha_{i-1}(x_i, \ldots, x_{i-1}, \hat{\theta}), \quad \text{for } i = 2, \ldots, n,$$
$$z_{n+1} = u - \alpha_n(x_i, \ldots, x_n, \hat{\theta}),$$
$$\varphi_1 = \phi_1,$$
$$\varphi_i = \phi_i - \sum_{j=1}^{i-1} \frac{\partial\alpha_{i-1}}{\partial x_j}\phi_j, \quad \text{for } i = 2, \ldots, n,$$

where α_{i-1}, for $i = 2, \ldots, n$, are stabilising functions obtained in the adaptive beackstepping design, and $\hat{\theta}$ denotes an estimate of θ.

We start the adaptive backstepping from the dynamics of z_1

$$\dot{z}_1 = z_2 + \alpha_1 + \varphi_1^T\theta.$$

Design α_1 as

$$\alpha_1 = -c_1 z_1 - \varphi_1^T\hat{\theta}, \tag{9.57}$$

where c_i, for $i = 1, \ldots, n$, are set of positive design parameters. The closed-loop dynamics are obtained as

$$\dot{z}_1 = -c_1 z_1 + z_2 + \varphi_1^T\tilde{\theta},$$

where $\tilde{\theta} = \theta - \hat{\theta}$.

For the dynamics of z_2, we have

$$\dot{z}_2 = \dot{x}_2 - \dot{\alpha}_1$$

$$= x_3 + \phi_2^T \theta - \frac{\partial \alpha_1}{\partial x_1} \dot{x}_1 - \frac{\partial \alpha_1}{\partial \hat{\theta}} \dot{\hat{\theta}}$$

$$= z_3 + \alpha_2 + \varphi_2^T \theta - \frac{\partial \alpha_1}{\partial x_1} x_2 - \frac{\partial \alpha_1}{\partial \hat{\theta}} \dot{\hat{\theta}}.$$

The dynamics of z_2 involve the adaptive law $\dot{\hat{\theta}}$. Even though the adaptive law has not been designed, it is surely known to the control design, and can be used in the control input. However, the dynamics of z_i with $i > 2$ will also affect the design of the adaptive law. For this reason, we would like to leave the design of the adaptive law to the end. Inevitably, the adaptive law will include z_i for $i > 2$. This causes a problem, if we use α_2 to cancel $\dot{\hat{\theta}}$ at this step, because z_3 depends on α_2. Instead, we only deal with the part of the adaptive law that depends on z_1 and z_2 at this step, and we denote that as τ_2, which is a function of z_1, z_2 and $\hat{\theta}$. In the subsequent steps, we use notations $\tau_i = \tau_i(z_1, \ldots, z_i, \hat{\theta})$ for $i = 3, \ldots, n$, which are often referred to as tuning functions.

Based on the above discussion, we design α_2 as

$$\alpha_2 = -z_1 - c_2 z_2 - \varphi_2^T \hat{\theta} + \frac{\partial \alpha_1}{\partial x_1} x_2 + \frac{\partial \alpha_1}{\partial \hat{\theta}} \tau_2. \tag{9.58}$$

The closed-loop dynamics are obtained as

$$\dot{z}_2 = -z_1 - c_2 z_2 + z_3 + \varphi_2^T \tilde{\theta} - \frac{\partial \alpha_1}{\partial \hat{\theta}} (\dot{\hat{\theta}} - \tau_2).$$

Then the adaptive backstepping can be carried on for z_i with $2 < i \leq n$. The dynamics of z_i can be written as

$$\dot{z}_i = \dot{x}_i - \dot{\alpha}_{i-1}$$

$$= x_{i+1} + \phi_i^T \theta - \sum_{j=1}^{i-1} \frac{\partial \alpha_{i-1}}{\partial x_j} \dot{x}_j - \frac{\partial \alpha_{i-1}}{\partial \hat{\theta}} \dot{\hat{\theta}}$$

$$= z_{i+1} + \alpha_i + \varphi_i^T \theta - \sum_{j=1}^{i-1} \frac{\partial \alpha_{i-1}}{\partial x_j} x_{j+1} - \frac{\partial \alpha_{i-1}}{\partial \hat{\theta}} \dot{\hat{\theta}}.$$

The stabilising function α_i, for $2 < i \leq n$, are designed as

$$\alpha_i = -z_{i-1} - c_i z_i - \varphi_i^T \hat{\theta} + \sum_{j=1}^{i-1} \frac{\partial \alpha_{i-1}}{\partial x_j} x_{j+1} + \frac{\partial \alpha_{i-1}}{\partial \hat{\theta}} \tau_i + \beta_i, \tag{9.59}$$

where $\beta_i = \beta_i(z_1, \ldots, z_i, \hat{\theta})$, for $i = 3, \ldots, n$, are functions to be designed later to tackle the terms $(\dot{\hat{\theta}} - \tau_i)$ in stability analysis. The closed-loop dynamics are obtained as, for $2 < i \leq n$,

$$\dot{z}_i = -z_{i-1} - c_i z_i + z_{i+1} + \varphi_i^T \tilde{\theta} - \frac{\partial \alpha_{i-1}}{\partial \hat{\theta}} (\dot{\hat{\theta}} - \tau_i) + \beta_i.$$

The control input u appears in the dynamics z_n, in the term z_{n+1}. When $i = n$, we have $\tau_i = \dot{\hat{\theta}}$ by definition. We obtain the control input by setting $z_{n+1} = 0$, which gives

$$
\begin{aligned}
u &= \alpha_n \\
&= -z_{n-1} - c_n z_n - \varphi_n^T \hat{\theta} \\
&\quad + \sum_{j=1}^{n-1} \frac{\partial \alpha_{n-1}}{\partial x_j} x_{j+1} + \frac{\partial \alpha_{n-1}}{\partial \hat{\theta}} \dot{\hat{\theta}} + \beta_n.
\end{aligned} \tag{9.60}
$$

We need to design the adaptive law, tuning functions and β_i to complete the control design. We will do it based on Lyapunov analysis. For notational convenience, we set $\beta_1 = \beta_2 = 0$.

Let

$$
V = \frac{1}{2} \sum_{i=1}^{n} z_i^2 + \frac{1}{2} \tilde{\theta}^T \Gamma^{-1} \tilde{\theta}, \tag{9.61}
$$

where $\Gamma \in \mathbb{R}^{p \times p}$ is a positive definite matrix. From the closed-loop dynamics of z_i, for $i = 1, \ldots, n$, we obtain

$$
\begin{aligned}
\dot{V} &= \sum_{i=1}^{n} \left(-c_i z_i^2 + z_i \varphi_i^T \tilde{\theta} + z_i \beta_i \right) - \sum_{i=2}^{n} z_i \frac{\partial \alpha_{i-1}}{\partial \hat{\theta}} (\dot{\hat{\theta}} - \tau_i) + \dot{\tilde{\theta}}^T \Gamma^{-1} \tilde{\theta} \\
&= -\sum_{i=1}^{n} c_i z_i^2 + \sum_{i=2}^{n} \left(z_i \beta_i - z_i \frac{\partial \alpha_{i-1}}{\partial \hat{\theta}} (\dot{\hat{\theta}} - \tau_i) \right) \\
&\quad + \left(\sum_{i=1}^{n} z_i \varphi_i - \Gamma^{-1} \dot{\hat{\theta}} \right)^T \tilde{\theta}.
\end{aligned}
$$

We set the adaptive law as

$$
\dot{\hat{\theta}} = \Gamma \sum_{i=1}^{n} z_i \varphi_i. \tag{9.62}
$$

We can conclude the design by setting the tuning functions to satisfy

$$
\sum_{i=2}^{n} \left(z_i \beta_i - z_i \frac{\partial \alpha_{i-1}}{\partial \hat{\theta}} (\dot{\hat{\theta}} - \tau_i) \right) = 0.
$$

Substituting the adaptive law in the above equation, and with some manipulation of index, we obtain that

$$
\begin{aligned}
0 &= \sum_{i=2}^{n} \left(z_i \beta_i - z_i \frac{\partial \alpha_{i-1}}{\partial \hat{\theta}} \left(\Gamma \sum_{j=1}^{n} z_j \varphi_j - \tau_i \right) \right) \\
&= \sum_{i=2}^{n} z_i \beta_i - \sum_{i=2}^{n} \sum_{j=2}^{n} z_i z_j \frac{\partial \alpha_{i-1}}{\partial \hat{\theta}} \Gamma \varphi_j + \sum_{i=2}^{n} z_i \frac{\partial \alpha_{i-1}}{\partial \hat{\theta}} (\tau_i - z_1 \Gamma \varphi_1)
\end{aligned}
$$

$$= \sum_{i=2}^{n} z_i \beta_i - \sum_{i=2}^{n} \sum_{j=i+1}^{n} z_i z_j \frac{\partial \alpha_{i-1}}{\partial \hat{\theta}} \Gamma \varphi_j$$

$$- \sum_{i=2}^{n} \sum_{j=2}^{i} z_i z_j \frac{\partial \alpha_{i-1}}{\partial \hat{\theta}} \Gamma \varphi_j + \sum_{i=2}^{n} z_i \frac{\partial \alpha_{i-1}}{\partial \hat{\theta}} (\tau_i - z_1 \Gamma \varphi_1)$$

$$= \sum_{i=2}^{n} z_i \beta_i - \sum_{j=3}^{n} \sum_{i=2}^{j-1} z_i z_j \frac{\partial \alpha_{i-1}}{\partial \hat{\theta}} \Gamma \varphi_j$$

$$+ \sum_{i=2}^{n} z_i \frac{\partial \alpha_{i-1}}{\partial \hat{\theta}} \left(\tau_i - z_1 \Gamma \varphi_1 - \sum_{j=2}^{i} z_j \Gamma \varphi_j \right)$$

$$= \sum_{i=3}^{n} z_i \left(\beta_i - \sum_{j=2}^{i-1} z_j \frac{\partial \alpha_{j-1}}{\partial \hat{\theta}} \Gamma \varphi_i \right)$$

$$+ \sum_{i=2}^{n} z_i \frac{\partial \alpha_{i-1}}{\partial \hat{\theta}} \left(\tau_i - z_1 \Gamma \varphi_1 - \sum_{j=2}^{i} z_j \Gamma \varphi_j \right).$$

Hence, we obtain

$$\beta_i = \sum_{j=2}^{i-1} z_j \frac{\partial \alpha_{j-1}}{\partial \hat{\theta}} \Gamma \varphi_i, \quad \text{for } i = 3, \ldots, n, \tag{9.63}$$

$$\tau_i = \sum_{j=1}^{i} z_j \Gamma \varphi_j, \quad \text{for } i = 2, \ldots, n. \tag{9.64}$$

With the complete design of u and $\dot{\hat{\theta}}$, we finally obtain that

$$\dot{V} = - \sum_{i=1}^{n} c_i z_i^2.$$

from which we can deduce that $z_i \in L_2 \cap L_\infty$, $i = 1, \ldots, n$, and $\hat{\theta}$ is bounded. Since all the variables are bounded, we have $\dot{z}_i$ bounded. From Barbalat's Lemma, we have $\lim_{t \to \infty} z_i(t) = 0$, $i = 1, \ldots, n$. Noticing that $x_1 = z_1$ and $\phi_1(0) = 0$, we can conclude that α_1 converges to zero, which further implies that $\lim_{t \to \infty} x_2(t) = 0$. Repeating the same process, we can show that $\lim_{t \to \infty} x_i(t) = 0$ for $i = 1, \ldots, n$.

Theorem 9.7. *For a system in the form of (9.57), the control input (9.60) and adaptive law (9.62) designed by adaptive backstepping ensure the boundedness of all the variables and $\lim_{t \to \infty} x_i(t) = 0$ for $i = 1, \ldots, n$.*

Example 9.2. Consider a second-order system

$$\dot{x}_1 = x_2 + (e^{x_1} - 1)\theta$$
$$\dot{x}_2 = u,$$

where $\theta \in \mathbb{R}$ is the only unknown parameter. We will follow the presented design procedure to design an adaptive control input to stabilise the system.

Let $z_1 = x_1$ and $z_2 = x_2 - \alpha_1$. We design the stabilising function α_1 as

$$\alpha_1 = -c_1 z_1 - (e^{x_1} - 1)\hat{\theta}.$$

The resultant dynamics of z_1 are given by

$$\dot{z}_1 = -c_1 z_1 + z_2 + (e^{x_1} - 1)\tilde{\theta}.$$

The dynamics of z_2 are obtained as

$$\dot{z}_2 = u - \frac{\partial \alpha_1}{\partial x_1}(x_2 + (e^{x_1} - 1)\theta) - \frac{\partial \alpha_1}{\partial \hat{\theta}}\dot{\hat{\theta}},$$

where

$$\frac{\partial \alpha_1}{\partial x_1} = -c_1 - e^{x_1}\hat{\theta}, \quad \frac{\partial \alpha_1}{\partial \hat{\theta}} = -(e^{x_1} - 1).$$

Therefore, we design the control input u as

$$u = -z_1 - c_2 z_2 + \frac{\partial \alpha_1}{\partial x_1}(x_2 + (e^{x_1} - 1)\hat{\theta}) + \frac{\partial \alpha_1}{\partial \hat{\theta}}\dot{\hat{\theta}}.$$

Note that the control input u contains $\dot{\hat{\theta}}$, which is to be designed later, as a function of the state variables and $\hat{\theta}$. The resultant dynamics of z_2 are obtained as

$$\dot{z}_2 = -z_1 - c_2 z_2 - \frac{\partial \alpha_1}{\partial x_1}(e^{x_1} - 1)\tilde{\theta}.$$

Let

$$V = \frac{1}{2}\left(z_1^2 + z_2^2 + \frac{\tilde{\theta}^2}{\gamma}\right).$$

We obtain that

$$\dot{V} = -c_1 z_1^2 - c_2 z_2^2 + \left(z_1(e^{x_1} - 1) - z_2\frac{\partial \alpha_1}{\partial x_1}(e^{x_1} - 1)\right)\tilde{\theta} - \frac{1}{\gamma}\tilde{\theta}\dot{\hat{\theta}}.$$

Therefore, an obvious choice of the adaptive law is

$$\dot{\hat{\theta}} = \gamma z_1(e^{x_1} - 1) - \gamma z_2\frac{\partial \alpha_1}{\partial x_1}(e^{x_1} - 1)$$

which gives

$$\dot{V} = -c_1 z_1^2 - c_2 z_2^2.$$

The rest part of the stability analysis follows Theorem 9.7.   ◁

Example 9.3. In Example 9.2, the nonlinear system is in the standard format as shown in (9.56). In this example, we show adaptive control design for a system which is slightly different from the standard form (9.56), but the same design procedure can be applied with some modifications.

Consider a second-order nonlinear system

$$\dot{x}_1 = x_2 + x_1^3\theta + x_1^2$$
$$\dot{x}_2 = (1 + x_1^2)u + x_1^2\theta,$$

where $\theta \in \mathbb{R}$ is the only unknown parameter.

Let $z_1 = x_1$ and $z_2 = x_2 - \alpha_1$. The stabilising function α_1 is designed as

$$\alpha_1 = -c_1 z_1 - x_1^3\hat{\theta} - x_1^2,$$

which results in the dynamics of z_1 as

$$\dot{z}_1 = -c_1 z_1 + z_2 + x_1^3\tilde{\theta}.$$

The dynamics of z_2 are obtained as

$$\dot{z}_2 = (1 + x_1^2)u + x_1^2\theta - \frac{\partial\alpha_1}{\partial x_1}(x_2 + x_1^3\theta + x_1^2) - \frac{\partial\alpha_1}{\partial\hat{\theta}}\dot{\hat{\theta}},$$

where

$$\frac{\partial\alpha_1}{\partial x_1} = -c_1 - 3x_1^2\hat{\theta} - 2x_1, \quad \frac{\partial\alpha_1}{\partial\hat{\theta}} = -x_1^3.$$

Therefore, we design the control input u as

$$u = \frac{1}{1 + x_1^2}\left(-z_1 - c_2 z_2 - x_1^2\hat{\theta} + \frac{\partial\alpha_1}{\partial x_1}(x_2 + x_1^3\hat{\theta} + x_1^2) + \frac{\partial\alpha_1}{\partial\hat{\theta}}\dot{\hat{\theta}}\right).$$

The resultant dynamics of z_2 are obtained as

$$\dot{z}_2 = -z_1 - c_2 z_2 + x_1^2\tilde{\theta} - \frac{\partial\alpha_1}{\partial x_1}x_1^3\tilde{\theta}.$$

Let

$$V = \frac{1}{2}\left(z_1^2 + z_2^2 + \frac{\tilde{\theta}^2}{\gamma}\right).$$

We obtain that

$$\dot{V} = -c_1 z_1^2 - c_2 z_2^2 + \left(z_1 x_1^3 + z_2\left(x_1^2 - \frac{\partial\alpha_1}{\partial x_1}x_1^3\right)\right)\tilde{\theta} - \frac{1}{\gamma}\tilde{\theta}\dot{\hat{\theta}}.$$

We can set the adaptive law as

$$\dot{\hat{\theta}} = \gamma z_1 x_1^3 + \gamma z_2\left(x_1^2 - \frac{\partial\alpha_1}{\partial x_1}x_1^3\right)$$

to obtain

$$\dot{V} = -c_1 z_1^2 - c_2 z_2^2.$$

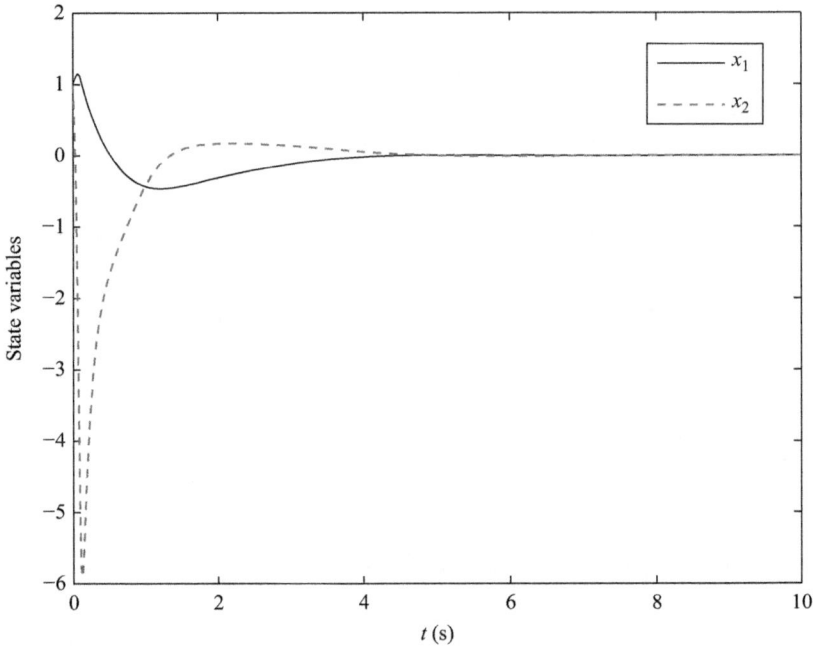

Figure 9.1   State variables

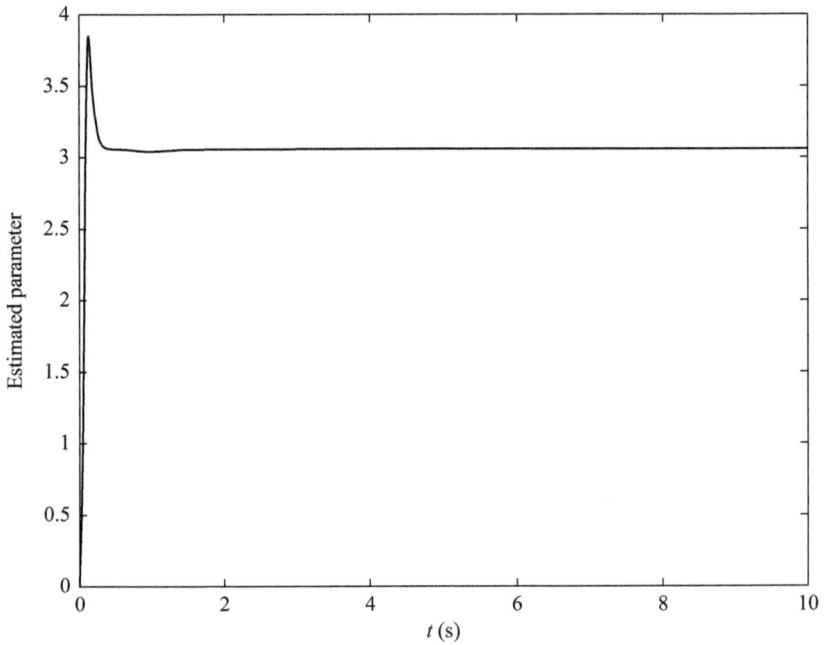

Figure 9.2   Estimated parameter $\hat{\theta}$

The rest part of the stability analysis follows Theorem 9.7. Simulation results are shown in Figures 9.1 and 9.2 with $x(0) = [1, 1]^T$, $c_1 = c_2 = \gamma = \theta = 1$. The state variables converge to zero as expected, and the estimated parameter converges to a constant, but not to the correct value $\theta = 1$. In general, estimated parameters in adaptive control are not guaranteed to converge to their actual values.   ◁

9.6 Adaptive observer backstepping

Backstepping can also be used to design control input for a class of nonlinear systems with unknown parameters using output feedback. We consider a system which can be transformed to the output feedback form with unknown parameters

$$\dot{x} = A_c x + b\sigma(y)u + \phi_0(y) + \sum_{i=1}^{p} \phi_i(y)a_i$$

$$:= A_c x + b\sigma(y)u + \phi_0(y) + \Phi(y)a \tag{9.65}$$

$$y = Cx$$

with

$$A_c = \begin{bmatrix} 0 & 1 & 0 & \cdots & 0 \\ 0 & 0 & 1 & \cdots & 0 \\ \vdots & \vdots & \vdots & \ddots & \vdots \\ 0 & 0 & 0 & \cdots & 1 \\ 0 & 0 & 0 & \cdots & 0 \end{bmatrix}, \quad C = \begin{bmatrix} 1 \\ 0 \\ \vdots \\ 0 \end{bmatrix}^T,$$

$$a = \begin{bmatrix} a_1 \\ a_2 \\ \vdots \\ a_p \end{bmatrix}, \quad b = \begin{bmatrix} 0 \\ \vdots \\ 0 \\ b_\rho \\ \vdots \\ b_n \end{bmatrix} := \begin{bmatrix} 0_{(\rho-1)\times 1} \\ \bar{b} \end{bmatrix},$$

where $x \in \mathbb{R}^n$ is the state vector, $u \in \mathbb{R}$ is the control and $b \in \mathbb{R}^n$ is an unknown constant Hurwitz vector with $b_\rho \neq 0$, which implies that the relative degree of the system is ρ, a denotes unknown system parameters, $\phi_i : \mathbb{R} \to \mathbb{R}^n$ for $0 \leq i \leq p$ are nonlinear functions with their elements being differentiable up to the ρth order, $\sigma : \mathbb{R} \to \mathbb{R}$ is a continuous function and $\sigma(y) \neq 0$, $\forall y \in \mathbb{R}$.

For the system (9.65), we assume that only the system output is available for control design. Both observer backstepping and backstepping with filtered transformation design methods presented earlier can be applied for the control design of (9.65) if there are no unknown parameters. Even with the unknown parameters, the same filtered transformation introduced earlier in (9.34) can still be applied to the system (9.65), and adaptive backstepping can then be applied to design the control input. Here we take an approach similar to observer backstepping. Because there are unknown parameters, we will not be able to design an observer in the same way as in the observer backstepping. Instead, we can design filters similar to observers, and obtain an expression of state estimation which contains unknown parameters. The unknown parameters in the state estimation are then tackled by adaptive backstepping.

For the state estimation, we re-arrange the system as

$$\dot{x} = A_c x + \phi_0(y) + F^T(y, u)\theta, \tag{9.66}$$

where the vector $\theta \in \mathbb{R}^q$, with $q = n - \rho + 1 + p$, is defined by

$$\theta = \begin{bmatrix} \bar{b} \\ a \end{bmatrix}$$

and

$$F(y, u)^T = \left[\begin{bmatrix} 0_{(\rho-1)\times(n-\rho+1)} \\ I_{n-\rho+1} \end{bmatrix} \sigma(y)u, \ \Phi(y) \right].$$

Similar to observer design, we design the following filters:

$$\dot{\xi} = A_0 \xi + Ly + \phi_0(y), \tag{9.67}$$

$$\dot{\Omega}^T = A_0 \Omega^T + F(y, u)^T, \tag{9.68}$$

where $\xi \in \mathbb{R}^n$, $\Omega^T \in \mathbb{R}^{n\times q}$ and

$$L = [l_1, \ldots, l_n]^T, \quad A_0 = A_c - LC$$

with L being chosen so that A_0 is Hurwitz. An estimate of the state is then given by

$$\hat{x} = \xi + \Omega^T \theta. \tag{9.69}$$

Let

$$\epsilon = x - \hat{x}$$

and from direct evaluation, we have

$$\dot{\epsilon} = A_0 \epsilon. \tag{9.70}$$

Therefore, the state estimate shown in (9.69) is an exponentially convergent one. Notice that this estimate contains unknown parameter vector θ, and therefore $\hat{x}$ cannot be directly used in control design. The relationship between an convergent estimate $\hat{x}$ and the unknown parameter vector can be used in adaptive backstepping control design.

We can reduce the order of the filters. Let us partition Ω^T as $\Omega^T = [v, \Xi]$ with $v \in \mathbb{R}^{n\times(n-\rho+1)}$ and $\Xi \in \mathbb{R}^{n\times p}$. We then obtain, from (9.68), that

$$\dot{\Xi} = A_0 \Xi + \Phi(y),$$

$$v_j = A_0 v_j + e_j \sigma(y)u, \quad \text{for } j = \rho, \ldots, n,$$

where e_j denotes jth column of identity matrix I in $\mathbb{R}^n$. For $1 < j < n$, we have

$$A_0 e_j = (A_c - LC)e_j = A_c e_j = e_{j+1}.$$

This implies that

$$v_j = A_0^{n-j} v_n.$$

Finally, we summarise the filters for Ω^T as

$$
\begin{aligned}
\Omega^T &= [v_\rho, \ldots, v_n, \Xi], \\
\dot{\Xi} &= A_0 \Xi + \Phi(y), \\
\dot{\lambda} &= A_0 \lambda + e_n \sigma(y) u, \\
v_j &= A_0^{n-j} \lambda, \quad \text{for } j = \rho, \ldots, n.
\end{aligned}
\tag{9.71}
$$

With the filters being designed, control design can be carried out using adaptive backstepping in a similar way as the observer backstepping, by combining the design of tuning functions in parameter adaptation. During the control design, the state variable x will be replaced by

$$x = \xi + \Omega^T \theta + \epsilon,$$

and in particular, whenever we encounter x_2, we will replace it by

$$x_2 = \xi_2 + \Omega_{(2)}^T \theta + \epsilon_2,$$

where the subscript (i) denotes the ith row of a matrix. In the following control design, we will consider the tracking control instead of stabilisation. The output y is designed to track a trajectory y_r with its derivatives available for the control design.

Let us define a number of notations:

$$
\begin{aligned}
z_1 &= y - y_r, \\
z_i &= v_{\rho,i} - \hat{\varrho} y_r^{(i-1)} - \alpha_{i-1}, \quad i = 2, \ldots, \rho, \\
z_{\rho+1} &= \sigma(y) u + v_{\rho,\rho+1} - \hat{\varrho} y_r^{(\rho)} - \alpha_\rho, \\
\varphi_0 &= \xi_2 + \phi_{0,1}, \\
\varphi &= [v_{\rho,2}, \ldots, v_{n,2}, \Phi_{(1)} + \Xi_{(2)}]^T, \\
\bar{\varphi} &= [0, v_{\rho+1,2}, \ldots, v_{n,2}, \Phi_{(1)} + \Xi_{(2)}]^T, \\
\bar{\lambda}_i &= [\lambda_1, \ldots, \lambda_i]^T, \\
\bar{y}_i &= [y_r, \dot{y}_r, \ldots, y_r^{(i)}]^T, \\
X_i &= [\xi^T, \text{vec}(\Xi)^T, \hat{\varrho}, \bar{\lambda}_i^T, \bar{y}_i^T]^T, \\
\sigma_{j,i} &= \frac{\partial \alpha_{j-1}}{\partial \hat{\theta}} \Gamma \frac{\partial \alpha_{i-1}}{\partial y} \varphi,
\end{aligned}
$$

where $\hat{\varrho}$, an estimate of $\varrho = 1/b_\rho$, and α_i, $i = 1, \ldots, \rho$, are the stabilising functions to be designed.

Consider the dynamics of z_1

$$
\begin{aligned}
\dot{z}_1 &= x_2 + \phi_{0,1}(y) + \Phi_{(1)}\theta - \dot{y}_r \\
&= \xi_2 + \Omega_{(2)}^T \theta + \epsilon_2 + \phi_{0,1}(y) + \Phi_{(1)}\theta - \dot{y}_r \\
&= b_\rho v_{\rho,2} + \varphi_0 + \bar{\varphi}^T \theta + \epsilon_2 - \dot{y}_r.
\end{aligned}
\tag{9.72}
$$

Note that b_ρ is unknown. To deal with unknown control coefficient, we often estimate its reciprocal, instead of itself, to avoid using the reciprocal of an estimate. This is the reason why we define

$$z_2 = v_{\rho,2} - \alpha_1 - \hat{\varrho}\dot{y}_r$$
$$:= v_{\rho,2} - \hat{\varrho}\bar{\alpha}_1 - \hat{\varrho}\dot{y}_r.$$

From $b_\rho\varrho = 1$, we have

$$b_\rho\hat{\varrho} = 1 - b_\rho\tilde{\varrho},$$

where $\tilde{\varrho} = \varrho - \hat{\varrho}$. Then from (9.72), we have

$$\begin{aligned}
\dot{z}_1 &= b_\rho(z_2 + \hat{\varrho}\bar{\alpha}_1 + \hat{\varrho}\dot{y}_r) + \varphi_0 + \bar{\varphi}^T\theta + \epsilon_2 - \dot{y}_r \\
&= b_\rho z_2 - b_\rho\tilde{\varrho}(\bar{\alpha}_1 + \dot{y}_r) + \bar{\alpha}_1 + \varphi_0 + \bar{\varphi}^T\theta + \epsilon_2.
\end{aligned} \tag{9.73}$$

Hence, we design

$$\bar{\alpha}_1 = -c_1 z_1 - k_1 z_1 - \varphi_0 - \bar{\varphi}^T\hat{\theta}, \tag{9.74}$$

where c_i and k_i for $i = 1, \ldots, \rho$ are positive real design parameters. Note that with $\alpha = \hat{\varrho}\bar{\alpha}_1$, we have $\alpha_1 = \alpha_1(y, X_1, \hat{\theta})$.

The resultant closed-loop dynamics are obtained as

$$\begin{aligned}
\dot{z}_1 &= -c_1 z_1 - k_1 z_1 + b_\rho z_2 - b_\rho\tilde{\varrho}(\bar{\alpha}_1 + \dot{y}_r) + \bar{\varphi}^T\tilde{\theta} + \epsilon_2 \\
&= -c_1 z_1 - k_1 z_1 + \hat{b}_\rho z_2 + \tilde{b}_\rho z_2 - b_\rho\tilde{\varrho}(\bar{\alpha}_1 + \dot{y}_r) + \bar{\varphi}^T\tilde{\theta} + \epsilon_2 \\
&= -c_1 z_1 - k_1 z_1 + \hat{b}_\rho z_2 - b_\rho\tilde{\varrho}(\bar{\alpha}_1 + \dot{y}_r) \\
&\quad + (\varphi - \hat{\varrho}(\bar{\alpha}_1 + \dot{y}_r)e_1)^T\tilde{\theta} + \epsilon_2,
\end{aligned} \tag{9.75}$$

where $\tilde{\theta} = \theta - \hat{\theta}$. Note that $b_\rho = \theta^T e_1$.

As shown in the observer backstepping, the term $-k_1 z_1$ is used to tackle the error term ϵ_2. In the subsequent steps, the terms headed by k_i are used to deal with the terms caused by ϵ_2 in stability analysis. Indeed, the method of tackling observer errors is exactly the same as in the observer backstepping when all the parameters are known. The adaptive law for $\hat{\varrho}$ can be designed in this step, as it will not appear in the subsequent steps,

$$\dot{\hat{\varrho}} = -\gamma \text{sgn}(b_\rho)(\bar{\alpha}_1 + \dot{y}_r))z_1, \tag{9.76}$$

where γ is a positive real constant.

Similar to adaptive backstepping shown in the previous section, the unknown parameter vector θ will appear in the subsequent steps, and tuning functions can be introduced in a similar way. We define the tuning functions τ_i as

$$\begin{aligned}
\tau_1 &= \Gamma(\varphi - \hat{\varrho}(\dot{y}_r + \bar{\alpha}_1)e_1)z_1, \\
\tau_i &= \tau_{i-1} - \Gamma\frac{\partial\alpha_{i-1}}{\partial y}\varphi z_i, \quad i = 2, \ldots, \rho,
\end{aligned} \tag{9.77}$$

where α_i are the stabilising functions to be designed, and $\Gamma \in \mathbb{R}^{q \times q}$ is a positive definite matrix, as the adaptive gain.

For the dynamics of z_2, we have

$$\dot{z}_2 = \dot{v}_{\rho,2} - \dot{\alpha}_1 - \dot{\hat{\varrho}}\ddot{y}_r - \hat{\varrho}\dddot{y}_r$$

$$= v_{\rho,3} - l_2 v_{\rho,1} - \frac{\partial \alpha_1}{\partial X_1}\dot{X}_1 - \frac{\partial \alpha_1}{\partial y}\dot{y} - \frac{\partial \alpha_1}{\partial \hat{\theta}}\dot{\hat{\theta}} - \dot{\hat{\varrho}}\ddot{y}_r - \hat{\varrho}\dddot{y}_r.$$

With

$$\dot{y} = \varphi_0 + \varphi^T \theta + \epsilon_2,$$

$$z_3 = v_{\rho,3} - \alpha_2 - \hat{\varrho}\ddot{y}_r,$$

we obtain that

$$\dot{z}_2 = z_3 + \alpha_2 - l_2 v_{\rho,1} - \frac{\partial \alpha_1}{\partial X_1}\dot{X}_1 - \frac{\partial \alpha_1}{\partial y}(\varphi_0 + \varphi^T \theta + \epsilon_2) - \frac{\partial \alpha_1}{\partial \hat{\theta}}\dot{\hat{\theta}} - \hat{\varrho}\dddot{y}_r,$$

from which the stabilising function α_2 is defined as

$$\alpha_2 = -\hat{b}_\rho z_1 - c_2 z_2 - k_2 \left(\frac{\partial \alpha_1}{\partial y}\right)^2 z_2 + l_2 v_{\rho,1} + \hat{\varrho}\dddot{y}_r$$

$$+ \frac{\partial \alpha_1}{\partial X_1}\dot{X}_1 + \frac{\partial \alpha_1}{\partial y}(\varphi_0 + \varphi^T \hat{\theta}) + \frac{\partial \alpha_1}{\partial \hat{\theta}}\tau_2. \tag{9.78}$$

The resultant dynamics of z_2 are obtained as

$$\dot{z}_2 = -\hat{b}_\rho z_1 - c_2 z_2 - k_2 \left(\frac{\partial \alpha_1}{\partial y}\right)^2 z_2 + z_3$$

$$- \frac{\partial \alpha_1}{\partial y}\varphi^T \tilde{\theta} - \frac{\partial \alpha_1}{\partial y}\epsilon_2 - \frac{\partial \alpha_1}{\partial \hat{\theta}}(\dot{\hat{\theta}} - \tau_2). \tag{9.79}$$

For the dynamics of z_i, $2 < i \le \rho$, we have

$$\dot{z}_i = z_{i+1} + \alpha_i - l_i v_{\rho,1} - \frac{\partial \alpha_{i-1}}{\partial X_{i-1}}\dot{X}_{i-1} - \frac{\partial \alpha_{i-1}}{\partial y}(\varphi_0 + \varphi^T \theta + \epsilon_2)$$

$$- \frac{\partial \alpha_{i-1}}{\partial \hat{\theta}}\dot{\hat{\theta}} - \hat{\varrho}\dddot{y}_r^{(i-1)}.$$

We design α_i, $2 < i \le \rho$, as

$$\alpha_i = -z_{i-1} - c_i z_i - k_i \left(\frac{\partial \alpha_{i-1}}{\partial y}\right)^2 z_i + l_i v_{\rho,1}$$

$$+ \hat{\varrho}\dddot{y}_r^{(i-1)} + \frac{\partial \alpha_{i-1}}{\partial X_{i-1}}\dot{X}_{i-1} + \frac{\partial \alpha_{i-1}}{\partial y}(\varphi_0 + \varphi^T \hat{\theta})$$

$$+ \frac{\partial \alpha_{i-1}}{\partial \hat{\theta}}\tau_i - \sum_{j=2}^{i-1} \sigma_{j,i} z_j, \quad i = 3, \dots, \rho, \tag{9.80}$$

where the last term $-\sum_{j=2}^{i-1} \sigma_{j,i} z_j$ is similar to the term β_i in the adaptive backstepping with tuning functions. The resultant dynamics of z_i are obtained as

$$\dot{z}_i = -z_{i-1} - c_i z_i - k_i \left(\frac{\partial \alpha_{i-1}}{\partial y}\right)^2 z_i + z_{i+1}$$

$$- \frac{\partial \alpha_{i-1}}{\partial y} \varphi^T \tilde{\theta} - \frac{\partial \alpha_{i-1}}{\partial y} \epsilon_2 - \sum_{j=2}^{i-1} \sigma_{j,i} z_j$$

$$- \frac{\partial \alpha_{i-1}}{\partial \hat{\theta}} (\dot{\hat{\theta}} - \tau_i) \quad i = 3, \dots, \rho. \tag{9.81}$$

Now we design the adaptive law for $\hat{\theta}$ as

$$\dot{\hat{\theta}} = \tau_\rho. \tag{9.82}$$

The control input is obtained by setting $z_{\rho+1} = 0$ as

$$u = \frac{1}{\sigma(y)}(\alpha_\rho - v_{\rho,\rho+1} + \hat{\varrho} y_r^{(\rho)}). \tag{9.83}$$

For the adaptive observer backstepping, we have the following stability result.

Theorem 9.8. *For a system in the form of (9.65), the control input (9.83) and adaptive laws (9.76) and (9.82) designed by adaptive observer backstepping ensure the boundedness of all the variables and $\lim_{t \to \infty} (y(t) - y_r(t)) = 0$.*

Proof. With the control design and adaptive laws presented earlier for the case $\rho > 1$, the dynamics of z_i, for $i = 1, \dots, \rho$ can be written as

$$\dot{z}_1 = -c_1 z_1 - k_1 z_1 + \epsilon_2 + (\varphi - \hat{\varrho}(\dot{y}_r + \bar{\alpha}_1)e_1)^T \tilde{\theta}$$
$$- b_\rho(\dot{y}_r + \bar{\alpha}_1)\tilde{\varrho} + \hat{b}_\rho z_2, \tag{9.84}$$

$$\dot{z}_2 = -\hat{b}_\rho z_1 - c_2 z_2 - k_2 \left(\frac{\partial \alpha_1}{\partial y}\right)^2 z_2 + z_3$$

$$- \frac{\partial \alpha_1}{\partial y} \varphi^T \tilde{\theta} - \frac{\partial \alpha_1}{\partial y} \epsilon_2 + \sum_{j=3}^{\rho} \sigma_{2,j} z_j, \tag{9.85}$$

$$\dot{z}_i = -z_{i-1} - c_i z_i - k_i \left(\frac{\partial \alpha_{i-1}}{\partial y}\right)^2 z_i + z_{i+1}$$

$$- \frac{\partial \alpha_{i-1}}{\partial y} \varphi^T \tilde{\theta} - \frac{\partial \alpha_{i-1}}{\partial y} \epsilon_2 + \sum_{j=i+1}^{\rho} \sigma_{i,j} z_j$$

$$- \sum_{j=2}^{i-1} \sigma_{j,i} z_j \quad i = 3, \dots, \rho. \tag{9.86}$$

Let

$$V_\rho = \frac{1}{2}\sum_{i=1}^{\rho} z_i^2 + \frac{1}{2}\tilde\theta^T\Gamma^{-1}\tilde\theta + \frac{|b_\rho|}{2\gamma}\tilde\varrho^2 + \sum_{i=1}^{\rho}\frac{1}{4k_i}\epsilon^T P\epsilon, \tag{9.87}$$

where P is a positive definite matrix that satisfies

$$A_0^T P + PA_0 = -I.$$

From (9.84)–(9.86), (9.82) and (9.76), it can be shown that

$$\dot V_\rho = -\sum_{i=1}^{\rho}\left(c_i + k_i\left(\frac{\partial\alpha_{i-1}}{\partial y}\right)^2\right)z_i^2 - \sum_{i=1}^{\rho} z_i\frac{\partial\alpha_{i-1}}{\partial y}\epsilon_2 - \sum_{i=1}^{\rho}\frac{1}{4k_i}\|\epsilon\|^2, \tag{9.88}$$

where we set $\frac{\partial\alpha_0}{\partial y} = -1$. Noting

$$\left|z_i\frac{\partial\alpha_{i-1}}{\partial y}\epsilon_2\right| \le \frac{1}{4k_i}\|\epsilon_2\|^2 + k_i\left(\frac{\partial\alpha_{i-1}}{\partial y}\right)^2 z_i^2, \tag{9.89}$$

we have

$$\dot V_\rho \le -\sum_{i=1}^{\rho} c_i z_i^2. \tag{9.90}$$

This implies that z_i, $i = 1,\ldots,\rho$, $\tilde\theta$, $\tilde\varrho$ and ϵ are bounded. The boundedness of $\tilde\theta$ further implies that $\hat\theta$ is bounded. The boundedness of $\hat\varrho$ follows from the fact that $\varrho = 1/b_\rho$ is a constant. Since $y = z_1 + y_r$, the boundedness of ξ and Ξ follows the boundedness of y. The boundedness of λ can be established from the minimum phase property of the system. Therefore, u is bounded and we can conclude that all the variables of the feedback control system are bounded. From the above analysis, we can deduce that $z_i \in L_2 \cap L_\infty$, $i = 1,\ldots,\rho$ and $\hat\theta$ are bounded. Since all the variables are bounded, we have $\dot z_i$ bounded. From Barbalat's Lemma, we have $\lim_{t\to\infty} z_i(t) = 0$, $i = 1,\ldots,\rho$, of which the result for z_1 means the asymptotic output tracking, $\lim_{t\to\infty}(y(t) - y_r(t)) = 0$. □

Chapter 10
Disturbance rejection and output regulation

Disturbances are often inevitable in control system design. A well-performed controller is expected to suppress undesirable effects of the disturbances in the system. There are various types of disturbances in physical systems, from random disturbances, wide-band, narrow-band disturbances, in terms of disturbance power spectra, to deterministic disturbances that include harmonic disturbances, i.e., sinusoidal functions, general periodic disturbances and other deterministic signals generated from nonlinear dynamic systems such as limit cycles. The spectral information of random disturbances may be considered in loop-shaping and other conventional design methods. In this chapter, we concentrate on suppression and rejection of deterministic periodic disturbances. One control design objective is to track a specific signal. If we take the tracking error as the state variable, the tracking problem could be converted to a stabilisation problem. Indeed, we can formulate both the disturbance rejection and output tracing problems in terms of output regulation. This will be discussed later in this chapter.

Disturbance rejection and output regulation are big topics in control systems, design. We have to be selective to limit the contents in one chapter. In this chapter, we will concentrate on rejection of deterministic disturbances in a class of nonlinear output feedback systems. As for disturbances, we will start from sinusoidal disturbances with unknown frequencies, then disturbances generated from nonlinear exosystems and then general periodical disturbances, etc. Adaptive control techniques are used to deal with the unknown disturbance frequencies and unknown parameters in the system. The presented design concepts can be applied to other classes of nonlinear systems.

10.1 Asymptotic rejection of sinusoidal disturbances

In this section, we consider asymptotic rejection of sinusoidal disturbances with unknown frequencies for dynamic systems in the output feedback form.

We consider a SISO nonlinear system which can be transformed into the output feedback form

$$\dot{\zeta} = A_c\zeta + bu + \phi(y) + Ew$$
$$y = C\zeta,$$

(10.1)

with

$$
A_c = \begin{bmatrix} 0 & 1 & 0 & \cdots & 0 \\ 0 & 0 & 1 & \cdots & 0 \\ \vdots & \vdots & \vdots & \ddots & \vdots \\ 0 & 0 & 0 & \cdots & 1 \\ 0 & 0 & 0 & \cdots & 0 \end{bmatrix}, \quad C = \begin{bmatrix} 1 \\ 0 \\ \vdots \\ 0 \end{bmatrix}^T, \quad b = \begin{bmatrix} 0 \\ \vdots \\ 0 \\ b_\rho \\ \vdots \\ b_n \end{bmatrix},
$$

where $\zeta \in \mathbb{R}^n$ is the state vector; $u \in \mathbb{R}$ is the control; $\phi : \mathbb{R} \to \mathbb{R}^n$ with $\phi(0) = 0$ is a nonlinear function with element ϕ_i being differentiable up to the $(n - i)$th order; $b \in \mathbb{R}^n$ is a known constant Hurwitz vector with $b_\rho \neq 0$, which implies the relative degree of the system is ρ; $E \in \mathbb{R}^{n \times m}$ is a constant matrix; and $w \in \mathbb{R}^m$ are disturbances, and they are generated from an unknown exosystem

$$
\dot{w} = Sw
$$

of which, S is a constant matrix with distinct eigenvalues of zero real parts.

Remark 10.1. For the system (10.1), if the disturbance $w = 0$, it is exactly same as (9.17), of which observer backstepping can be used to design a control input. Due to the unknown disturbance, although the observer backstepping presented in Chapter 9 cannot be applied directly for control design, a similar technique can be developed using an observer and an adaptive internal model. Also note that the linear system characterised by (A_c, b, C) is minimum phase.   ◁

Remark 10.2. The dynamic model $\dot{w} = Sw$ is referred to as an exosystem, because w is a disturbance, not a part of the system state. This is a convention adopted for disturbance rejection and output regulation. Of course, one could argue that w could be considered as a part of the system state, or at least, of an augmented system. In such a case, w is not controllable.   ◁

Remark 10.3. With the assumption that S has distinct eigenvalues with zero real parts, w is restrict to sinusoidal signals (sinusoidal disturbances) with a possible constant bias. This is a common assumption for disturbance rejection. Roughly speaking, all the periodic signal can be approximated by finite number of sinusoidal functions.   ◁

The disturbance rejection problem to be solved here is to design a control input that ensures the boundedness of the variables in the closed-loop system, and the convergence to zero of the system output.

To solve the problem, we start from state transformation, based on an invariant manifold. The basic idea for disturbance suppression or output regulation is the internal model principle. A controller for disturbance rejection or output regulation should generate a feedforward input term to cancel the influence caused by the disturbance or to track a desired trajectory. This feedforward term is also referred to as the equivalent input disturbance. For nonlinear systems, asymptotic disturbance rejection depends on the existence of an invariant manifold. We shall show that there exists an invariant manifold for any exosystem specified in (10.1). This invariant manifold is then used in the state transformation. The following lemma summarises the results.

Lemma 10.1. *For the system (10.1) with an exosystem whose eigenvalues are distinct and with zero real parts, there exist $\pi(w) \in \mathbb{R}^n$ with $\pi_1(w) = 0$ and $\alpha(w) \in \mathbb{R}$ such that*

$$\frac{\partial \pi(w)}{\partial w} Sw = A_c \pi(w) + Ew + b\alpha(w). \tag{10.2}$$

Proof. Asymptotic disturbance rejection aims at $y = 0$, which implies that $\pi_1 = 0$. The manifold π is invariant and it should satisfy the system equation (10.1) with $y \equiv 0$. From the first equation of (10.1), we have

$$\pi_2 = -E_1 w \tag{10.3}$$

where E_1 denotes the first row of E. Furthermore, we have, for $2 \le i \le \rho$,

$$\pi_i = \frac{d}{dt}\pi_{i-1} - E_{i-1}w. \tag{10.4}$$

From equations ρ to n of (10.1), we obtain

$$\sum_{i=\rho}^{n} \frac{d^{n-i}}{dt^{n-i}} b_i \alpha(w) = \frac{d^{n-\rho+1}}{dt^{n-\rho+1}}\pi_\rho - \sum_{i=\rho}^{n} \frac{d^{n-i}}{dt^{n-i}} E_i w. \tag{10.5}$$

A solution of $\alpha(w)$ can always be found from (10.5). With $\alpha(w)$, we can write, for $\rho < i \le n$,

$$\pi_i = \frac{d}{dt}\pi_{i-1} - E_{i-1}w - b_{i-1}\alpha(w). \tag{10.6}$$

$\square$

With the invariant manifold $\pi(w)$, we define a transformation of state as

$$x = \zeta - \pi(w). \tag{10.7}$$

It can be easily shown from (10.1) and (10.2) that

$$\dot{x} = A_c x + b(u - \alpha) + \phi(y)$$
$$y = Cx. \tag{10.8}$$

The stabilisation and disturbance suppression problem of (10.1) degenerates to the stabilisation problem of (10.8).

For dynamic output feedback control, state variables need to be estimated for the control design directly or indirectly. The difficulty for designing a state estimator for (10.8) is that $\alpha(w)$, the feedforward control input for disturbance suppression, is unknown. Let us consider the observers

$$\dot{p} = (A_c - LC)p + \phi(y) + bu + Ly \tag{10.9}$$
$$\dot{q} = (A_c - LC)q + b\alpha(w), \tag{10.10}$$

where $L \in \mathbb{R}^n$ is chosen so that $A_c - LC$ is Hurwitz, and q denotes the steady-state contribution of $\alpha(w)$ to the state variable. Notice that the observer (10.10) cannot be implemented because $\alpha(w)$ is unknown, due to the unknown exosystem. Nevertheless, if we define

$$\hat{x} = p - q \tag{10.11}$$

then the error of the observer defined by $\epsilon = x - \hat{x}$ is an exponentially decaying signal with the dynamics

$$\dot{\epsilon} = (A_c - LC)\epsilon, \tag{10.12}$$

which can be obtained by a direct evaluation from (10.8) to (10.10).

Observe from (10.5) that $\alpha(w)$ is a linear combination of w. Since the filter (10.10) is just a stable linear system, the steady-state solution of state variables is linear combinations of w as well. Therefore, there exists an $l \in \mathbb{R}^m$ for q_2, and we can write

$$\dot{w} = Sw$$
$$q_2 = l^T w. \tag{10.13}$$

We re-parameterise (10.13) for state estimation here for q_2. For any known controllable pair (F, G) with $F \in \mathbb{R}^{m \times m}$ being Hurwitz and $G \in \mathbb{R}^m$, there exists a $\psi \in \mathbb{R}^m$ so that

$$\dot{\eta} = (F + G\psi^T)\eta$$
$$q_2 = \psi^T \eta, \tag{10.14}$$

with the initial value $\eta(0)$ dependent on exogenerous variables.

Remark 10.4. The importance of (10.14) compared with (10.13) is the re-formulation of the uncertainty caused by the unknown exosystem. The uncertainty in (10.13) parameterised by unknown S and l is represented by a single vector ψ in (10.14). The relation between the two parameterisations is discussed here. Suppose that $M \in \mathbb{R}^{m \times m}$ is the unique solution of

$$MS - FM = Gl^T. \tag{10.15}$$

The existence of a non-singular M is ensured by the fact that S and F have exclusively different eigenvalues, and (S, l) and (F, G) are observable and controllable respectively. From (10.15), we have

$$MSM^{-1} = F + Gl^T M^{-1}, \tag{10.16}$$

which implies $\eta = Mw$ and $\psi^T = l^T M^{-1}$

To achieve asymptotic rejection with global stability using output feedback, an asymptotic state estimator, probably depending on unknown parameters, is instrumental. In the control design shown later for the problem considered, the estimate of state variable x_2 is crucial, and therefore we introduce the internal model (10.14) to describe the influence of disturbance w on this state.   ◁

The following lemma summarises the results on state observation.

Lemma 10.2. *The state variable x can be expressed as*

$$x = p - q + \epsilon, \tag{10.17}$$

where p is generated from (10.9) with q and ϵ satisfying (10.10) and (10.12) respectively. In particular,

$$x_2 = p_2 - \psi^T \eta + \epsilon_2, \tag{10.18}$$

where η satisfies (10.14).

Remark 10.5. The expression (10.18) cannot be directly implemented, because ψ and η are not available. The relation shown in (10.18) is very useful in the control design, and it allows adaptive control technique to be introduced to deal with the unknown parameter ψ later.   ◁

Based on the parameterisation (10.14) of the internal model, we design an estimator of η as

$$\begin{aligned}
\dot{\xi} &= (F + G\hat{\psi}^T)\xi + \iota(y) \\
\hat{q}_2 &= \hat{\psi}^T \xi,
\end{aligned} \tag{10.19}$$

where $\iota(y)$ is an interlace function to be designed later.

With the state estimation introduced in previous chapter, in particular, the filter (10.9), the relation (10.18) and estimator (10.19), the control design can be carried out using adaptive observer backstepping technique described in previous chapter. In the backstepping design, c_i, for $i = 1, \ldots, \rho$, denote constant design parameters which can be set in the design, while k_i, for $i = 1, \ldots, \rho$, denote the unknown parameters depending on an upper bound of $\|\psi\|$. The estimates, $\hat{k}_i$, for $i = 1, \ldots, \rho$, are then used. To apply the observer backstepping through p in (10.9), we define

$$z_1 = y, \tag{10.20}$$

$$z_i = p_i - \alpha_{i-1}, \quad i = 2, \ldots, \rho, \tag{10.21}$$

$$z_{\rho+1} = b_\rho u + p_{\rho+1} - \alpha_\rho, \tag{10.22}$$

where α_i, for $i = 1, \ldots, \rho$, are stabilising functions decided in the control design. Consider the dynamics of z_1

$$\dot{z}_1 = x_2 + \phi_1(y). \tag{10.23}$$

We use (10.18) to replace the unmeasurable x_2 in (10.23), resulting at

$$\begin{aligned}
\dot{z}_1 &= p_2 - \psi^T \eta + \epsilon_2 + \phi_1(y) \\
&= z_2 + \alpha_1 - \psi^T \eta + \epsilon_2 + \phi_1(y).
\end{aligned} \tag{10.24}$$

We design α_1 as

$$\alpha_1 = -c_1 z_1 - \hat{k}_1 z_1 - \phi_1(y) + \hat{\psi}^T \xi. \tag{10.25}$$

Then from (10.25) and (10.24), we have

$$\dot{z}_1 = z_2 - c_1 z_1 - \hat{k}_1 z_1 + \epsilon_2 + (\hat{\psi}^T \xi - \psi^T \eta). \tag{10.26}$$

After the design of α_1, the remaining stabilising functions can be designed in a similar way to the standard adaptive backstepping with tuning functions shown in the previous chapter. Omitting the deriving procedures, we briefly describe the final results

$$\alpha_i = -z_{i-1} - c_i z_i - \hat{k}_i \left(\frac{\partial \alpha_{i-1}}{\partial y}\right)^2 z_i - l_i(y - p_1) - \phi_i(y)$$

$$+ \frac{\partial \alpha_{i-1}}{\partial y}(p_2 - \hat{\psi}^T \xi + \phi_1) + \sum_{j=1}^{i-1} \frac{\partial \alpha_{i-1}}{\partial p_j}\dot{p}_j + \sum_{j=1}^{i-1} \frac{\partial \alpha_{i-1}}{\partial \hat{k}_j}\dot{\hat{k}}_j + \frac{\partial \alpha_{i-1}}{\partial \xi}\dot{\xi}$$

$$+ \frac{\partial \alpha_{i-1}}{\partial \hat{\psi}}\tau_i + \sum_{j=2}^{i-1} \frac{\partial \alpha_{j-1}}{\partial \hat{\psi}}\Gamma\frac{\partial \alpha_{i-1}}{\partial y}\xi z_j, \quad i = 2, \ldots, \rho, \tag{10.27}$$

where l_i are the ith element of the observer gain L in (10.9), Γ is a positive definite matrix, τ_i, $i = 2, \ldots, \rho$, are the tuning functions defined by

$$\tau_i = \sum_{j=1}^{i} \Gamma\frac{\partial \alpha_{j-1}}{\partial y}\xi z_j, \quad i = 2, \ldots, \rho, \tag{10.28}$$

where we set $\frac{\partial \alpha_0}{\partial y} = -1$. The adaptive law for $\hat{\psi}$ is set as

$$\dot{\hat{\psi}} = \tau_\rho. \tag{10.29}$$

The adaptive laws for $\hat{k}_i$, $i = 1, \ldots, \rho$, are given by

$$\dot{\hat{k}}_i = \gamma_i \left(\frac{\partial \alpha_{i-1}}{\partial y}\right)^2 z_i^2, \tag{10.30}$$

where γ_i is a positive real design parameter. The control input is obtained by setting $z_{\rho+1} = 0$ as

$$u = \frac{1}{b_\rho}(\alpha_\rho - p_{\rho+1}). \tag{10.31}$$

Considering the stability, we set a restriction on c_2 by

$$c_2 > \|PG\|^2, \tag{10.32}$$

where P is a positive definite matrix, satisfying

$$F^T P + PF = -2I. \tag{10.33}$$

To end the control design, we set the interlace function in (10.19) by

$$\iota(y) = -(FG + c_1 G + \hat{k}_1 G)y. \tag{10.34}$$

Remark 10.6. The adaptive coefficients $\hat{k}_i$, $i = 1, \ldots, \rho$, are introduced to allow the exosystem to be truly unknown. It implies that the proposed control design can reject disturbances completely at any frequency. If an upper bound of $\|\psi\|$ is known, we can replace $\hat{k}_i$, $i = 1, \ldots, \rho$, by constant positive real parameters. ◁

Remark 10.7. The final control u in (10.31) does not explicitly contain $\alpha(w)$, the feedforward control term for disturbance rejection. Instead, the proposed control design considers q_2, the contribution of the influence of $\alpha(w)$ to x_2, from the first step in α_1 throughout to the final step in α_ρ. ◁

Theorem 10.3. *For the system (10.1), the control input (10.31) ensures the boundedness of all the variables, and asymptotically rejects the unknown disturbances in the sense that $\lim_{t \to \infty} y(t) = 0$. Furthermore, if $w(0) \in \mathbb{R}^m$ is such that $w(t)$ contains the components at $m/2$ distinct frequencies, then $\lim_{t \to \infty} \hat{\psi}(t) = \psi$.*

Proof. We start from the analysis of the internal model. Define

$$e = \xi - \eta - Gy. \tag{10.35}$$

From (10.14), (10.19), (10.34) and (10.26), it can be obtained that

$$\dot{e} = Fe - G(z_2 + \epsilon_2). \tag{10.36}$$

Based on the stabilising functions shown in (10.27), the dynamics of z_i, for $i = 2, \ldots, \rho$, can be written as

$$\dot{z}_i = -z_{i-1} - c_i z_i - \hat{k}_i \left(\frac{\partial \alpha_{i-1}}{\partial y} \right)^2 z_i + z_{i+1} - \frac{\partial \alpha_{i-1}}{\partial y} (\hat{\psi}^T \xi - \psi^T \eta)$$

$$- \frac{\partial \alpha_{i-1}}{\partial y} \epsilon_2 - \sum_{j=i+1}^{\rho} \frac{\partial \alpha_{i-1}}{\partial \hat{\psi}} \Gamma \frac{\partial \alpha_j}{\partial y} \xi z_j$$

$$+ \sum_{j=2}^{i-1} \frac{\partial \alpha_{j-1}}{\partial \hat{\psi}} \Gamma \frac{\partial \alpha_{i-1}}{\partial y} \xi z_j, \quad i = 2, \ldots, \rho, \tag{10.37}$$

where the term $\hat{\psi}^T \xi - \psi^T \eta$ can be rephrased by

$$\hat{\psi}^T \xi - \psi^T \eta = \psi^T e - \tilde{\psi}^T \xi + \psi^T G z_1 \tag{10.38}$$

with $\tilde{\psi} = \psi - \hat{\psi}$.

In the following analysis, we denote κ_0, $\kappa_{i,j}$, $i = 1, \ldots, 3$, and $j = 1, \ldots, \rho$, as constant positive reals, which satisfy

$$\kappa_0 + \sum_{j=1}^{\rho} \kappa_{3,j} < \frac{1}{2}. \tag{10.39}$$

Furthermore, we set constant positive reals k_i, $i = 1, \ldots, \rho$, satisfying the following conditions:

$$k_1 > |\psi^T G| + \kappa_{1,1} + \sum_{j=2}^{\rho} \kappa_{2,i} |\psi^T G|^2 + \frac{\|\psi\|^2}{4\kappa_{3,1}}, \tag{10.40}$$

$$k_i > \kappa_{1,i} + \frac{1}{4\kappa_{2,i}} + \frac{\|\psi\|^2}{4\kappa_{3,i}} \quad i = 2, \ldots, \rho. \tag{10.41}$$

Define a Lyapunov function candidate

$$V = \frac{1}{2} \left(e^T P e + \sum_{i=1}^{\rho} z_i^2 + \sum_{i=1}^{\rho} \gamma_i^{-1} \tilde{k}_i^2 + \tilde{\psi}^T \Gamma^{-1} \tilde{\psi} + \beta \epsilon^T P_\epsilon \epsilon \right), \tag{10.42}$$

where $\tilde{k}_i = k_i - \hat{k}_i$, $i = 1, \ldots, \rho$, P_ϵ is a positive definite matrix satisfying

$$(A_c - LC)^T P_\epsilon + P_\epsilon (A_c - LC) = -2I,$$

and β is a constant positive real satisfying

$$\beta > \frac{\|PG\|^2}{4\kappa_0} + \sum_{j=1}^{\rho} \frac{1}{4\kappa_{1,i}}. \tag{10.43}$$

Evaluating the derivative of V along the dynamics in (10.12), (10.26), (10.36) and (10.37) together with adaptive laws in (10.29) and (10.30), we have

$$\dot{V} = -\sum_{i=1}^{\rho} \left(c_i + k_i \left(\frac{\partial \alpha_{i-1}}{\partial y} \right)^2 \right) z_i^2 - \sum_{i=1}^{\rho} z_i \frac{\partial \alpha_{i-1}}{\partial y} \epsilon_2$$

$$- \sum_{i=1}^{\rho} z_i \frac{\partial \alpha_{i-1}}{\partial y} (\psi^T e + \psi^T G z_1)$$

$$- e^T e - e^T PG(z_2 + \epsilon_2) - \beta \epsilon^T \epsilon. \tag{10.44}$$

For the cross-terms in (10.44), we have

$$\left| e^T PG \epsilon_2 \right| < \kappa_0 e^T e + \frac{1}{4\kappa_0} \|PG\|^2 \epsilon_2^2,$$

$$\left| z_i \frac{\partial \alpha_{i-1}}{\partial y} \epsilon_2 \right| < \kappa_{1,i} \left(\frac{\partial \alpha_{i-1}}{\partial y} \right)^2 z_i^2 + \frac{1}{4\kappa_{1,i}} \epsilon_2^2, \quad i = 1, \ldots, \rho,$$

$$\left| \frac{\partial \alpha_{i-1}}{\partial y} \psi^T G z_1 z_i \right| < \kappa_{2,i} |\psi^T G|^2 z_1^2 + \frac{1}{4\kappa_{2,i}} \left(\frac{\partial \alpha_{i-1}}{\partial y} \right)^2 z_i^2, \quad i = 2, \ldots, \rho,$$

$$\left| \frac{\partial \alpha_{i-1}}{\partial y} \psi^T e z_i \right| < \kappa_{3,i} e^T e + \frac{\|\psi\|^2}{4\kappa_{3,i}} \left(\frac{\partial \alpha_{i-1}}{\partial y} \right)^2 z_i^2, \quad i = 1, \ldots, \rho,$$

$$\left| e^T PG z_2 \right| < e^T \frac{e}{2} + \|PG\|^2 z_2^2 / 2.$$

Then based on the conditions specified in (10.32), (10.39), (10.40), (10.41) and (10.43), we can conclude that there exist positive real constants δ_i, for $i = 1, 2, 3$, such that

$$\dot{V} \le -\delta_1 \sum_{i=1}^{\rho} z_i^2 - \delta_2 e^T e - \delta_3 \epsilon^T \epsilon. \tag{10.45}$$

We conclude $z_i \in L_2 \cap L_\infty$, $i = 1, \ldots, \rho$, and $\|e\| \in L_2 \cap L_\infty$ and the boundedness of the variables $\hat{\psi}$, and $\hat{k}_i$, $i = 1, \ldots, \rho$. With $y = z_1 \in L_\infty$, the boundedness of p can be established from the minimum phase property of the system and the neutral stability of w. Therefore, we can conclude that all the variables in the proposed closed-loop system are bounded. Since the derivatives of z_i and e are bounded, from $z_i \in L_2 \cap L_\infty, i = 1, \ldots, \rho$, $\|e\| \in L_2 \cap L_\infty$ and Babalat's lemma, we further conclude that $\lim_{t \to \infty} z_i = 0$, $i = 1, \ldots, \rho$, and $\lim_{t \to \infty} \|e\| = 0$.

From (10.35), we have $\lim_{t \to \infty} \|\xi(t) - \eta(t)\| = 0$, which means that ξ asymptotically converges to η. From the boundedness of all the variables and $\lim_{t \to \infty} z_i = 0$, $i = 1, \ldots, \rho$, we can conclude that $\lim_{t \to \infty} \dot{\hat{\psi}} = 0$ and $\lim_{t \to \infty} \dot{\hat{k}}_i = 0$, $i = 1, \ldots, \rho$, which implies that the adaptive controller will converge to a fixed-parameter type.

We now establish the convergence of $\hat{\psi}$. Since $\lim_{t \to \infty} \dot{\hat{\psi}} = 0$, there exists a $\psi_\infty = \lim_{t \to \infty} \hat{\psi}(t)$. From (10.14), (10.19) and (10.35), we obtain

$$\frac{d}{dt}(\eta - \xi) = F(\eta - \xi) + G(\psi - \psi_\infty)^T \eta + \varepsilon(t), \tag{10.46}$$

where

$$\varepsilon(t) = G(\psi_\infty - \hat{\psi})^T \xi + G\psi_\infty^T(e + Gy) - \iota.$$

From $\lim_{t \to \infty} (\eta - \xi) = 0$ and $\lim_{t \to \infty} \varepsilon(t) = 0$, we can conclude that

$$(\psi - \psi_\infty)^T \eta = 0, \tag{10.47}$$

because, if otherwise, $(\psi - \psi_\infty)^T \eta$ would be a persistently excited signal, and we could never have $\lim_{t \to \infty} (\eta - \xi) = 0$. If the disturbance $w(t)$ contains components at $m/2$ distinct frequencies, the signal $\eta(t)$ is persistently excited, and from (10.47) we have $\psi - \psi_\infty = 0$, i.e., $\lim_{t \to \infty} \hat{\psi}(t) = \psi$. $\qquad \Box$

Remark 10.8. The condition imposed on $w(0)$ does not really affect the convergence of $\hat{\psi}$. It is added to avoid the situation that $w(t)$ degenerates to have less independent frequency components. In that situation, we can reform the exosystem and E in (10.1) with a smaller dimension $\bar{m}$ such that the reformed $w(t)$ is persistently excited in reduced space $\mathbb{R}^{\bar{m}}$ and accordingly we have $\eta, \psi \in \mathbb{R}^{\bar{m}}$. With the estimate $\hat{\psi}$, together with (F, G), the disturbance frequencies can be estimated. $\qquad \triangleleft$

Example 10.1. Consider the nonlinear system

$$\dot{\zeta}_1 = \zeta_2 + (e^y - 1) + w_1$$
$$\dot{\zeta}_2 = u + w_1$$
$$y = \zeta_1,$$

where $w_1 \in \mathbb{R}$ is a disturbance with two sinusoidal components at unknown frequencies ω_1 and ω_2. An augmented disturbance vector $w = [w_1, w_2, w_3, w_4]^T$ satisfies $\dot{w} = Sw$ with eigenvalues of S at $\{\pm j\omega_1, \pm j\omega_2\}$. It is easy to obtain that $\pi = [0, -w_1]^T$ and $\alpha = -w_1 - \dot{w}_1$. Using the state transform $x = \zeta - \pi$, we have

$$\begin{aligned}
\dot{x}_1 &= x_2 + (e^y - 1) \\
\dot{x}_2 &= u - \alpha \\
y &= x_1.
\end{aligned} \tag{10.48}$$

A transform $w = T\bar{w}$ can also be introduced to the disturbance model so that the disturbance is generated by

$$\dot{\bar{w}} = \begin{bmatrix} \begin{bmatrix} 0 & \omega_1 \\ -\omega_1 & 0 \end{bmatrix} & 0_{2\times 2} \\ 0_{2\times 2} & \begin{bmatrix} 0 & \omega_2 \\ -\omega_2 & 0 \end{bmatrix} \end{bmatrix} \bar{w} := \bar{S}\bar{w}$$

$$w_1 = [t_{11}, t_{12}, t_{13}, t_{14}]\bar{w},$$

where T, ω_1 and ω_2 are unknown. With the coordinate $\bar{w}$, we have

$$\alpha(w) = [-t_{11} + \omega_1 t_{12}, -t_{12} - \omega_1 t_{11}, -t_{13} + \omega_2 t_{14}, -t_{14} - \omega_2 t_{13},]\bar{w}.$$

The steady-state contribution of w to x_2 is given by $q_2 = Q_2\bar{w}$, i.e., $l^T = Q_2 T^{-1}$, where Q_2 is the second row of Q which satisfies

$$Q\bar{S} = (A_c + kC)Q + [0, 1]^T[-t_{11} + \omega_1 t_{12}, -t_{12} - \omega_1 t_{11}, -t_{13} + \omega_2 t_{14}, -t_{14} - \omega_2 t_{13}].$$

The system (10.48) is of relative degree 2, and the control input and parameter estimator can be designed by following the steps presented earlier.

In the simulation study, we chose the pair

$$F = \begin{bmatrix} 0 & 1 & 0 & 0 \\ 0 & 0 & 1 & 0 \\ 0 & 0 & 0 & 1 \\ -1 & -4 & -6 & -4 \end{bmatrix}, \quad G = \begin{bmatrix} 0 \\ 0 \\ 0 \\ 1/10 \end{bmatrix},$$

which is controllable with the eigenvalues of F at $\{-1, -1, -1, -1\}$ and $\|PG\| < 1$. We set $c_1 = c_2 = 10$, with the condition (10.32) being satisfied. Other parameters in the control design were set as $\gamma_1 = \gamma_2 = 1$, $\Gamma = 1000I$, $L_1 = 3$ and $L_2 = 2$. The disturbance was set as

$$w_1 = 4\sin\omega_1 t + 4\sin\omega_2 t$$

with $\omega_1 = 1$ and

$$\omega_2 = \begin{cases} 2, & 250 > t \geq 0, \\ 1.5, & t \geq 250. \end{cases}$$

A set of simulation results are presented here. Figure 10.1 shows the system output together with the control input. Figure 10.2 shows the estimates of ψ. The ideal values of ψ are $[-30, 40, 10, 40]^T$ and $[-12.5, 40, 27.5, 40]^T$ for $\omega_1 = 1, \omega_2 = 2$ and $\omega_1 = 1$,

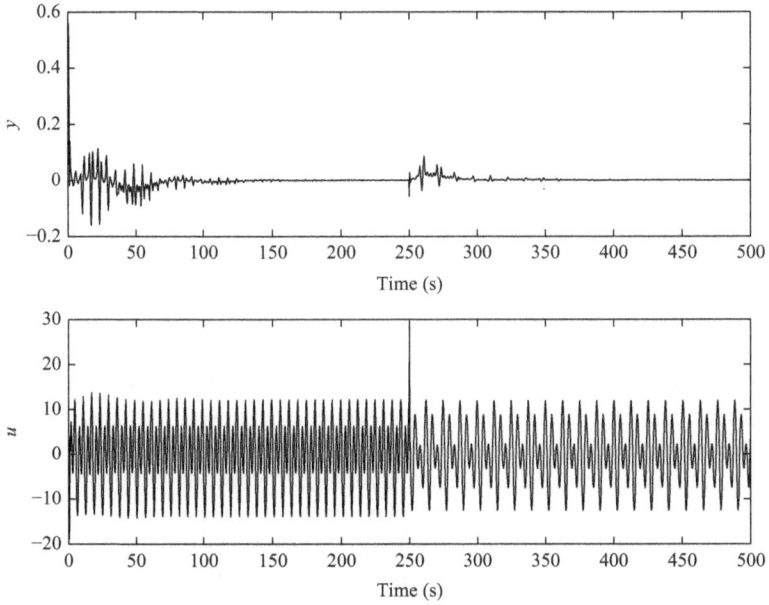

Figure 10.1   System output and control input

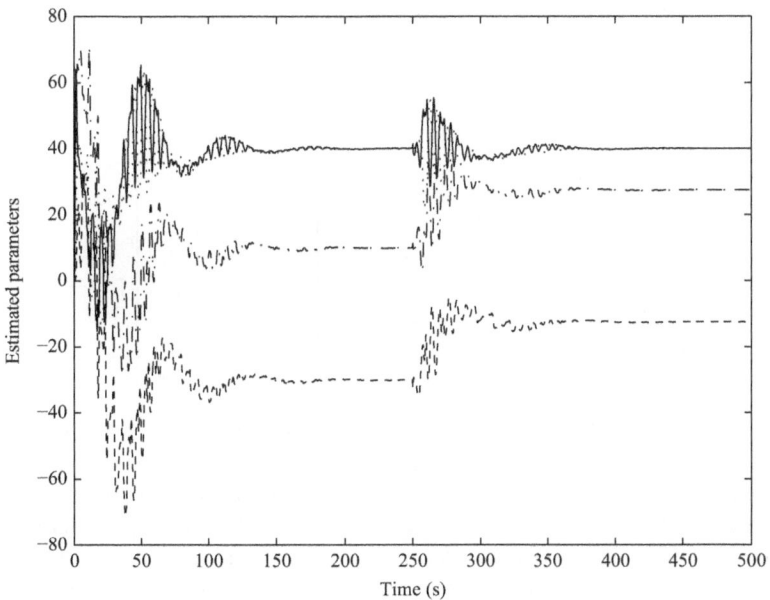

Figure 10.2   *Estimated parameters* $\hat{\psi}_1$ *(dashed),* $\hat{\psi}_2$ *(dotted),* $\hat{\psi}_3$ *(dashdot) and* $\hat{\psi}_4$ *(solid)*

$\omega_2 = 1.5$ respectively. Under both sets of the frequencies, $\hat{\psi}$ converged to the ideal values. It can be seen through this example that the disturbance of two unknown frequencies has been rejected completely.

◁

10.2 Adaptive output regulation

In the previous section, we presented a control design for asymptotic rejection of harmonic disturbances in nonlinear systems when the disturbance frequencies are unknown. When the measurement, or the output, does not explicitly contain the disturbances, the control design is referred to as asymptotic rejection problem for the boundedness of the signals and the asymptotic convergence to zero. When the measurement contains the disturbance or exogenous signals, the control design to ensure the measurement converge to zero asymptotically is often referred to as the output regulation problem. When the measurement contains the exogenous signal, the output is normally different from the measurement, and in such a case, the output can be viewed as to track an exogenous signal. In this section, we will present a control design for output regulation when the measurement is different from the controlled output.

We consider a SISO nonlinear system which can be transformed into the output feedback form

$$
\begin{aligned}
\dot{x} &= A_c x + \phi(y, w, a) + bu \\
y &= Cx \\
e &= y - q(w),
\end{aligned}
\tag{10.49}
$$

with

$$
A_c = \begin{bmatrix} 0 & 1 & 0 & \cdots & 0 \\ 0 & 0 & 1 & \cdots & 0 \\ \vdots & \vdots & \vdots & \ddots & \vdots \\ 0 & 0 & 0 & \cdots & 1 \\ 0 & 0 & 0 & \cdots & 0 \end{bmatrix}, \quad C = \begin{bmatrix} 1 \\ 0 \\ \vdots \\ 0 \end{bmatrix}^T, \quad b = \begin{bmatrix} 0 \\ \vdots \\ 0 \\ b_\rho \\ \vdots \\ b_n \end{bmatrix},
$$

where $x \in \mathbb{R}^n$ is the state vector; $u \in \mathbb{R}$ is the control; $y \in \mathbb{R}$ is the output; e is the measurement output; $a \in \mathbb{R}^q$ and $b \in \mathbb{R}^n$ are vectors of unknown parameters, with b being a Hurwitz vector with $b_\rho \neq 0$, which implies the relative degree of the system is ρ; $\phi : \mathbb{R} \times \mathbb{R}^m \times \mathbb{R}^q \to \mathbb{R}^n$ is a smooth vector field with each element being polynomials with a known upper order of its variables and satisfying $\phi(0, w, a) = 0$; q is an unknown polynomial of w; and $w \in \mathbb{R}^m$ are disturbances, and they are generated from an unknown exosystem

$$
\dot{w} = S(\sigma)w
$$

with unknown $\sigma \in \mathbb{R}^s$, of which, $S \in \mathbb{R}^{m \times m}$ is a constant matrix with distinct eigenvalues of zero real parts.

The control design problem considered in this section is to design a control input using the measurement feedback to ensure that the regulated measurement converges to zero asymptotically while keeping all other variables bounded.

The system (10.1) has unknown parameters in the system and in the exosystem, and therefore adaptive control techniques are used to tackle the uncertainties. For this reason, the problem under consideration is referred to as adaptive output regulation.

Remark 10.9. Different from (10.1), the system (10.49) has a measurement e that is perturbed by an unknown polynomial of the unknown disturbance. This is the reason why the problem to be solved is an output regulation problem, rather than a disturbance rejection problem.   ◁

Remark 10.10. The nonlinear functions in (10.49) are restricted to be polynomials of its variables, to guarantee the existence of invariant manifold for the solution of output regulation problem. Comparing with (10.1), the nonlinear term $\phi(y, w, a)$ has a more complicated structure, containing the output, disturbance, and unknown parameters all together. Assuming the nonlinear functions to be polynomials also allows that the nonlinear terms in the control design can be bounded by polynomials of the measurement.   ◁

Remark 10.11. In the system, all the parameters are assumed unknown, including the sign of the high-frequency gain, b_ρ, and the parameters of the exosystem. A Nussbaum gain is used to deal with the unknown sign of the high-frequency gain. The adaptive control techniques presented in this control design can be easily applied to other adaptive control schemes introduced in this book. The nonlinear functions $\phi(y, w, a)$ and $q(w)$ are only assumed to have known upper orders. This class of nonlinear systems perhaps remains as the largest class of uncertain nonlinear systems of which global output regulation problem can be solved with unknown disturbance frequencies.   ◁

When we set the unknown disturbance w and unknown parameter a to zero, the system (10.49) is in exactly the same format as (9.32), to which backstepping with filtered transformation has been applied. With unknown parameters, adaptive backstepping with filtered transformation can be applied to solve the problem here. One reason for us to use backstepping with filtered transformation is due to the uncertainty in the nonlinear function $\phi(y, w, a)$, which prevents the application of adaptive observer backstepping. In the following design, we only consider the case for relative degree greater than 1.

For the system (10.49) with relative degree $\rho > 1$, we introduce the same filtered transformation as in Section 9.3 with the filter

$$\dot{\xi}_1 = -\lambda_1 \xi_1 + \xi_2$$
$$\cdots$$
$$\dot{\xi}_{\rho-1} = -\lambda_{\rho-1}\xi_{\rho-1} + u, \tag{10.50}$$

where $\lambda_i > 0$, for $i = 1, \ldots, \rho - 1$, are the design parameters, and the filtered transformation

$$\bar{z} = x - [\bar{d}_1, \ldots, \bar{d}_{\rho-1}]\xi, \tag{10.51}$$

where $\xi = [\xi_1, \ldots, \xi_{\rho-1}]^T$ and $\bar{d}_i \in \mathbb{R}^n$ for $i = 1, \ldots, \rho - 1$, and they are generated recursively by $\bar{d}_{\rho-1} = b$ and $\bar{d}_i = (A_c + \lambda_{i+1}I)\bar{d}_{i+1}$ for $i = \rho - 2, \ldots, 1$. The system (10.49) is then transformed to

$$\dot{\bar{z}} = A_c\bar{z} + \phi(y, w, a) + d\xi_1$$
$$y = C\bar{z}, \tag{10.52}$$

where $d = (A_c + \lambda_1 I)\bar{d}_1$. It has been shown in (9.36) that $d_1 = b_\rho$ and

$$\sum_{i=1}^{n} d_i s^{n-i} = \prod_{i=1}^{\rho-1}(s + \lambda_i) \sum_{\rho}^{n} b_i s^{n-i}. \tag{10.53}$$

With ξ_1 as the input, the system (10.52) is with relative degree 1 and minimum phase. We introduce another state transform to extract the internal dynamics of (10.52) with $z \in \mathbb{R}^{n-1}$ given by

$$z = \bar{z}_{2:n} - \frac{d_{2:n}}{d_1}y, \tag{10.54}$$

where $(\cdot)_{2:n}$ refers to the vector or matrix formed by the 2nd row to the nth row. With the coordinates (z, y), (10.52) is rewritten as

$$\dot{z} = Dz + \psi(y, w, \theta)$$
$$\dot{y} = z_1 + \psi_y(y, w, \theta) + b_\rho\xi_1, \tag{10.55}$$

where the unknown parameter vector $\theta = [a^T, b^T]^T$, and D is the left companion matrix of d given by

$$D = \begin{bmatrix} -d_2/d_1 & 1 & \cdots & 0 \\ -d_3/d_1 & 0 & \ddots & 0 \\ \vdots & \vdots & \vdots & \vdots \\ -d_{n-1}/d_1 & 0 & \cdots & 1 \\ -d_n/d_1 & 0 & \cdots & 0 \end{bmatrix}, \tag{10.56}$$

and

$$\psi(y, w, \theta) = D\frac{d_{2:n}}{d_1}y + \phi_{2:n}(y, w, a) - \frac{d_{2:n}}{d_1}\phi_1(y, w, a),$$

$$\psi_y(y, w, \theta) = \frac{d_2}{d_1}y + \frac{d_{2:n}}{d_1}\phi_1(y, w, a).$$

Notice that D is Hurwitz, from (9.36), and that the dependence of d on b is reflected in the parameter θ in $\psi(y, w, \theta)$ and $\psi_y(y, w, \theta)$, and it is easy to check that $\psi(0, w, \theta) = 0$ and $\psi_y(0, w, \theta) = 0$.

The solution of the output regulation problem depends on the existence of certain invariant manifold and feedforward input. For this problem, we have the following result.

Proposition 10.4. *Suppose that an invariant manifold $\pi(w) \in \mathbb{R}^{n-1}$ satisfies*

$$\frac{\partial \pi(w)}{\partial w}S(\sigma)w = D\pi(w) + \psi(q(w), w, \theta). \tag{10.57}$$

Then there exists an immersion for the feedforward control input

$$\frac{\partial \tau(w, \theta, \sigma)}{\partial w}S(\sigma)w = \Phi(\sigma)\tau(w, \theta, \sigma)$$

$$\alpha(w, \theta, \sigma) = \Gamma\tau(w, \theta, \sigma),$$

where

$$\alpha(w, \theta, \sigma) = b_\rho^{-1}\left(\frac{\partial q(w)}{\partial w}S(\sigma)w - \pi_1(w) - \psi_y(q(w), w, \theta)\right).$$

Furthermore, this immersion can be re-parameterised as

$$\dot{\eta} = (F + Gl^T)\eta$$

$$\alpha = l^T\eta, \tag{10.58}$$

where (F, G) is a controllable pair with compatible dimensions, $\eta = M\tau$ and $l = \Gamma M^{-1}$ with M satisfying

$$M(\sigma)\Phi(\sigma) - FM(\sigma) = G\Gamma. \tag{10.59}$$

Proof. With ξ_1 being viewed as the input, α is the feedforward term used for output regulation to tackle the disturbances, and from the second equation of (10.55), we have

$$\alpha(w, \theta, \sigma) = b_\rho^{-1}\left(\frac{\partial q(w)}{\partial w}S(\sigma)w - \pi_1(w) - \psi_y(q(w), w, \theta)\right).$$

From the structure of the exosystem, the disturbances are sinusoidal functions. Polynomials of sinusoidal functions are still sinusoidal functions, but with some high-frequency terms. Since all the nonlinear functions involved in the system (10.49) are polynomials of their variables, the immersion in (10.58) always exists. For a controllable pair (F, G), M is an invertible solution of (10.59) if (Φ, Γ) is observable, which is guaranteed by the immersion. □

We now introduce the last transformation based on the invariant manifold with

$$\tilde{z} = z - \pi \tag{10.60}$$

Finally we have the model for the control design

$$\dot{z} = D\tilde{z} + \tilde{\psi}$$
$$\dot{e} = \tilde{z}_1 + \tilde{\psi}_y + b_\rho(\xi - l^T\eta)$$
$$\dot{\xi}_1 = -\lambda_1\xi_1 + \xi_2 \qquad\qquad (10.61)$$

$$\cdots$$

$$\dot{\xi}_{\rho-1} = -\lambda_{\rho-1}\xi_{\rho-1} + u,$$

where

$$\tilde{\psi} = \psi(y, w, \theta) - \psi(q(w), w, \theta)$$

and

$$\tilde{\psi}_y = \psi_y(y, w, \theta) - \psi_y(q(w), w, \theta).$$

Since the state in the internal model η is unknown, we design the adaptive internal model

$$\dot{\hat{\eta}} = F\hat{\eta} + G\xi_1. \qquad\qquad (10.62)$$

If we define the auxiliary error

$$\tilde{\eta} = \eta - \hat{\eta} + b_\rho^{-1}Ge, \qquad\qquad (10.63)$$

it can be shown that

$$\dot{\tilde{\eta}} = F\tilde{\eta} - FGb_\rho^{-1}e + b_\rho^{-1}G\tilde{z}_1 + b_\rho^{-1}G\tilde{\psi}_y. \qquad\qquad (10.64)$$

If the system (10.49) is of relative degree 1, then ξ_1 in (10.61) is the control input. For the systems with higher relative degrees, adaptive backstepping will be used to find the final control input u from the desirable value of ξ_1. Supposing that $\hat{\xi}_1$ is desirable value for ξ_1, we introduce a Nussbaum gain $N(\kappa)$ such that

$$\hat{\xi}_1 = N(\kappa)\bar{\xi}_1$$
$$\dot{\kappa} = e\bar{\xi}_1, \qquad\qquad (10.65)$$

where the Nussbaum gain N is a function (e.g. $N(\kappa) = \kappa^2 \cos\kappa$) which satisfies the two-sided Nussbaum properties

$$\lim_{\kappa\to\pm\infty} \sup \frac{1}{\kappa} \int_0^\kappa N(s)ds = +\infty, \qquad\qquad (10.66)$$

$$\lim_{\kappa\to\pm\infty} \inf \frac{1}{\kappa} \int_0^\kappa N(s)ds = -\infty, \qquad\qquad (10.67)$$

where $\kappa \to \pm\infty$ denotes $\kappa \to +\infty$ and $\kappa \to -\infty$ respectively. From (10.61) and the definition of the Nussbaum gain, we have

$$\dot{e} = \tilde{z}_1 + (b_\rho N - 1)\bar{\xi}_1 + \bar{\xi}_1 + \tilde{b}_\rho\xi_1 + \hat{b}_\rho\bar{\xi}_1 - l_b^T\eta + \tilde{\psi}_y,$$

where $l_b = b_\rho l$, $\hat{b}_\rho$ is an estimate of b_ρ and $\tilde{b}_\rho = b_\rho - \hat{b}_\rho$, and $\tilde{\xi}_1 = \xi_1 - \hat{\xi}_1$. Sine the nonlinear functions involved in $\tilde{\psi}$ and $\tilde{\psi}_y$ are polynomials with $\tilde{\psi}(0, w, \theta, \sigma) = 0$ and $\tilde{\psi}_y(0, w, \theta, \sigma) = 0$, w is bounded, and the unknown parameters are constants, it can be shown that

$$|\tilde{\psi}| < \bar{r}_z(|e| + |e|^p),$$
$$|\tilde{\psi}_y| < \bar{r}_y(|e| + |e|^p),$$

where p is a known positive integer, depending on the polynomials in $\tilde{\psi}$ and $\tilde{\psi}_y$, and $\bar{r}_z$ and $\bar{r}_y$ are unknown positive real constants. We now design the virtual control $\hat{\xi}_1$ as, with $c_0 > 0$,

$$\bar{\xi}_1 = -c_0 e - \hat{k}_0(e + e^{2p-1}) + \hat{l}_b^T \hat{\eta}.$$

Using (10.63), we have the resultant error dynamics

$$\dot{e} = -c_0 e - \hat{k}_0(e + e^{2p-1}) + \tilde{z}_1 + (b_\rho N - 1)\bar{\xi}_1 + \tilde{b}_\rho \tilde{\xi}_1 + \hat{b}_\rho \tilde{\xi}_1$$
$$- l_b^T \tilde{\eta} - \tilde{l}_b^T \hat{\eta} + l^T Ge + \tilde{\psi}_y. \tag{10.68}$$

The adaptive laws are given by

$$\dot{\hat{k}}_0 = e^2 + e^{2p},$$
$$\tau_{b,0} = \tilde{\xi}_1 e, \tag{10.69}$$
$$\tau_{l,0} = -\hat{\eta} e,$$

where $\tau_{b,0}$ and $\tau_{l,0}$ denote the first tuning functions in adaptive backstepping design for the final adaptive laws for $\hat{b}_\rho$ and $\hat{l}_b$. If the relative degree $\rho = 1$, we set $u = \hat{\xi}_1$. For $\rho > 1$, adaptive backstepping can be used to obtain the following results:

$$\hat{\xi}_2 = -\hat{b}_\rho e - c_1 \tilde{\xi}_1 - k_1 \left(\frac{\partial \hat{\xi}_1}{\partial e}\right)^2 \tilde{\xi}_1$$

$$+ \frac{\partial \hat{\xi}_1}{\partial e}(\hat{b}_\rho \xi_1 - \hat{l}_b^T \hat{\eta}) + \frac{\partial \hat{\xi}_1}{\partial \hat{\eta}}\dot{\hat{\eta}}$$

$$+ \frac{\partial \hat{\xi}_1}{\partial \hat{k}_0}\dot{\hat{k}}_0 + \frac{\partial \hat{\xi}_1}{\partial \hat{l}_b}\tau_{l,1}, \tag{10.70}$$

$$\hat{\xi}_i = -\tilde{\xi}_{i-2} - c_{i-1}\tilde{\xi}_{i-1} - k_{i-1}\left(\frac{\partial\hat{\xi}_{i-1}}{\partial e}\right)^2 \tilde{\xi}_{i-1}$$

$$+ \frac{\partial\hat{\xi}_{i-1}}{\partial e}(\hat{b}_\rho\xi_1 - \hat{l}_b^T\hat{\eta}) + \frac{\partial\hat{\xi}_{i-1}}{\partial\hat{\eta}}\dot{\hat{\eta}}$$

$$+ \frac{\partial\hat{\xi}_{i-1}}{\partial\hat{k}_0}\dot{\hat{k}}_0 + \frac{\partial\hat{\xi}_{i-1}}{\partial\hat{b}_\rho}\tau_{b,i-1} + \frac{\partial\hat{\xi}_{i-1}}{\partial\hat{l}_b}\tau_{l,1}$$

$$- \sum_{j=4}^{i} \frac{\partial\hat{\xi}_{i-1}}{\partial e}\frac{\partial\hat{\xi}_{j-2}}{\partial\hat{b}_\rho}\xi_1\tilde{\xi}_{j-2}$$

$$+ \sum_{j=3}^{i} \frac{\partial\hat{\xi}_{i-1}}{\partial e}\frac{\partial\hat{\xi}_{j-2}}{\partial\hat{l}_b}\hat{\eta}\tilde{\xi}_{j-2} \quad \text{for } i = 2,\ldots,\rho, \tag{10.71}$$

where $\tilde{\xi}_i = \xi_i - \hat{\xi}_i$ for $i = 1,\ldots,\rho - 2$, c_i and k_i, $i = 2,\ldots,\rho - 1$, are positive real design parameters, and $\tau_{b,i}$ and $\tau_{l,i}$, for $i = 1,\ldots,\rho - 2$, are tuning functions. The adaptive law and tuning functions are given by

$$\tau_{b,i} = \tau_{b,i-1} - \frac{\partial\hat{\xi}_i}{\partial e}\xi_1\tilde{\xi}_i, \quad \text{for } i = 1,\ldots,\rho - 1,$$

$$\tau_{l,i} = \tau_{l,i-1} + \frac{\partial\hat{\xi}_i}{\partial e}\hat{\eta}\tilde{\xi}_i, \quad \text{for } i = 1,\ldots,\rho - 1,$$

$$\dot{\hat{b}}_\rho = \tau_{b,\rho-1}, \tag{10.72}$$

$$\dot{\hat{l}}_b = \tau_{l,\rho-1}. \tag{10.73}$$

Finally we design the control input as

$$u = \hat{\xi}_\rho. \tag{10.74}$$

For the proposed control design, we have the following result for stability.

Theorem 10.5. *For a system (10.49) satisfying the invariant manifold condition (10.57), the adaptive output regulation problem is globally solved by the feedback control system consisting the ξ-filters (10.50), the adaptive internal model (10.62), Nussbaum gain parameter (10.65), the parameter adaptive laws (10.69), (10.72), (10.73) and the feedback control (10.74), which ensures the convergence to zero of the regulated measurement, and the boundedness of all the variables in the closed-loop system.*

Proof. Define a Lyapunov function candidate

$$V = \beta_1\tilde{\eta}^T P_\eta\tilde{\eta} + \beta_2\tilde{z}^T P_z\tilde{z}$$

$$+ \frac{1}{2}\left(e^2 + \sum_{i=1}^{\rho-1}\tilde{\xi}_i^2 + (k_0 - \hat{k}_0)^2 + \tilde{b}_\rho^2 + \tilde{l}_b^T\tilde{l}_b\right),$$

where β_1 and β_2 are two positive reals and P_z and P_η are positive definite matrices satisfying

$$P_z D + D^T P_z = -I,$$

$$P_\eta F + F^T P_\eta = -I.$$

With the design of $\hat{\xi}_i$, for $i = 1, \ldots, \rho$, the dynamics of $\tilde{\xi}_i$ can be easily evaluated. From the dynamics of $\tilde{z}$ in (10.61) and the dynamics of $\tilde{\eta}$ in (10.64), virtual controls and adaptive laws designed earlier, we have the derivative of V as

$$\dot{V} = \beta_1(-\tilde{\eta}^T\tilde{\eta} - 2\tilde{\eta}^T P_\eta F b_\rho^{-1} Ge + 2\tilde{\eta}^T P_\eta b_\rho^{-1} G\tilde{z}_1 + 2\tilde{\eta}^T P_\eta b_\rho^{-1} G\tilde{\psi}_y)$$

$$+ \beta_2(-\tilde{z}^T\tilde{z} + 2\tilde{z}^T P_z \tilde{\psi}) + (\hat{k}_0 - k_0)(e^2 + e^{2p})$$

$$- c_0 e^2 - \hat{k}_0(e^2 + e^{2p}) + (b_\rho N - 1)e\bar{\xi}_1$$

$$+ e\tilde{z}_1 + e\tilde{\psi}_y - el_b^T\tilde{\eta} + l^T Ge^2$$

$$+ \sum_{i=1}^{\rho-1}\left(-c_i\tilde{\xi}_i^2 - k_i\left(\frac{\partial\hat{A}\xi_i}{\partial e}\right)^2\tilde{\xi}_i^2 - \tilde{\xi}_i\frac{\partial\hat{\xi}_i}{\partial e}\tilde{z}_1\right.$$

$$\left. - \tilde{\xi}_i\frac{\partial\hat{\xi}_i}{\partial e}\tilde{\psi}_y + \tilde{\xi}_i\frac{\partial\hat{\xi}_i}{\partial e}l_b^T\tilde{\eta} - \tilde{\xi}_i\frac{\partial\hat{\xi}_i}{\partial e}l^T Ge\right).$$

The stability analysis can be proceeded by using the inequalities $2xy < rx^2 + y^2/r$ or $xy < rx^2 + y^2/(4r)$ for $x > 0$, $y > 0$ and r being any positive real, to tackle the cross-terms between the variables $\tilde{z}$, $\tilde{\eta}$, e, $\tilde{\xi}_i$, for $i = 1, \ldots, \rho - 1$. It can be shown that there exist sufficiently big positive real β_1, and then sufficiently big positive real β_2, and finally the sufficient big k_0 such that the following result holds:

$$\dot{V} \leq (b_\rho N(\kappa) - 1)\dot{\kappa} - \frac{1}{3}\beta_1\tilde{\eta}^T\tilde{\eta} - \frac{1}{4}\beta_2\tilde{z}^T\tilde{z} - c_0 e^2 - \sum_{i=1}^{\rho-1}c_i\tilde{\xi}_i^2. \qquad (10.75)$$

The boundedness of V can be established based on the Nussbaum gain properties (10.66) and (10.67) via an argument of contradiction. In fact, integrating (10.75) gives

$$V(t) + \int_0^t\left(\frac{1}{3}\beta_1\tilde{\eta}^T\tilde{\eta} + \frac{1}{4}\beta_2\tilde{z}^T\tilde{z} + c_0 e^2 + \sum_{i=1}^{\rho-1}c_i\tilde{\xi}_i^2\right)dt$$

$$\leq b_\rho\int_0^{\kappa(t)}N(s)ds - \kappa(t) + V(0). \qquad (10.76)$$

If $\kappa(t)$, $\forall t \in \mathbb{R}^+$, is not bounded from above or below, then from (10.66) and (10.67) it can be shown that the right-hand side of (10.76) will be negative at some instances of time, which is a contradiction, since the left-hand side of (10.76) is non-negative.

Therefore, κ is bounded, which implies the boundedness of V. The boundedness of V further implies $\tilde{\eta}$, $\tilde{z}$, e, $\tilde{\xi}_i \in \mathcal{L}_2 \cap \mathcal{L}_\infty$ for $i = 1, \ldots, \rho - 1$, and the boundedness of $\hat{k}_0$, $\hat{b}_\rho$ and $\hat{l}_b$. Since the disturbance w is bounded, e, $\tilde{z}$, $\tilde{\eta} \in \mathcal{L}_\infty$ implies the boundedness of y, z and $\hat{\eta}$, which further implies the boundedness of $\hat{\xi}_1$ and then the boundedness of ξ_1. The boundedness of $\hat{\xi}_1$ and ξ_1, together with the boundedness of e, $\hat{\eta}$, $\hat{k}_0$, $\hat{b}_\rho$ and $\hat{l}_b$, implies the boundedness of $\hat{\xi}_2$, and then the boundedness of $\tilde{\xi}_2$ follows the boundedness of $\tilde{\xi}_2$. Applying the above reasoning recursively, we can establish the boundedness of $\hat{\xi}_i$ for $i > 2$ to $i = \rho - 1$. We then conclude that all the variables are bounded.

The boundedness of all the variables implies the boundedness of $\dot{\tilde{\eta}}$, $\dot{\tilde{z}}$, $\dot{e}$ and $\dot{\tilde{\xi}}_i$, which further implies, together with $\tilde{\eta}$, $\tilde{z}$, e, $\tilde{\xi}_i \in \mathcal{L}_2 \cap \mathcal{L}_\infty$ and Barbalat's lemma, $\lim_{t \to \infty} \tilde{\eta} = 0$, $\lim_{t \to \infty} \tilde{z} = 0$, $\lim_{t \to \infty} e(t) = 0$ and $\lim_{t \to \infty} \tilde{\xi}_i = 0$ for $i = 1, \ldots, \rho - 1$. $\qquad\qquad\square$

10.3   Output regulation with nonlinear exosystems

In the previous sections, the disturbances are sinusoidal functions which are generated from linear exosystems with the restriction that the eigenvalues of the exosystem matrix are distinct and with zero real parts. Sinusoidal disturbances are important as disturbances in practical systems can often be approximated by a finite number of sinusoidal functions. However, there are situations where disturbances are generated from nonlinear exosystems, such as nonlinear vibration, etc. Such disturbances can still be approximated by a finite number of sinusoidal functions, possibly with a big number of sinusoidal functions for a good accuracy. The more sinusoidal functions involved in the approximation of a periodic signal, the higher order will be the corresponding system matrix for the linear exosystem. If an internal model can be designed directly based on the nonlinear exosystem, it is possible to achieve asymptotic rejection of the disturbances, which cannot be achieved by approximation using sinusoidal functions, and the order of the internal model can also remain much lower. Of course, it is expected to be a very difficult problem to directly design an internal model to deal with nonlinear exosystems, even though there exists an invariant manifold for output regulation. In this section, we will show nonlinear internal model design for a class of nonlinear exosystem to achieve output regulation. For the internal model design, we exploit a technique for nonlinear observer design based on conditions similar to circle criteria. The dynamic model considered for output regulation is still the class of output feedback form.

We consider a SISO nonlinear system

$$\dot{x} = A_c x + \phi(y)a + E(w) + bu$$

$$y = Cx \qquad\qquad\qquad\qquad (10.77)$$

$$e = y - q(w),$$

with

$$A_c = \begin{bmatrix} 0 & 1 & 0 & \cdots & 0 \\ 0 & 0 & 1 & \cdots & 0 \\ \vdots & \vdots & \vdots & \ddots & \vdots \\ 0 & 0 & 0 & \cdots & 1 \\ 0 & 0 & 0 & \cdots & 0 \end{bmatrix}, \quad C = \begin{bmatrix} 1 \\ 0 \\ \vdots \\ 0 \end{bmatrix}^T, \quad b = \begin{bmatrix} 0 \\ \vdots \\ 0 \\ b_\rho \\ \vdots \\ b_n \end{bmatrix},$$

where $x \in \mathbb{R}^n$ is the state vector; $u \in \mathbb{R}$ is the control; $y \in \mathbb{R}$ is the output; e is the measurement output; $a \in \mathbb{R}^q$ and $b \in \mathbb{R}^n$ are vectors of unknown parameters, with b being a Hurwitz vector with $b_\rho \neq 0$, which implies the relative degree of the system is ρ, and with known sign of b_ρ, $E : \mathbb{R}^m \to \mathbb{R}^n$, $\phi : \mathbb{R} \to \mathbb{R}^{n \times q}$ with $\phi(0) = 0$ and $|\phi(y_1) - \phi(y_2)| \leq \Delta_1(|y_1|)\delta_1(|y_1 - y_2|)$ and $\delta_1(\cdot) \in \mathcal{K}$ and $\Delta_1(\cdot)$ is non-decreasing and the function $\delta_1(\cdot)$ is a known smooth function; and $w \in \mathbb{R}^m$ are disturbances, and they are generated from a nonlinear exosystem

$$\dot{w} = s(w) \tag{10.78}$$

of which the flows are bounded and converge to periodic solutions.

Remark 10.12. The assumption about the function ϕ is satisfied for many kinds of functions, for example polynomial functions. ◁

Remark 10.13. The nonlinear exosystem (10.78) includes nonlinear systems that have limit cycles. ◁

Remark 10.14. The system (10.78) is very similar to the system (10.49) considered in the previous section. The main difference is that the exosystem is nonlinear. ◁

The system (10.78) has the same structure of A_c, *bu* and C as in (10.49), and therefore the same filtered transformation as in the previous section can be used here. We can use the same filtered transformation, and the transformation for extracting the zero dynamics of the system as in the previous section. Using the transformations (10.51) and (10.54), the system (10.78) is put in the coordinate (z, y) as

$$\dot{z}_i = -\frac{d_{i+1}}{d_1}z_1 + z_{i+1} + \left(\frac{d_{i+2}}{d_1} - \frac{d_{i+1}d_2}{d_1^2} \right) y + (\phi_{i+1}(y) - \frac{d_{i+1}}{d_1}\phi_1(y))a$$

$$+ E_{i+1}(w) - \frac{d_{i+1}}{d_1}E_1(w), \quad i = 1, \ldots, n-2, \tag{10.79}$$

$$\dot{z}_{n-1} = -\frac{d_n}{d_1}z_1 - \frac{d_n d_2}{d_1^2}y + \left(\phi_n(y) - \frac{d_n}{d_1}\phi_1(y) \right) a + E_n(w) - \frac{d_n}{d_1}E_1(w),$$

$$\dot{y} = z_1 + \frac{d_2}{d_1}y + \phi_1(y)a + E_1(w) + b_\rho \xi_1$$

where d_i are defined in (10.53).

It is necessary to have the existence of certain invariant manifolds for a solution to the output regulation problem. When the exosystem is nonlinear, it is even more challenging to have the necessary conditions for invariant manifold to exist.

Proposition 10.6. *Suppose that there exist $\varpi(w) \in \mathbb{R}^n$ and $\iota(w)$ with $\varpi_1(w) = q(w)$ for each a, b such that*

$$\frac{\partial \varpi}{\partial w} s(w) = A_c \varpi + \phi(q(w))a + E(w) + b\iota(w). \tag{10.80}$$

Then there exists $\pi(w) \in \mathbb{R}^{n-1}$ along the trajectories of exosystem satisfying

$$\frac{\partial \pi_i(w)}{\partial w} s(w) = -\frac{d_{i+1}}{d_1} \pi_1(w) + \pi_{i+1}(w) + q(w) \left(\frac{d_{i+2}}{d_1} - \frac{d_{i+1}d_2}{d_1^2} \right)$$

$$+ E_{i+1}(w) - E_1(w)\frac{d_{i+1}}{d_1} + (\phi_{i+1}(q(w)))$$

$$- \frac{d_{i+1}}{d_1} \phi_1(q(w)))a, \quad i = 1, \ldots, n-2,$$

$$\frac{\partial \pi_{n-1}(w)}{\partial w} s(w) = -\frac{d_n}{d_1} \pi_1(w) - \frac{d_n d_2}{d_1^2} q(w) + (\phi_n(q(w)))$$

$$- \frac{d_n}{d_1} \phi_1(q(w)))a + E_n(w) - \frac{d_n}{d_1} E_1(w).$$

Proof. Since the last equation of input filter (10.50) used for the filtered transformation is an asymptotically stable linear system, there is a static response for every external input $u(w)$, i.e., there exists a function $\chi_{\rho-1}(w)$ such that

$$\frac{\partial \chi_{\rho-1}(w)}{\partial w} s(w) = -\lambda_{\rho-1} \chi_{\rho-1}(w) + \iota(w).$$

Recursively, if there exists $\chi_i(w)$ such that

$$\frac{\partial \chi_i(w)}{\partial w} s(w) = -\lambda_i \chi_i(w) + \chi_{i+1}(w),$$

then there exists $\chi_{i-1}(w)$ such that

$$\frac{\partial \chi_{i-1}(w)}{\partial w} s(w) = -\lambda_{i-1} \chi_{i-1}(w) + \chi_i(w).$$

Define

$$\begin{bmatrix} \pi(w) \\ q(w) \end{bmatrix} = D_a(\varpi(w) - [\bar{d}_1, \ldots, \bar{d}_{\rho-1}]\chi),$$

where $\chi = [\chi_1, \ldots, \chi_{\rho-1}]^T$ and

$$D_a = \begin{bmatrix} -d_2/d_1 & 1 & \cdots & 0 \\ \vdots & \vdots & \ddots & \vdots \\ -d_n/d_1 & 0 & \cdots & 1 \\ 1 & 0 & \cdots & 0 \end{bmatrix}.$$

It can be seen that $\pi(w)$ satisfies the dynamics of z along the trajectories of (10.78) as shown in (10.80), and hence the proposition is proved. $\qquad\square$

Based on the above lemma, we have

$$\frac{\partial q(w)}{\partial w}s(w) = \pi_1(w) + \frac{d_2}{d_1}q(w) + \phi_1(q(w))a + E_1(w) + b_\rho\alpha(w),$$

where $\alpha(w) = \chi_1(w)$. With ξ_1 being viewed as the input, $\alpha(w)$ is the feedforward term used for output regulation to tackle the disturbances, and it is given by

$$\alpha = b_\rho^{-1}\left(\frac{\partial q(w)}{\partial w}s(w) - \pi_1(w) - \frac{d_2}{d_1}q(w) - \phi_1(q(w))a - E_1(w)\right).$$

We now introduce the last transformation based on the invariant manifold with

$$\tilde{z} = z - \pi(w(t)).$$

Finally we have the model for the control design

$$\dot{\tilde{z}}_i = -\frac{d_{i+1}}{d_1}\tilde{z}_1 + \tilde{z}_{i+1} + \left(\frac{d_{i+2}}{d_1} - \frac{d_{i+1}d_2}{d_1^2}\right)e$$
$$+ (\phi_{i+1}(y) - \phi_{i+1}(q(w)))a$$
$$- \frac{d_{i+1}}{d_1}(\phi_1(y) - \phi_1(q(w)))a, \quad i = 1, \ldots, n-2$$

$$\dot{\tilde{z}}_{n-1} = -\frac{d_n}{d_1}\tilde{z}_1 - \frac{d_nd_2}{d_1^2}e + (\phi_n(y) - \phi_n(q(w)))a$$
$$- \frac{d_n}{d_1}(\phi_1(y) - \phi_1(q(w)))a$$

$$\dot{e} = \tilde{z}_1 + \frac{d_2}{d_1}e + (\phi_1(y) - \phi_1(q(w)))a + b_\rho(\xi_1 - \alpha(w)),$$

i.e., the system can be represented as

$$\dot{\tilde{z}} = D\tilde{z} + \Xi e + \Omega(y, w, d)a$$
$$\dot{e} = \tilde{z}_1 + \frac{d_2}{d_1}e + (\phi_1(y) - \phi_1(q(w)))a + b_\rho(\xi_1 - \alpha(w)),$$

$$(10.81)$$

where D is a companion matrix of d shown in (10.56), and

$$\Xi = \left(\frac{d_3}{d_1} - \frac{d_2^2}{d_1^2}, \ldots, \frac{d_n}{d_1} - \frac{d_{n-1}d_2}{d_1^2}, -\frac{d_nd_2}{d_1^2}\right)^T,$$

$$\Omega(y, w, d) = \begin{pmatrix} \phi_2(y) - \phi_2(q(w)) - \frac{d_2}{d_1}(\phi_1(y) - \phi_1(q(w))) \\ \vdots \\ \phi_n(y) - \phi_n(q(w)) - \frac{d_n}{d_1}(\phi_1(y) - \phi_1(q(w))) \end{pmatrix}.$$

Lemma 10.7. *There exists a known function $\zeta(\cdot)$ which is non-decreasing and an unknown constant Δ, which is dependent on the initial state w_0 of exosystem, such that*

$$|\Omega(y, w, d)| \leq \Delta|e|\zeta(|e|),$$
$$|\phi_1(y) - \phi_1(q(w))| \leq \Delta|e|\zeta(|e|).$$

Proof. From the assumption of ϕ we can see that

$$|\phi(y) - \phi(q(w))| \leq \Delta_1(|q(w)|)\delta_1(|e|).$$

Since the trajectories of exosystem are bounded and $\delta_1(\cdot)$ is smooth there exist smooth nondecreasing known function $\zeta(\cdot)$ and a nondecreasing known function $\Delta_2(|w_0|)$, such that

$$\delta_1(|e|) \leq |e|\zeta(|e|),$$
$$\Delta_1(|q(w)|) \leq \Delta_2(|w_0|).$$

From previous discussion the result of the lemma is obtained. □

Let

$$V_z = \tilde{z}^T P_d \tilde{z},$$

where

$$P_d D + D^T P_d = -I.$$

Then using $2ab \leq ca^2 + c^{-1}b^2$ and $\zeta^2(|e|) \leq \zeta^2(1 + e^2)$, there exist unknown positive real constants Λ_1 and Λ_2 such that

$$\dot{V}_z = -\tilde{z}^T\tilde{z} + 2\tilde{z}^T P_d(\Xi e + \Omega(y, w, d)a)$$
$$\leq -\frac{3}{4}\tilde{z}^T\tilde{z} + \Lambda_1 e^2 + \Lambda_2 e^2\zeta^2(1 + e^2), \tag{10.82}$$

noting that

$$2\tilde{z}^T P_d \Xi e \leq \frac{1}{8}\tilde{z}^T\tilde{z} + 8e^T \Xi^T P_d^2 \Xi e$$
$$\leq \frac{1}{8}\tilde{z}^T\tilde{z} + \Lambda_1 e^2,$$

and

$$2\tilde{z}^T P_d \Omega(y, w, d)a \leq \frac{1}{8}\tilde{z}^T\tilde{z} + 8a^T \Omega^T P_d^2 \Omega a$$
$$\leq \frac{1}{8}\tilde{z}^T\tilde{z} + \Lambda_2^1|\Omega|^2$$
$$\leq \frac{1}{8}\tilde{z}^T\tilde{z} + \Lambda_2^1\Delta^2|e|^2\zeta^2(|e|)$$
$$\leq \frac{1}{8}\tilde{z}^T\tilde{z} + \Lambda_2 e^2\zeta^2(1 + e^2),$$

where Λ_2^1 is an unknown positive real constant.

Now let us consider the internal model design. We need an internal model to produce a feedforward input that converges to the ideal feedforward control term $\alpha(w)$, which can be viewed as the output of the exosystem as

$$\dot{w} = s(w)$$
$$\alpha = \alpha(w).$$

Suppose that there exists an immersion of the exosystem

$$\dot{\eta} = F\eta + G\gamma(J\eta)$$
$$\alpha = H\eta, \tag{10.83}$$

where $\eta \in \mathbb{R}^r$, $H = [1, 0, \ldots, 0]$, (H, F) is observable

$$(v_1 - v_2)^T(\gamma(v_1) - \gamma(v_2)) \geq 0,$$

and G and J are some appropriate dimensional matrices. We then design an internal model as

$$\dot{\hat{\eta}} = (F - KH)(\hat{\eta} - b_\rho^{-1}Ke) + G\gamma(J(\hat{\eta} - b_\rho^{-1}Ke)) + K\xi_1, \tag{10.84}$$

where $K \in \mathbb{R}^r$ is chosen such that $F_0 = F - KH$ is Hurwitz and there exist a positive definite matrix P_f and a semi-positive definite matrix Q satisfying

$$\begin{cases} P_f F_0 + F_0^T P_f = -Q \\ P_f G + J^T = 0 \\ \eta^T Q\eta \geq \gamma_0 |\eta_1|^2, \quad \gamma_0 > 0, \quad \eta \in \mathbb{R}^r \\ \operatorname{span}(P_F K) \subseteq \operatorname{span}(Q). \end{cases} \tag{10.85}$$

Remark 10.15. It reminds us a challenging problem to design internal models for output regulation with nonlinear exosystems. It is not clear at the moment what general conditions are needed to guarantee the existence of an internal model for output regulation of nonlinear systems with nonlinear internal models. Here we use the condition of the existence of an immersion (10.83) for the internal model design. Also note that even when the exosystem is linear, an internal model can be nonlinear for a nonlinear dynamic system.   ◁

Remark 10.16. Note the condition specified in (10.85) is weaker than the condition that there exist $P_f > 0$ and $Q > 0$ satisfying

$$\begin{cases} P_F F_0 + F_0^T P_F = -Q \\ P_F G + J^T = 0, \end{cases} \tag{10.86}$$

which can be checked by LMI. This will be seen in the example later in this section. In particular, if G and J^T are two column vectors, (F_0, G) controllable, (J, F_0) observable and $Re[-J(j\omega I - F_0)^{-1}G] > 0$, $\forall \omega \in \mathbb{R}$, then there exists a solution of (10.86) from Kalman–Yacubovich lemma.   ◁

If we define the auxiliary error

$$\tilde{\eta} = \eta - \hat{\eta} + b_\rho^{-1} Ke,$$

it can be shown that

$$\dot{\tilde{\eta}} = F_0\tilde{\eta} + G(\gamma(J\eta) - \gamma(J(\hat{\eta} - b_\rho^{-1}Ke)))$$
$$+ b_\rho^{-1}K\left(\tilde{z}_1 + \frac{d_2}{d_1}e + (\phi_1(y) - \phi_1(q(w)))a\right).$$

Let

$$V_\eta = \tilde{\eta} P_F \tilde{\eta}.$$

Then following the spirit of (10.82), there exist unknown positive real constants Θ_1 and Θ_2 such that

$$\dot{V}_\eta = -\tilde{\eta}^T Q\tilde{\eta} + 2\tilde{\eta}^T P_F b_\rho^{-1} K\left(\tilde{z}_1 + \frac{d_2}{d_1}e\right)$$
$$+ 2\tilde{\eta}^T P_F b_\rho^{-1} K(\phi_1(y) - \phi_1(q(w)))a$$
$$+ 2\tilde{\eta}^T P_F G(\gamma(J\eta) - \gamma(J(\hat{\eta} - b_\rho^{-1}Ke)))$$
$$\le -\frac{3}{4}\gamma_0|\tilde{\eta}_1|^2 + \frac{12}{\gamma_0}b_\rho^{-2}\tilde{z}_1^2 + \Theta_1 e^2 + \Theta_2 e^2\varsigma^2(1 + e^2). \tag{10.87}$$

Let us proceed with the control design. From (10.81) and

$$\alpha = \eta_1 = \hat{\eta}_1 + \tilde{\eta}_1 - b_\rho^{-1}K_1e,$$

we have

$$\dot{e} = \tilde{z}_1 + \frac{d_2}{d_1}e + (\phi_1(y) - \phi_1(q(w)))a + \bar{\xi}_1 + b_\rho(\tilde{\xi}_1 - \tilde{\eta}_1 - \hat{\eta}_1 + b_\rho^{-1}K_1e),$$

where $\tilde{\xi}_1 = \xi_1 - \hat{\xi}_1$ and

$$\hat{\xi}_1 = b_\rho^{-1}\bar{\xi}_1. \tag{10.88}$$

For the virtual control $\hat{\xi}_1$, we design $\bar{\xi}_1$ as, with $c_0 > 0$,

$$\bar{\xi}_1 = -c_0 e + b_\rho\hat{\eta}_1 - K_1 e - \hat{l}e(1 + \varsigma^2(1 + e^2)), \tag{10.89}$$

where $\hat{l}$ is an adaptive coefficient. Then we have the resultant error dynamics

$$\dot{e} = \tilde{z}_1 - c_0 e + \frac{d_2}{d_1}e - \hat{l}e(1 + \varsigma^2(1 + e^2)) + (\phi_1(y) - \phi_1(q(w)))a + b_\rho(\tilde{\xi}_1 - \tilde{\eta}_1).$$

Then for

$$V_e = \frac{1}{2}e^2,$$

there exist unknown positive real constants Ψ_1 and Ψ_2, and a sufficiently large unknown positive constant β such that

$$
\begin{aligned}
\dot{V}_e &= -c_0 e^2 + e\tilde{z}_1 + \frac{d_2}{d_1} e^2 + eb_\rho(\tilde{\xi}_1 - \tilde{\eta}_1) \\
&\quad + e(\phi_1(y) - \phi_1(q(w)))a - \hat{l}e^2(1 + \zeta^2(1 + e^2)) \\
&\leq -c_0 e^2 + \frac{1}{8}\beta\tilde{z}_1^2 + \frac{1}{4}\gamma_0\tilde{\eta}_1^2 + \Psi_1 e^2 + \Psi_2 e^2 \zeta(1 + e^2) \\
&\quad -\hat{l}e^2(1 + \zeta^2(1 + e^2)) + b_\rho e\tilde{\xi}_1.
\end{aligned}
\tag{10.90}
$$

Let

$$
V_0 = \beta V_z + V_\eta + V_e + \frac{1}{2}\gamma^{-1}(\hat{l} - l)^2,
$$

where $\beta \geq \frac{96}{\gamma_0}b_\rho^{-2}$ is chosen and $l = \Psi_1 + \Psi_2 + \Theta_1 + \Theta_2 + \beta(\Lambda_1 + \Lambda_2)$ is an unknown constant. Let

$$
\dot{\hat{l}} = \gamma e^2(1 + \zeta^2(1 + e^2)).
$$

Then, it can be obtained that

$$
\dot{V}_0 \leq -\frac{1}{2}\beta\tilde{z}^T\tilde{z} - \frac{1}{2}\gamma_0|\tilde{\eta}_1|^2 - c_0 e^2 + b_\rho e\tilde{\xi}_1.
$$

If the system (10.78) has relative degree 1, the virtual control $\hat{\xi}_1$ shown in (10.88) together with $\bar{\xi}_1$ in (10.88) gives the input, i.e., $u = \hat{\xi}_1$. For the system with higher relative degrees, the control design can be proceeded with backstepping using (10.50) in the same way as the adaptive backstepping with filtered transformation shown in the previous section for adaptive output regulation with linear exosystems. We summarise the stability result in the following theorem.

Theorem 10.8. *For the system (10.78) with the nonlinear exosystem (10.78), if there exists an invariant manifold (10.80) and an immersion (10.83), then there exists $K \in \mathbb{R}^r$ such that $F_0 = F - KH$ is Hurwitz and there exist a positive definite matrix P_F and a semi-positive definite matrix Q satisfying (10.85), and there exists a controller to solve the output regulation in the sense the regulated measurement converges to zero asymptotically while other variables remain bounded.*

We use an example to illustrate the proposed control design, concentrating on the design of nonlinear internal model.

Example 10.2. Consider a first-order system

$$
\begin{aligned}
\dot{y} &= 2y + \theta \sin y - y^3 - \theta \sin w_1 + w_2 + u \\
e &= y - w_1,
\end{aligned}
$$

where θ is an unknown parameter, and the disturbance w is generated by

$$\dot{w}_1 = w_1 + w_2 - w_1^3$$
$$\dot{w}_2 = -w_1 - w_2^3.$$

It is easy to see that $V(w) = \frac{1}{2}w_1^2 + \frac{1}{2}w_2^2$ satisfies

$$\frac{dV}{dt} = w_1^2 - w_1^4 - w_2^4 \leq 0, \quad \text{when } |w_1| \geq 1,$$

and that

$$q(w) = w_1,$$
$$\pi = w_1,$$
$$\alpha(w) = -w_1.$$

From the exosystem and the desired feedforward input α, it can be seen that the condition specified in (10.85) is satisfied with $\eta = -w$ and

$$\begin{cases} F = \begin{pmatrix} 1 & 1 \\ -1 & 0 \end{pmatrix}, \quad G = \begin{pmatrix} -1 & 0 \\ 0 & -1 \end{pmatrix} \\ \gamma_1(s) = \gamma_2(s) = s^3, \quad J = \begin{pmatrix} 1 & 0 \\ 0 & 1 \end{pmatrix}. \end{cases}$$

Let $K = [2, 0]^T$. Then with

$$F_0 = \begin{pmatrix} -1 & 1 \\ -1 & 0 \end{pmatrix}, \quad P_F = I, \quad Q = \text{diag}(2, 0),$$

the internal model is designed as the following:

$$\dot{\hat{\eta}}_1 = -(\hat{\eta}_1 - 2e) + \hat{\eta}_2 - (\hat{\eta}_1 - 2e)^3 + 2u$$
$$\dot{\hat{\eta}}_2 = -(\hat{\eta}_1 - 2e) - \hat{\eta}_2^3.$$

The control input and the adaptive law are given by

$$u = -ce + \hat{\eta}_1 - \hat{l}e(1 + (e^2 + 1)^2),$$
$$\dot{\hat{l}} = \gamma e^2(1 + (e^2 + 1)^2).$$

For simulation study, we set $c = 1$, $\theta = 1$, $\gamma = 1$, and the initial states are $y(0) = 1$, $w_1(0) = 2$ and $w_2(0) = 2$. The initial state of dynamic controller is zero. The system output and input are shown in Figure 10.3, while the feedforward term and its estimation are shown in Figure 10.4 and the portrait of the exosystem is shown in Figure 10.5. As shown in the figures, the internal model successfully reproduces the feedforward control needed after a transient period, and the system output measurement is regulated to zero, as required. ◁

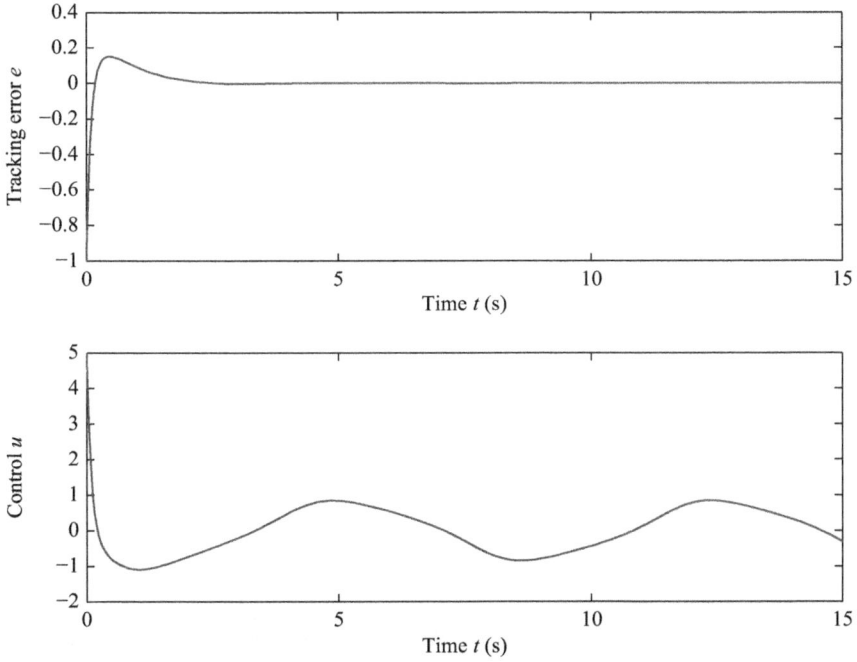

Figure 10.3   The system's output e and input u

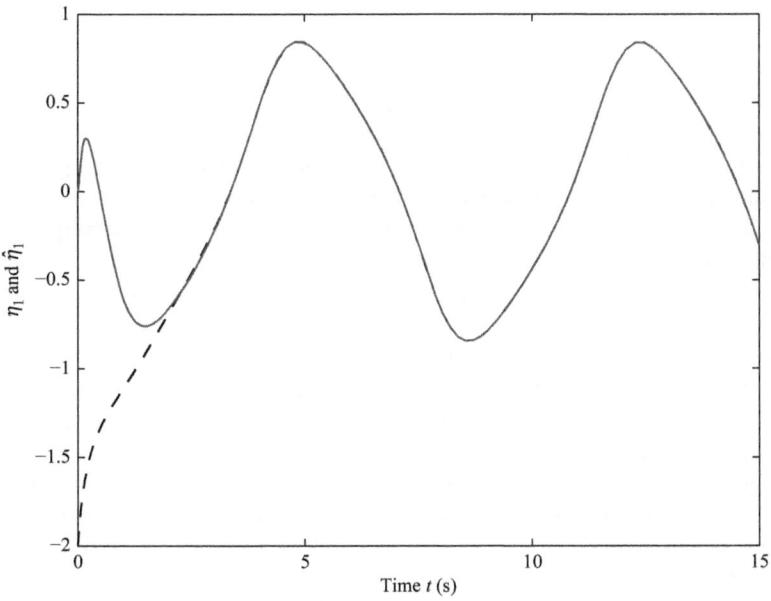

Figure 10.4   The systems's feedforward control η_1 and its estimation $\hat{\eta}_1$

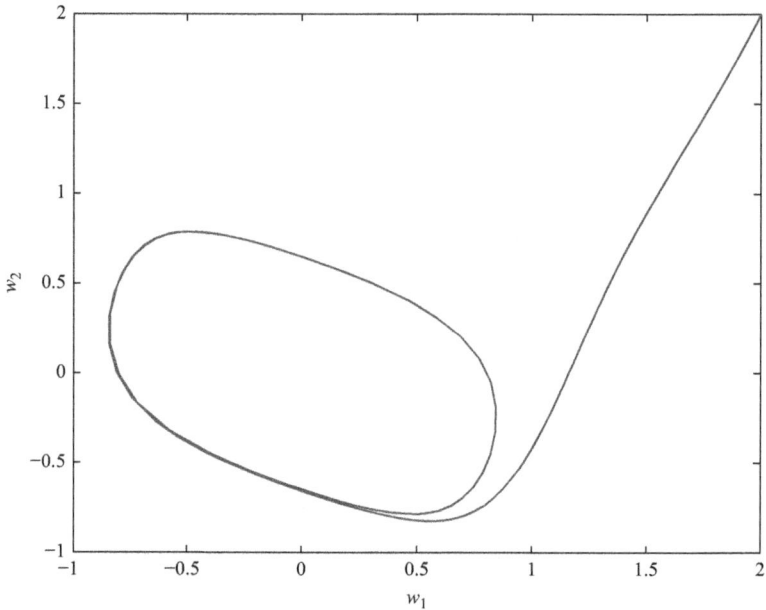

Figure 10.5   The portrait of exosystems

10.4   Asymptotic rejection of general periodic disturbances

We have presented design methods for asymptotic rejection and output regulation of disturbances generated from linear exosystems, i.e., sinusoidal functions and disturbances from a specific class of nonlinear exosystems, which generally produce non-harmonic but still periodic disturbances. For disturbances from linear exosystem, internal models can normally be designed under some mild assumptions, while for nonlinear exosystems, the conditions are more restrictive for designing an internal model for asymptotic rejection and output regulation. The difficulty lies in the guaranteed existence of the invariant manifold, and then the nonlinear internal model design, which is often more involved even than nonlinear observers for nonlinear systems.

   We consider some more general periodic disturbances than harmonic disturbances in this section. These general periodic disturbances can be modelled as outputs of nonlinear systems, and in particular, as the outputs of linear dynamic systems with nonlinear output functions. For the systems with Lipschitz nonlinearities, nonlinear observers can be designed as shown in Section 8.4 and other results in literature. Of course, the problem addressed in this chapter cannot be directly solved by nonlinear observer design, not even the state estimation of the disturbances system, as the disturbance is not measured. However, there is an intrinsic relationship between observer design and internal model design, as evidenced in the previous results of disturbance rejection and output regulation in the earlier sections in this chapter.

With the formulation of general periodic disturbances as the nonlinear outputs of a linear dynamic system, the information of the phase and amplitude of a general periodic disturbance is then embedded in the state variables, and the information of the wave profile in the nonlinear output function. The nonlinear output functions are assumed to be Lipschitz. By formulating the general periodic disturbances in this way, we are able to explore nonlinear observer design of nonlinear systems with output Lipschitz nonlinearities shown in Section 8.4. We will show that general periodic disturbances can be modelled as nonlinear outputs of a second-order linear system with a pair of pure imaginary poles which depend on the frequencies. For this specific system with Lipschitz nonlinear output, a refined condition on the Lipschitz constant will be given by applying the proposed method in this section, and observer gain will be explicitly expressed in terms of the Lipschitz constant and the period or frequency of the disturbance.

An internal model design is then introduced based on the proposed Lipschitz output observer for a class of nonlinear systems. Conditions are identified for the nonlinear system, and control design is carried out using the proposed internal model. Two examples are included to demonstrate the proposed internal model and control design procedures. These examples also demonstrate that some other problems can be converted to the problem addressed in this section.

We consider a nonlinear system

$$
\begin{aligned}
\dot{y} &= a(z) + \psi_0(y) + \psi(y, v) + b(u - \mu(v)) \\
\dot{z} &= f(z, v, y),
\end{aligned}
\tag{10.91}
$$

where $y \in \mathbb{R}$ is the output; a and $\psi_0 : \mathbb{R}^n \to \mathbb{R}$ are continuous functions; $v \in \mathbb{R}^m$ denotes general periodic disturbances; $\mu : \mathbb{R}^m \to \mathbb{R}$ is a continuous function; $\psi :$ $\mathbb{R} \times \mathbb{R}^m \to \mathbb{R}$ is a continuous function and satisfies the condition that $|\psi(y, v)|^2 \leq y\bar{\psi}(y)$ with $\bar{\psi}$ being a continuous function; b is a known constant; $u \in \mathbb{R}$ is the input; $z \in \mathbb{R}^n$ is the internal state variable; and $f : \mathbb{R}^n \times \mathbb{R}^m \times \mathbb{R} \to \mathbb{R}^n$ is a continuous function.

Remark 10.17. For the convenience of presentation, we only consider the system with relative degree 1 as in (10.91). The systems with higher relative degrees can be dealt with similarly by invoking backstepping. The second equation in (10.91) describes the internal dynamics of the system states, and if we set $v = 0$ and $y = 0$, $\dot{z} = f(z, 0, 0)$ denotes the zero dynamics of this system. ◁

Remark 10.18. The system in (10.91) specifies a kind of standard form for asymptotic rejection of general periodic disturbances. For example, consider

$$
\begin{aligned}
\dot{x} &= Ax + \phi(y, v) + bu \\
y &= c^T x,
\end{aligned}
$$

with $b, c \in \mathbb{R}^n$ and

$$A = \begin{bmatrix} -a_1 & 1 & \cdots & 0 \\ -a_2 & 0 & \ddots & 0 \\ \vdots & \vdots & \ddots & \vdots \\ -a_n & \cdots & \cdots & 0 \end{bmatrix}, \quad b = \begin{bmatrix} b_1 \\ b_2 \\ \vdots \\ b_n \end{bmatrix}, \quad c = \begin{bmatrix} 1 \\ 0 \\ \vdots \\ 0 \end{bmatrix},$$

where $x \in \mathbb{R}^n$ is the state vector; y and $u \in \mathbb{R}$ are the output and input respectively of the system; $v \in \mathbb{R}^m$ denotes general periodic disturbances; and $\phi : \mathbb{R} \times \mathbb{R}^m \to \mathbb{R}^n$ is a nonlinear smooth vector field in $\mathbb{R}^n$ with $\phi(0, 0) = 0$. This system is similar to the system (10.49) with $q(w) = 0$. For this class of nonlinear systems, the asymptotic disturbance rejection depends on the existence of state transform to put the systems in the form shown in (10.91), and it has been shown in Section 10.2 that such a transformation exists under some mild assumptions.   ◁

The wave profile information of a general periodic disturbance is used to construct a nonlinear function as the output function for a linear exosystem, for the generation of the desired feedforward input. By doing that, an observer with nonlinear output function can be designed, viewing the feedforward input as the output, and an internal model can then be designed based on the nonlinear observer. The problem to solve in this section is to design a control scheme, using a nonlinear observer-based internal model, to asymptotically reject a class of general periodic disturbances for the system in (10.91).

We start with modelling general periodic disturbances as the outputs of a linear oscillator with nonlinear output functions, and then propose nonlinear observer design for such a system, for the preparation of internal model design.

Many periodic functions with period T can be modelled as outputs of a second-order system

$$\dot{w} = Aw, \quad \text{with } A = \begin{bmatrix} 0 & \omega \\ -\omega & 0 \end{bmatrix}$$

$$\mu(v) = h(w),$$

where $\omega = \frac{2\pi}{T}$. Here, the desired feedforward input $\mu(v)$ is modelled as the nonlinear output $h(w)$ of the second-order system. With

$$e^{At} = \begin{bmatrix} \cos \omega t & \sin \omega t \\ -\sin \omega t & \cos \omega t \end{bmatrix},$$

the linear part of the output $h(w)$, Hw, is always in the form of $a \sin(\omega t + \phi)$ where a and ϕ denote the amplitude and phase respectively. Hence, we can set $H = [1 \ 0]$ without loss of generality, as the amplitude and the phase can be decided by the initial value with

$$w(0) = [a \sin(\phi) \ a \cos(\phi)]^T.$$

Based on the above discussion, the dynamic model for general periodic disturbance is described by

$$\dot{w}_1 = \omega w_2$$
$$\dot{w}_2 = -\omega w_1 \qquad\qquad (10.92)$$
$$\mu = w_1 + h_1(w_1, w_2),$$

where $h_1(w_1, w_2)$ is a Lipschitz nonlinear function with Lipschitz constant γ.

Remark 10.19. General periodic disturbances can be modelled as $af(t + \phi)$ with a and ϕ for the amplitude and phase of a disturbance, and the wave profile is specified by a periodic function f. In the model shown in (10.93), the amplitude and phase of the disturbance are determined by the system state variables w_1 and w_2, and the profile is determined by the nonlinear output function. In some results shown in literature, the phase and amplitude are obtained by delay and half-period integral operations. Here, we use nonlinear observers for the estimation of phases and amplitudes of general periodic disturbances.   ◁

For the model shown in (10.93), the dynamics are linear, but the output function is nonlinear. Many results in literature on observer design for nonlinear Lipschitz systems are for the system with nonlinearities in the system dynamics while the output functions are linear. Here we need the results for observer design with nonlinear output functions. Similar techniques to the observer design of nonlinearities in dynamics can be applied to the case when the output functions are nonlinear.

We have shown the observer design for a linear dynamic system with a nonlinear Lipschitz output function in Section 8.4 with the observer format in (8.38) and gain in Theorem 8.11. Now we can apply this result to observer design for the model of general periodic disturbances. For the model shown in (10.93), the observer shown in (8.38) can be applied with $A = \begin{bmatrix} 0 & \omega \\ -\omega & 0 \end{bmatrix}$ and $H = [1\ 0]$. We have the following lemma for the stability of this observer.

Lemma 10.9. *An observer in the form of (8.38) can be designed to provide an exponentially convergent state estimate for the general periodic disturbance model (10.93) if the Lipschitz constant γ for h_1 satisfies $\gamma < \frac{1}{\sqrt{2}}$.*

Proof. Our proof is constructive. Let

$$P = \begin{bmatrix} p & -\dfrac{1}{4\gamma^2\omega} \\[2ex] -\dfrac{1}{4\gamma^2\omega} & p \end{bmatrix},$$

where $p > \frac{1}{4\gamma^2\omega}$. It is easy to see that P is positive definite. A direct evaluation gives

$$PA + A^T P - \frac{H^T H}{\gamma^2} = -\frac{1}{2\gamma^2} I.$$

Therefore, the second condition in (8.40) is satisfied. Following the first condition specified in (8.39), we set

$$L = \begin{bmatrix} \dfrac{4\omega(4p\gamma^2\omega)}{(4p\gamma^2\omega)^2 - 1} \\[4mm] \dfrac{4\omega}{(4p\gamma^2\omega)^2 - 1} \end{bmatrix}. \tag{10.93}$$

The rest part of the proof can be completed by invoking Theorem 8.11.   □

Hence from the above lemma, we design the observer for the general disturbance model as

$$\dot{\hat{x}} = A\hat{x} + L(y - h(\hat{x})), \tag{10.94}$$

where $A = \begin{bmatrix} 0 & \omega \\ -\omega & 0 \end{bmatrix}$ and $H = [1\ 0]$ with the observer gain L as shown in (10.93).

Before introducing the control design, we need to examine the stability issues of the $z-$subsystem, and hence introduce a number of functions that are needed later for the control design and stability analysis of the entire system.

Lemma 10.10. *Assuming that the subsystem*

$$\dot{z} = f(z, v, y)$$

is ISS with state z and input y, characterised by an ISS pair (α, σ), and furthermore, $\alpha(s) = \mathcal{O}(a^2(s))$ as $s \to 0$, there exist a differentiable positive definite function $\tilde{V}(z)$ and a $\mathcal{K}_\infty$ function β satisfying $\beta(\|z\|) \geq a^2(z)$ such that

$$\dot{\tilde{V}}(z) \leq -\beta(\|z\|) + \bar{\sigma}(y). \tag{10.95}$$

where $\bar{\sigma}$ is a continuous function.

Proof. From Corollary 5.10, there exists a Lyapunov function $V_z(z)$ that satisfies that

$$\begin{aligned} \alpha_1(\|z\|) &\leq V_z(z) \leq \alpha_2(\|z\|) \\ \dot{V}_z(z) &\leq -\alpha(\|z\|) + \sigma(|y|), \end{aligned} \tag{10.96}$$

where α, α_1 and α_2 are class $\mathcal{K}_\infty$ functions, and σ is a class $\mathcal{K}$ function. Let β be a $\mathcal{K}_\infty$ function such that $\beta(\|z\|) \geq a^2(z)$ and $\beta(s) = \mathcal{O}(a^2(s))$ as $s \to 0$. Since $\beta(s) = \mathcal{O}(a^2(s)) = \mathcal{O}(\alpha(s))$ as $s \to 0$, there exists a smooth nondecreasing ($\mathcal{SN}$) function $\tilde{q}$ such that, $\forall r \in \mathbb{R}^+$

$$\frac{1}{2}\tilde{q}(r)\alpha(r) \geq \beta(r).$$

Let us define two functions

$$q(r) := \tilde{q}(\alpha_1^{-1}(r)),$$

$$\rho(r) := \int_0^r q(t)dt.$$

Define

$$\tilde{V}(z) := \rho(V(z)),$$

and it can be obtained that

$$
\begin{aligned}
\dot{\tilde{V}}(z) &\leq -q(V(z))\alpha(z) + q(V(z))\sigma(|y|) \\
&\leq -\frac{1}{2}q(V(z))\alpha(z) + q(\theta(|y|))\sigma(|y|) \\
&\leq -\frac{1}{2}q(\alpha_1(\|z\|))\alpha(z) + q(\theta(|y|))\sigma(|y|) \\
&= -\frac{1}{2}\tilde{q}(\|z\|)\alpha(z) + q(\theta(|y|))\sigma(|y|),
\end{aligned}
$$

where θ is defined as

$$\theta(r) := \alpha_2(\alpha^{-1}(2\sigma(r)))$$

for $r \in \mathbb{R}^+$. Let us define a smooth function $\bar{\sigma}$ such that

$$\bar{\sigma}(r) \geq q(\theta(|r|))\sigma(|r|)$$

for $r \in \mathbb{R}$ and $\bar{\sigma}(0) = 0$, and then we have established (10.95). □

Based on observer design presented in (10.94), we design the following internal model:

$$\dot{\eta} = A\eta + b^{-1}L\psi_0(y) + Lu - b^{-1}ALy - Lh(\eta - b^{-1}Ly), \tag{10.97}$$

where L is designed as in (10.93).

The control input is then designed as

$$u = -b^{-1}\left(\psi_0(y) + k_0 y + k_1 y + k_2\frac{\bar{\sigma}(y)}{y} + k_3\bar{\psi}(y)\right) + h(\eta - b^{-1}Ly), \tag{10.98}$$

where k_0 is a positive real constant, and

$$
\begin{aligned}
k_1 &= \kappa^{-1}b^2(\gamma + \|H\|)^2 + \frac{3}{4}, \\
k_2 &= 4\kappa^{-1}\|b^{-1}PL\|^2 + 2, \\
k_3 &= 4\kappa^{-1}\|b^{-1}PL\|^2 + \frac{1}{2}.
\end{aligned}
$$

For the stability of the closed-loop system, we have the following theorem.

Theorem 10.11. *For a system in the form shown in (10.91), if*

- *feedforward term $\mu(v)$ can be modelled as the output of a system in the format shown in (10.93) and the Lipschitz constant of the output nonlinear function γ satisfies $\gamma < \frac{1}{\sqrt{2}}$*

- the subsystem $\dot{z} = f(z, v, y)$ is ISS with state z and input y, characterized by ISS pair (α, σ), and furthermore, $\alpha(s) = \mathcal{O}(a^2(s))$ as $s \to 0$

the output feedback control design with the internal model (10.97) and the control input (10.98) ensures the boundedness of all the variables of the closed-loop system and the asymptotic convergence to zero of the state variables z and y and the estimation error $(w - \eta + b^{-1}Ly)$.

Proof. Let

$$\xi = w - \eta + b^{-1}Ly.$$

It can be obtained from (10.97) that

$$\dot{\xi} = (A - LH)\xi + b^{-1}L(h_1(w) - h_1(w - \xi)) + b^{-1}La(z) + b^{-1}L\psi(y, v).$$

Let $V_w = \xi^T P \xi$. It can be obtained that

$$
\begin{aligned}
\dot{V}_w(\xi) &\leq -\kappa \|\xi\|^2 + 2|\xi^T b^{-1} PLa(z)| + 2|\xi^T b^{-1} PL\psi(y, v)| \\
&\leq -\frac{1}{2}\kappa \|\xi\|^2 + 2\kappa^{-1}\|b^{-1}PL\|^2(a^2(z) + |\psi(y, v)|^2) \\
&\leq -\frac{1}{2}\kappa \|\xi\|^2 + (k_2 - 2)\beta(\|z\|) + \left(k_3 - \frac{1}{2}\right)y\bar{\psi}(y) \qquad (10.99)
\end{aligned}
$$

where $\kappa = \frac{1}{2\gamma^2} - 1$.

Based on the control input (10.98), we have

$$\dot{y} = -k_0 y - k_1 y - k_2 \frac{\bar{\sigma}(y)}{y} - k_3 \bar{\psi}(y) + a(z) + \psi(y, v) + b(h(w - \xi) - h(w)).$$

Let $V_y = \frac{1}{2}y^2$. It follows from the previous equation that

$$
\begin{aligned}
\dot{V}_y &= -(k_0 + k_1)y^2 - k_2\bar{\sigma}(y) - k_3 y\bar{\psi}(y) + ya(z) + y\psi(y, v) + yb(h(w - \xi) - h(w)) \\
&\leq -k_0 y^2 - k_2\bar{\sigma}(y) - \left(k_3 - \frac{1}{2}\right)y\bar{\psi}(y) + \beta(\|z\|) + \frac{1}{4}\kappa\|\xi\|^2. \quad (10.100)
\end{aligned}
$$

Let us define a Lyapunov function candidate for the entire closed-loop system as

$$V = V_y + V_w + k_2 \tilde{V}_z.$$

Following the results shown in (10.96), (10.99) and (10.100), we have

$$\dot{V} \leq -k_0 y^2 - \frac{1}{4}\kappa\|\xi\|^2 - \beta(\|z\|).$$

Therefore, we can conclude that closed-loop system is asymptotically stable with respect to the state variables y, z and the estimation error ξ.

Several types of disturbance rejection and output regulation problems can be converted to the form (10.91). In this section, we show two examples. The first example deals with rejection of general periodic disturbances, and the second example demonstrates how the proposed method can be used for output regulation.

Example 10.3. Consider

$$
\begin{aligned}
\dot{x}_1 &= x_2 + \phi_1(x_1) + b_1 u \\
\dot{x}_2 &= \phi_2(x_1) + v(w) + b_2 u \\
\dot{w} &= Aw \\
y &= x_1,
\end{aligned}
\tag{10.101}
$$

where $y \in \mathbb{R}$ is the measurement output; $\phi_i : \mathbb{R} \to \mathbb{R}$, for $i = 1, 2$, are continuous nonlinear functions; $v : \mathbb{R}^2 \to \mathbb{R}$ is a nonlinear function which produces a periodic disturbance from the exosystem state w; and b_1 and b_2 are known constants with the same sign, which ensures that stability of the zero dynamics. The control objective is to design an output feedback control input to ensure the overall stability of the entire system, and the asymptotic convergence to zero of the measurement output. The system shown in (10.101) is not in the form of (10.91) and the disturbance is not matched. We will show that the problem can be transformed to the problem considered in the previous section.

Let

$$
\bar{z} = x_2 - \frac{b_2}{b_1} x_1.
$$

In the coordinates $(y, \bar{z})$, we have

$$
\dot{y} = \bar{z} + \frac{b_2}{b_1} y + \phi_1(y) + b_1 u
$$

$$
\dot{\bar{z}} = -\frac{b_2}{b_1} \bar{z} + \phi_2(y) - \frac{b_2}{b_1} \phi_1(y) - \left(\frac{b_2}{b_1}\right)^2 y + v(w).
$$

Consider

$$
\dot{\pi}_z = -\frac{b_2}{b_1} \pi_z + v(w).
$$

It can be shown that there exists a steady-state solution, and furthermore, we can express the solution as a nonlinear function of w, denoted by $\pi_z(w)$. Let us introduce another state transformation with $z = \bar{z} - \pi_z(w)$. We then have

$$
\dot{y} = z + \frac{b_2}{b_1} y + \phi_1(y) + b_1(u + b_1^{-1} \pi_z(w))
$$

$$
\dot{z} = -\frac{b_2}{b_1} z + \phi_2(y) - \frac{b_2}{b_1} \phi_1(y) - \left(\frac{b_2}{b_1}\right)^2 y.
\tag{10.102}
$$

Comparing (10.102) with (10.91), we have

$$a(z) = z,$$

$$\psi(y) = \frac{b_2}{b_1}y + \phi_1(y),$$

$$b = b_1,$$

$$h(w) = -b_1^{-1}\pi_z(w),$$

$$f(z, v, y) = \phi_2(y) - \frac{b_2}{b_1}\phi_1(y) - \left(\frac{b_2}{b_1}\right)^2 y.$$

From $a(z) = z$, we can set $\beta(\|z\|) = \|z\|^2 = z^2$.

It can be shown that the second condition of Theorem 10.11 is satisfied by (10.102). Indeed, let $V(z) = \frac{1}{2}z^2$, and we have

$$\dot{V}_z = -\frac{b_2}{b_1}z^2 + z\left(\phi_2(y) - \frac{b_2}{b_1}\phi_1(y) - \left(\frac{b_2}{b_1}\right)^2 y\right)$$

$$\leq -\frac{1}{2}\frac{b_2}{b_1}z^2 + \frac{1}{2}\frac{b_1}{b_2}\left(\phi_2(y) - \frac{b_2}{b_1}\phi_1(y) - \left(\frac{b_2}{b_1}\right)^2 y\right)^2.$$

Let

$$\tilde{V}_z = 2\frac{b_1}{b_2}V_z$$

and finally we have

$$\dot{\tilde{V}}_z \leq -\beta(|z|) + \left(\frac{b_1}{b_2}\right)^2\left(\phi_2(y) - \frac{b_2}{b_1}\phi_1(y) - \left(\frac{b_2}{b_1}\right)^2 y\right)^2. \tag{10.103}$$

It can be seen that there exists a class $\mathcal{K}$ function $\sigma(|y|)$ to dominate the second term on the right-hand side of (10.103), and the z-subsystem is ISS. For the control design, we can take

$$\bar{\sigma}(y) = \left(\frac{b_1}{b_2}\right)^2\left(\phi_2(y) - \frac{b_2}{b_1}\phi_1(y) - \left(\frac{b_2}{b_1}\right)^2 y\right)^2.$$

The rest part of the control design follows the steps shown earlier.

For the simulation study, we set the periodic disturbance as a square wave. For convenience, we abuse the notations of $v(w(t))$ and $h(w(t))$ as $v(t)$ and $h(t)$. For v with t in one period, we have

$$v = \begin{cases} d, & 0 \leq t < \dfrac{T}{2}, \\ -d, & \dfrac{T}{2} \leq t < T, \end{cases} \tag{10.104}$$

where d is an unknown positive constant, denoting the amplitude. It can be obtained that

$$h = d\bar{h}(t),$$

where

$$\bar{h}(t) = \begin{cases} -\dfrac{1}{b_2}(1 - e^{-\frac{b_2}{b_1}t}) + \dfrac{1}{b_2}e^{-\frac{b_2}{b_1}t}\tanh\left(\dfrac{T\,b_2}{4\,b_1}\right), & 0 \le t < \dfrac{T}{2}, \\[2ex] \dfrac{1}{b_2}(1 + e^{-\frac{b_2}{b_1}t} - 2e^{\frac{b_2}{b_1}(\frac{T}{2}-t)}) + \dfrac{1}{b_2}e^{-\frac{b_2}{b_1}t}\tanh\left(\dfrac{T\,b_2}{4\,b_1}\right), & \dfrac{T}{2} \le t < T. \end{cases} \quad (10.105)$$

Eventually we have the matched periodic disturbance $h(w)$ given by

$$h(w) = \sqrt{w_1^2 + w_2^2}\,\bar{h}\left(\arctan\left(\frac{w_2}{w_1}\right)\right).$$

Note that $\sqrt{w_1^2 + w_2^2}$ decides the amplitude, which can be determined by the initial state of w.

In the simulation study, we set $T = 1, d = 10, \phi_1 = y^3, \phi_2 = y^2$ and $b_1 = b_2 = 1$. The simulation results are shown in Figures 10.6–10.9. It can be seen from Figure 10.6 that the measurement output converges to zero and the control input converges to a periodic function. In fact, the control input converges to $h(w)$ as shown in Figure 10.7.

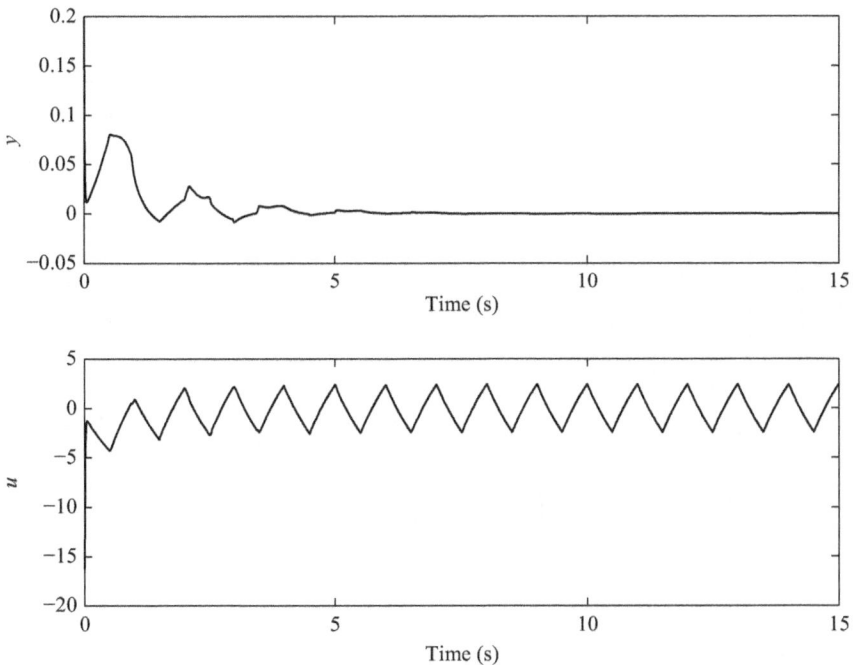

Figure 10.6 The system input and output

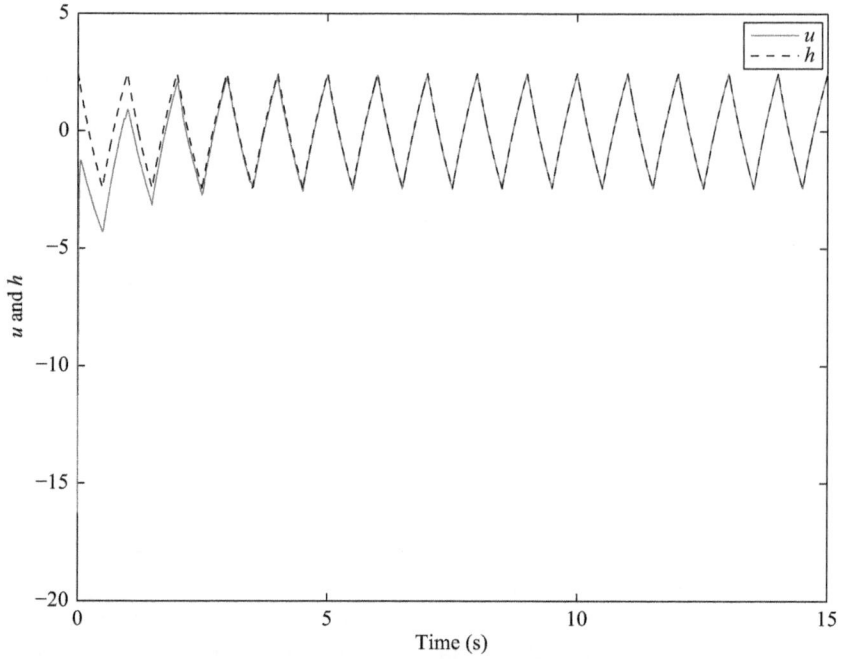

Figure 10.7   Control input and the equivalent input disturbance

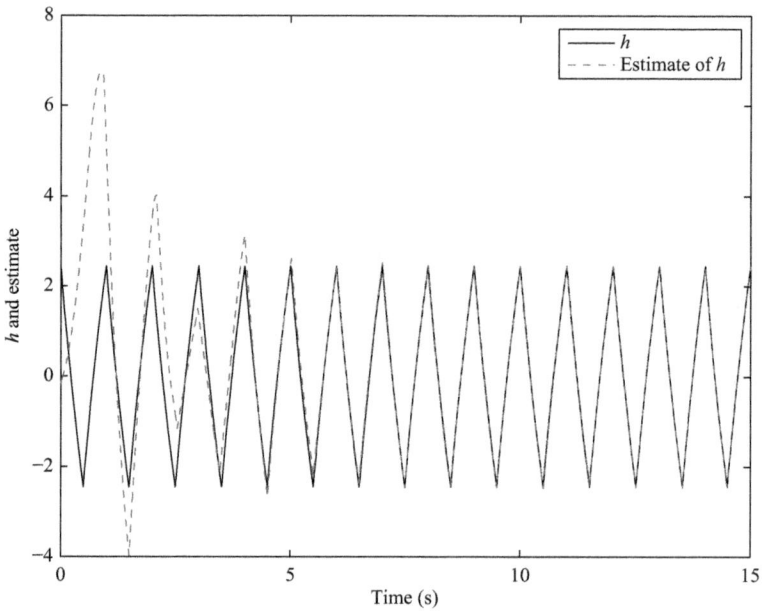

Figure 10.8   The equivalent input disturbance and its estimate

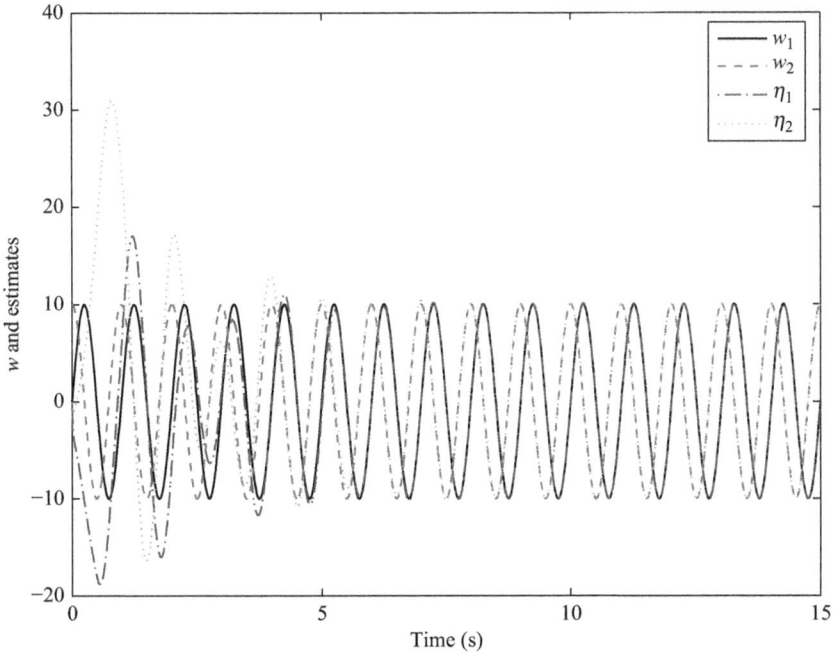

Figure 10.9 The exosystem states and the internal model states

As for the internal model and state estimation, it is clear from Figure 10.8 that the estimated equivalent input disturbance converges to $h(w)$, and η converges to w.

◁

Example 10.4. In this example, we briefly show that an output regulation problem can also be converted to the form in (10.91). Consider

$$
\begin{aligned}
\dot{x}_1 &= x_2 + (e^y - 1) + u \\
\dot{x}_2 &= (e^y - 1) + 2w_1 + u \\
\dot{w} &= Aw \\
y &= x_1 - w_1,
\end{aligned}
\tag{10.106}
$$

where $y \in \mathbb{R}$ is the measurement output and $w_1 = [1\ 0]w$. In this example, the measured output contains the unknown disturbance, unlike Example 10.3. The control objective remains the same, to design an output feedback control law to ensure the overall stability of the system and the convergence to zero of the measured output. The key step in the control design is to show that the system shown in (10.106) can be converted to the form as shown in (10.91).

Let

$$
\pi_z = \frac{1}{1 + \omega^2}[1\ -\omega]w,
$$

and it is easy to check that π_z satisfies

$$\dot{\pi}_z = -\pi_z + [1\ 0]w.$$

Let $z = x_2 - \pi_z - x_1$. It can be obtained that

$$\dot{y} = z + y + e^{w_1}(e^y - 1) + (u - h(w))$$
$$\dot{z} = -z - y + 2w_1,$$

where

$$h(w) = \frac{2 + \omega^2}{1 + \omega^2}[-1\ \omega]w - (e^{w_1} - 1).$$

It can be seen that we have transformed the system to the format as shown in (10.91) with $\psi(y, v) = e^{w_1}(e^y - 1)$.

To make $H = [1\ 0]$, we introduce a state transform for the disturbance model as

$$\zeta = \frac{2 + \omega^2}{1 + \omega^2}\begin{bmatrix} -1 & \omega \\ -\omega & -1 \end{bmatrix}w.$$

It can be easily checked that $\dot{\zeta} = A\zeta$. The inverse transformation is given as

$$w = \frac{1}{2 + \omega^2}\begin{bmatrix} -1 & -\omega \\ \omega & -1 \end{bmatrix}\zeta.$$

With ζ as the disturbance state, we can write the transformed system as

$$\dot{y} = z + y + e^{1/(2+\omega^2)[-1-\omega]\zeta}(e^y - 1) + (u - h(\zeta))$$
$$\dot{z} = -z - y + \frac{2}{2 + \omega^2}[-1\ -\omega]\zeta \tag{10.107}$$
$$\dot{\zeta} = A\zeta,$$

where

$$h(\zeta) = \zeta_1 - (e^{1/(2+\omega^2)[-1-\omega]\zeta} - 1).$$

Note that $\frac{e^y - 1}{y}$ is a continuous function, and we can take $\bar{\psi}(y) = d_0 y(\frac{e^y-1}{y})^2$ where d_0 is a positive real constant depending on the frequency and the knowledge of an upper limit of the disturbance amplitude. The control design presented in the previous section can then be applied to (10.108). Simulation studies were carried out with the results shown in Figures 10.10–10.13.

◁

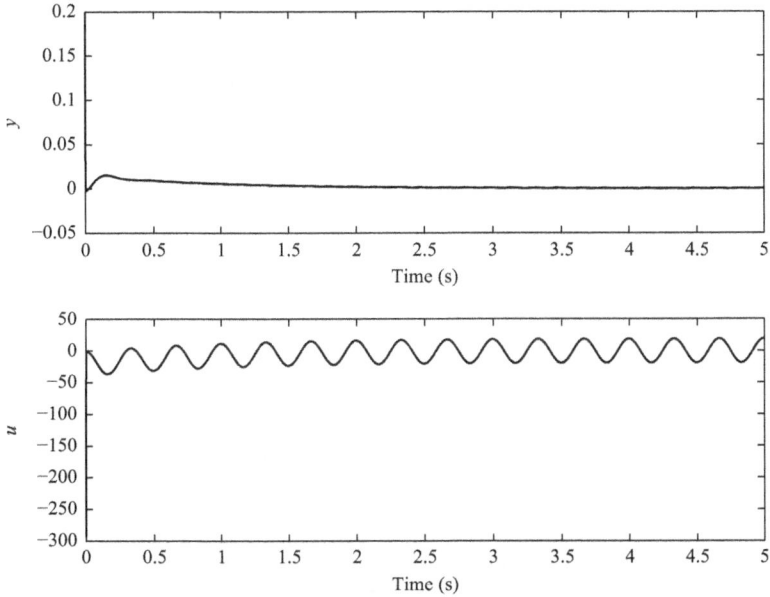

Figure 10.10 The system input and output

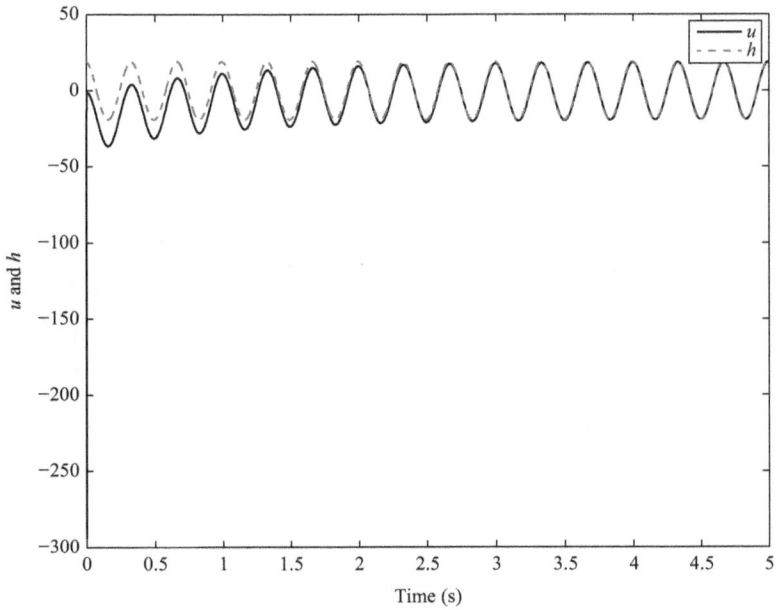

Figure 10.11 Control input and the equivalent input disturbance

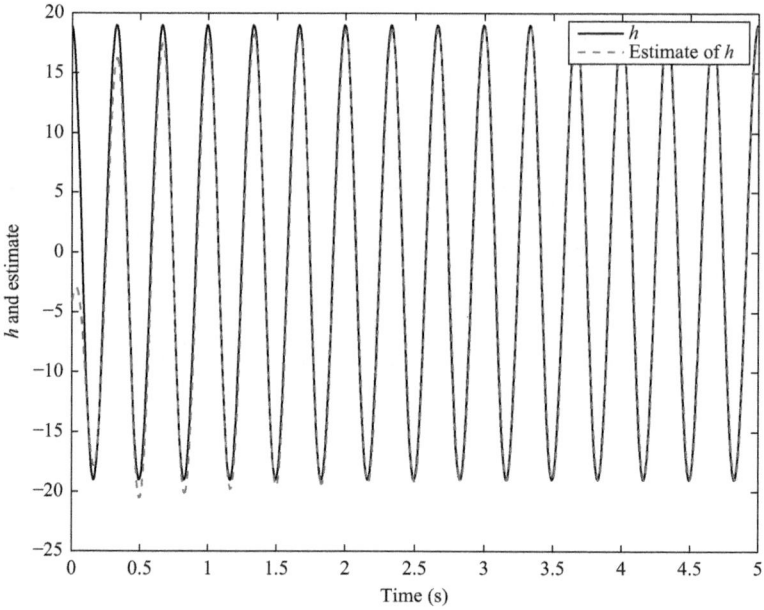

Figure 10.12   The equivalent input disturbance and its estimate

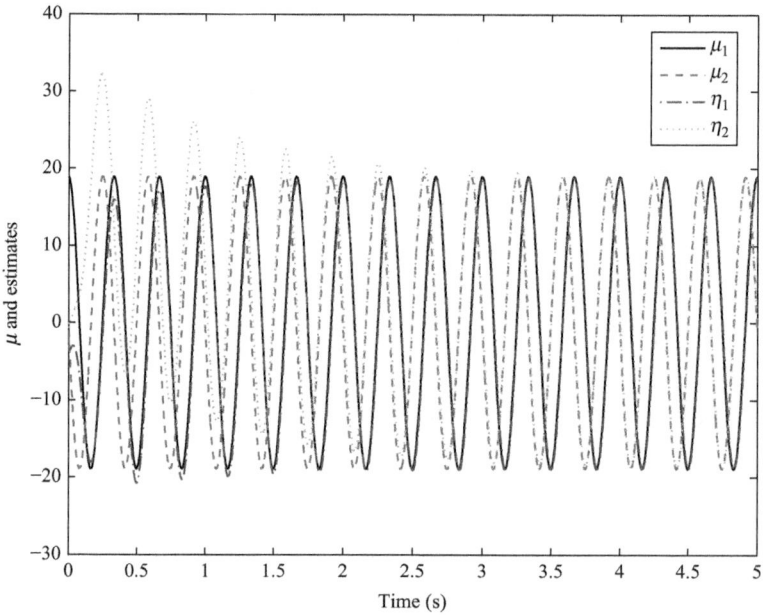

Figure 10.13   The exosystem states and the internal model states

Chapter 11

Control applications

In this chapter, we will address a few issues about control applications. Several methods of disturbance rejection are presented in Chapter 10, including rejection of general periodic disturbances. A potential application can be the estimation and rejection of undesirable harmonics in power systems. Harmonics, often referred to high-order harmonics in power systems, are caused by nonlinearities in power systems, and the successful rejection depends on accurate estimation of amplitudes and phase of harmonics. We will show an iterative estimation method based on a new observer design method.

There are tremendous nonlinearities in biological systems, and there have been some significant applications of nonlinear system analysis and control methods in system biology. We will show a case that nonlinear observer and control are applied to circadian rhythms. A Lipschitz observer is used to estimate unknown states, and backstepping control design is then applied to restore circadian rhythms.

Most of the control systems are implemented in computers or other digital devices which are in discrete-time in nature. Control implementation using digital devices inevitably ends with sample-data control. For linear systems, the sampled systems are still linear, and the stability of the sampled-date system can be resolved in stability analysis using standard tools of linear systems in discrete-time. However, when a nonlinear system is sampled, the system description may not have a closed form, and the structure cannot be preserved. The stability cannot be assumed for a sampled-data implementation of nonlinear control strategy. We will show that for certain nonlinear control schemes, the stability can be preserved by fast sampling in the last section.

11.1 Harmonics estimation and rejection in power distribution systems

There is a steady increase in nonlinear loading in power distribution networks due to the increase in the use of electrical cars, solar panels for electricity generation, etc. Nonlinear loading distorts the sinusoidal waveforms of voltage and current in networks. The distorted waveforms are normally still periodic, and they can be viewed as general periodic disturbances. Based on Fourier series, a general periodic signal can be decomposed into sinusoidal functions with multiple frequencies of the base frequency. The components with high frequencies are referred to as harmonics. In other words, they are individual frequency modes at frequencies which are multiples of the base frequency.

Harmonics are undesirable in power distribution networks for various reasons. They occupy the limited power capacity, and can be harmful to electrical and electronic devices. Often in power distribution networks, active and passive power filters are used to reduce the undesired harmonics. Effective rejection of harmonics depends on accurate phase and amplitude estimation of individual frequency modes. In a power distribution network, harmonics with certain frequencies are more critical for rejection than others. For example the double frequency harmonics normally disappear in the system due to a particular connection in power distribution networks, and third-order harmonics would be most important to reject, perhaps due to high-attenuation high frequencies of distribution networks. Based on this discussion, estimation and rejection of specific frequency modes will be of interest to power distribution networks.

As explained above, harmonics appear as general periodic signals. Rejection of general periodic disturbances is discussed in Chapter 10, for matched cases, i.e., the input is in exactly the same location as the disturbances, which is a restriction to certain applications, even though we may be able to convert an unmatched case to a matched one for a certain class of nonlinear systems. The method in Chapter 10 does not apply to rejection of individual frequency modes. We will show the conversion of an unmatched disturbance to a matched one, and show how individual frequency modes can be estimated and rejected.

We will consider a class of nonlinear systems that has a similar structure to the systems considered in the earlier chapters, but with unmatched general periodic disturbances. We will show how an equivalent input, also a periodic disturbance, can be obtained for this system, and then propose an estimation and rejection method for the system. The presentation at this stage is not exactly based on power systems, but the proposed methods can be directly applied to power distribution systems. We will show an example of estimation of harmonics using the proposed method for individual frequency modes.

11.1.1   System model

Consider a SISO nonlinear system which can be transformed into the output feedback form

$$\dot{\zeta} = A_c \zeta + \phi(y) + bu + dw$$
$$y = C\zeta, \tag{11.1}$$

with

$$A_c = \begin{bmatrix} 0 & 1 & 0 & \cdots & 0 \\ 0 & 0 & 1 & \cdots & 0 \\ \vdots & \vdots & \vdots & \ddots & \vdots \\ 0 & 0 & 0 & \cdots & 1 \\ 0 & 0 & 0 & \cdots & 0 \end{bmatrix}, \quad b = \begin{bmatrix} 0 \\ \vdots \\ 0 \\ b_\rho \\ \vdots \\ b_n \end{bmatrix}, \quad C = \begin{bmatrix} 1 \\ 0 \\ \vdots \\ 0 \end{bmatrix}^T, \quad d = \begin{bmatrix} d_1 \\ \vdots \\ d_n \end{bmatrix},$$

where $\zeta \in \mathbb{R}^n$ is the state vector; $u \in \mathbb{R}$ is the control input; $\phi : \mathbb{R} \to \mathbb{R}^n$ with $\phi(0) = 0$ is a nonlinear function with element ϕ_i being differentiable up to the $(n - i)$th order;

$b \in \mathbb{R}^n$ is a known constant Hurwitz vector, with $b_\rho \neq 0$, which implies the relative of the system is ρ; d is an unknown constant vector; and $w \in R$ is a bounded periodic disturbance, which has continuous derivative up to the order of $\max\{\rho - \iota, 0\}$ with ι being the index of the first non-zero element of vector d.

Remark 11.1. The disturbance-free system of (11.1) is in the output feedback form discussed in the previous chapters. With the disturbance, it is similar to the system (10.1), with the only difference that w is a general periodic disturbance. It is also different from the system (10.91) of which the disturbance is in the matched form. ◁

Remark 11.2. The continuity requirement specified for the general periodic disturbance w in (11.1) is for the existence of a continuous input equivalent disturbance and a continuous invariant manifold in the state space. For the case of $\rho < \iota$, we may allow disturbance to have finite discontinuous points within each period, and for each of the discontinuous points, the left and right derivatives exist. ◁

Remark 11.3. The minimum phase assumption is needed for the convenience of presentation of the equivalent input disturbance and the control design based on backstepping. It is not essential for control design, disturbance rejection or disturbance estimation. We could allow the system to be non-minimum phase, provided that there exists a control design for the disturbance-free system which renders the closed-loop system exponentially stable. ◁

The zero dynamics of (11.1) is linear. To obtain the equivalent input disturbance, we need a result for steady-state response for stable linear systems.

Lemma 11.1. *For a linear system*

$$\dot{x} = Ax + bw, \tag{11.2}$$

where $x \in \mathbb{R}^n$ is the system state, A is Hurwitz, $b \in \mathbb{R}^n$ is a constant vector and w is the periodic disturbance with period T, the steady state under the input of the periodic disturbance w is given by

$$x_s(t) = \int_0^t e^{A(t-\tau)}bw(\tau)d\tau + e^{At}(I - e^{AT})^{-1}e^{AT}W_T, \tag{11.3}$$

where W_T is a constant vector in R^n given by

$$W_T = \int_0^T e^{-A\tau}bw(\tau)d\tau.$$

Proof. The state response to the input w is given by

$$x(t) = e^{At}x(0) + \int_0^t e^{A(t-\tau)}bw(\tau)d\tau.$$

Considering the response after a number of periods, we have

$$x(NT + t) = e^{A(NT+t)}x(0) + e^{A(NT+t)} \int_0^{NT+t} e^{-A\tau} bw(\tau)d\tau$$

$$= e^{A(NT+t)}x(0) + e^{A(NT+t)} \sum_{i=0}^{N-1} \int_{iT}^{(i+1)T} e^{-A\tau} bw(\tau)d\tau$$

$$+ e^{A(NT+t)} \int_{NT}^{NT+t} e^{-A\tau} bw(\tau)d\tau,$$

where N is a positive integer. Since $w(t)$ is a periodic function, we have

$$\int_{iT}^{(i+1)T} e^{-A\tau} bw(\tau)d\tau = \int_{iT}^{(i+1)T} e^{-A\tau} bw(\tau - iT)d\tau$$

$$= \int_0^T e^{-A(iT+\tau)} bw(\tau)d\tau$$

$$= e^{-iAT} \int_0^T e^{-A\tau} bw(\tau)d\tau.$$

Therefore, we have

$$x(NT + t) = e^{A(NT+t)}x(0) + e^{At} \sum_{i=0}^{N-1} e^{A(N-i)T} \int_0^T e^{-A\tau} bw(\tau)d\tau$$

$$+ \int_0^t e^{A(t-\tau)} bw(\tau)d\tau. \tag{11.4}$$

The steady-state response in (11.3) is obtained by taking the limit of (11.4) for $t \to \infty$.   □

To obtain the equivalent input disturbance, we need to introduce state transformation to (11.1). To extract the zero dynamics, we introduce a partial state transformation for system (11.1) as

$$z = \begin{bmatrix} \zeta_{\rho+1} \\ \vdots \\ \zeta_n \end{bmatrix} - \sum_{i=1}^{\rho} B^{\rho-i}\bar{b}\zeta_i,$$

where

$$B = \begin{bmatrix} -b_{\rho+1}/b_\rho & 1 & \cdots & 0 \\ \vdots & \vdots & \ddots & \vdots \\ -b_{n-1}/b_\rho & 0 & \cdots & 1 \\ -b_n/b_\rho & 0 & \cdots & 0 \end{bmatrix}, \quad \bar{b} = \begin{bmatrix} b_{\rho+1}/b_\rho \\ \vdots \\ b_n/b_\rho \end{bmatrix}.$$

The dynamics with the coordinates $(\zeta_1, \ldots, \zeta_\rho, z)$ can be obtained as

$$\dot{\zeta}_i = \zeta_{i+1} + \phi_i(y) + d_i w, \quad i = 1, \ldots, \rho - 1$$

$$\dot{\zeta}_\rho = z_1 + \sum_{i=1}^{\rho} r_i \zeta_i + \phi_\rho(y) + d_\rho w + b_\rho u \quad (11.5)$$

$$\dot{z} = Bz + \phi_z(y) + d_z w,$$

where

$$r_i = (B^{\rho-i}\bar{b})_i, \quad \text{for } i = 1, \ldots, \rho,$$

$$\phi_z(y) = \begin{bmatrix} \phi_{\rho+1} \\ \vdots \\ \phi_n \end{bmatrix} - \sum_{i=1}^{\rho} B^{\rho-i}\bar{b}\phi_i + B^\rho \bar{b}y,$$

and

$$d_z = \begin{bmatrix} d_{\rho+1} \\ \vdots \\ d_n \end{bmatrix} - \sum_{i=1}^{\rho} B^{\rho-i}\bar{b}d_i.$$

The periodic trajectory and the equivalent input disturbance can be found using the system in the coordinate $(\zeta_1, \ldots, \zeta_\rho, z)$. Since the system output y does not contain the periodic disturbance, we have the invariant manifold for $\pi_1 = 0$. From Lemma 11.1 which is used for the result of the steady-state response of linear systems to the periodic input, we have, for $0 \leq t < T$,

$$\pi_z(t) = \int_0^t e^{B(t-\tau)} d_z w(\tau) d\tau + e^{Bt}(I - e^{BT})^{-1} e^{BT} W_T$$

with $W_T = \int_0^T e^{-B\tau} d_z w(\tau) d\tau$. From the first equation of (11.6), we have, for $i = 1, \ldots, \rho - 1$

$$\pi_{i+1}(t) = \frac{d\pi_i(t)}{dt} - d_i w.$$

Based on the state transformation introduced earlier, we can use its inverse transformation to obtain

$$\begin{bmatrix} \pi_{\rho+1} \\ \vdots \\ \pi_n \end{bmatrix} = \pi_z + \sum_{i=1}^{\rho} B^{\rho-i}\bar{b}\pi_i.$$

Therefore, the periodic trajectory in the state space is obtained as

$$\pi = [\pi, \ldots, \pi_n]^T. \quad (11.6)$$

Finally, the equivalent input disturbance μ is given by

$$\mu = \frac{1}{b_\rho} \left(\frac{d\pi_\rho(t)}{dt} - \pi_{z,1} - \sum_{i=1}^{\rho} r_i \pi_{z,i} - d_\rho w \right). \tag{11.7}$$

Let $x = \zeta - \pi$ denote the difference between the state variable ζ and the periodic trajectory.

The periodic trajectory, π, plays a similar role as the invariant manifold in the set-up for the rejection of disturbances generated from linear exosystems. For this, we have the following result.

Theorem 11.2. *For the general periodic disturbance w in (11.1), the periodic trajectory given in (11.6) and the equivalent input disturbance given in (11.7) are well defined and continuous, and the difference between the state variable (11.1) and the periodic trajectory, denoted by $x = \zeta - \pi$, satisfies the following equation:*

$$\begin{aligned} \dot{x} &= A_c x + \phi(y) + b(u - \mu) \\ y &= Cx. \end{aligned} \tag{11.8}$$

The control design and disturbance rejection will be based on (11.8) instead of (11.1).

Remark 11.4. The control design and disturbance rejection only use the output y, with no reference to any of other state of the system. Therefore, there is no difference whether we refer to (11.1) or (11.8) for the system, because they have the same output. The format in (11.8) shows that there exist an invariant manifold and an equivalent input disturbance. However, the proposed control design does not depend on any information of μ, other than its period, which is the same as the period of w. In other words, control design only relies on the form shown in (11.1). The form shown in (11.8) is useful for the analysis of the performance of the proposed control design, including the stability. In this section, we start our presentation from (11.1) rather than (11.8) in order to clearly indicate the class of the systems to which the proposed control design can be applied, without the restriction to the rejection of matched disturbances. ◁

11.1.2   Iterative observer design for estimating frequency modes in input

We will propose an iterative observer design method to estimate specific frequency modes in the input to a linear system from its output. For the convenience of discussion, we have the following definitions.

Definition 11.1. A T-periodic function f is said to be orthogonal to a frequency mode with frequency ω_k if

$$\int_0^T f(\tau)\sin\omega_k\tau d\tau = 0, \tag{11.9}$$

$$\int_0^T f(\tau)\cos\omega_k\tau d\tau = 0. \tag{11.10}$$

Definition 11.2. A function f is said to be asymptotically orthogonal to a frequency mode with frequency ω_k if

$$\lim_{t\to\infty}\int_t^{t+T} f(\tau)\sin\omega_k\tau d\tau = 0, \tag{11.11}$$

$$\lim_{t\to\infty}\int_t^{t+T} f(\tau)\cos\omega_k\tau d\tau = 0. \tag{11.12}$$

We consider a stable linear system

$$\begin{aligned}\dot{x} &= Ax + b\mu \\ y &= Cx,\end{aligned} \tag{11.13}$$

where $x \in \mathbb{R}^n$ is the state variable; and y and $u \in \mathbb{R}$ are output and input, the matrices A, b and C are with proper dimensions; and the system is stable and observable with the transfer function $\mathbf{Q}(s) = C(sI - A)^{-1}b$. The problem considered in this subsection is to create a signal such that specific frequency modes can then be removed from the input μ, which is a general periodic disturbance described by

$$\mu(t) = \sum_{k=1}^{\infty} a_k\sin(\omega_k t + \phi_k),$$

where $\omega_k = \frac{2\pi k}{T}$ and a_k and ϕ_k are the amplitude and phase angle of the mode for frequency ω_k. The dynamics for a single frequency mode can be described as

$$\dot{w}_k = S_k w_k, \tag{11.14}$$

where

$$S_k = \begin{bmatrix} 0 & \omega_k \\ -\omega_k & 0 \end{bmatrix}.$$

For a single frequency mode with the frequency ω_k as the input denoted by $\mu_k = a_k\sin(\omega_k t + \phi_k)$, its output in y, denoted by y_k, is a sinusoidal function with the same frequency, if we only consider the steady-state response. In fact, based on the frequency response, we have the following result.

Lemma 11.3. *Consider a stable linear system* (A, b, C) *with no zero at* $j\omega_k$ *for any integer* k. *For the output* y_k *of a single frequency mode with the frequency* ω_k, *there exists an initial state* $w_k(0)$ *such that*

$$y_k = g^T w_k, \tag{11.15}$$

where $g = [1 \ 0]^T$ and w_k is the state of (11.14). Furthermore, the input μ_k for this frequency mode can be expressed by

$$\mu_k = g_k^T w_k, \tag{11.16}$$

where

$$g_k = \frac{1}{m_k} \begin{bmatrix} \cos\theta_k & \sin\theta_k \\ -\sin\theta_k & \cos\theta_k \end{bmatrix} g := Q_k g \tag{11.17}$$

with $\theta_k = \angle Q(j\omega_k)$ and $m_k = |Q(j\omega_k)|$.

Proof. For the state model (11.14), we have

$$e^{S_k t} = \begin{bmatrix} 1 & 1 \\ j & -j \end{bmatrix} \begin{bmatrix} e^{j\omega_k t} & \\ & e^{j\omega_k t} \end{bmatrix} \begin{bmatrix} 1 & 1 \\ j & -j \end{bmatrix}^{-1}$$

$$= \begin{bmatrix} \cos(\omega_k t) & \sin(\omega_k t) \\ -\sin(\omega_k t) & \cos(\omega_k t) \end{bmatrix}.$$

For the single frequency mode ω_k, the output is given by

$$y_k = m_k a_k \sin(\omega_k t + \phi_k + \theta_k).$$

With $g = [1 \ 0]^T$, we have

$$w_k(0) = \begin{bmatrix} m_k a_k \sin(\phi_k + \theta_k) \\ m_k a_k \cos(\phi_k + \theta_k) \end{bmatrix}$$

and

$$w_k(t) = e^{S_k t} w_k(0) = \begin{bmatrix} m_k a_k \sin(\omega_k t + \phi_k + \theta_k) \\ m_k a_k \cos(\omega_k t + \phi_k + \theta_k) \end{bmatrix}$$

such that

$$y_k = g^T e^{S_k t} w_k(0) = g^T w_k(t).$$

Considering the gain and phase shift of $Q(s)$, we have

$$\mu_k = \frac{1}{m_k} g^T e^{S_k(t-(\theta_k/\omega_k))} w_k(0)$$

$$= \frac{1}{m_k} g^T e^{-(\theta_k/\omega_k)S_k} e^{S_k t} w_k(0).$$

Hence, we have

$$g_k = \frac{1}{m_k} e^{-(\theta_k/\omega_k)S_k^T} g$$

and therefore

$$Q_k = \frac{1}{m_k} e^{-(\theta_k/\omega_k)S_k^T} = \frac{1}{m_k} e^{(\theta_k/\omega_k)S_k} = \frac{1}{m_k} \begin{bmatrix} \cos\theta_k & \sin\theta_k \\ -\sin\theta_k & \cos\theta_k \end{bmatrix}.$$

For a stable single-input linear system, if the input is a T-periodic signal that is orthogonal to a frequency ω_k, the steady state, as shown earlier, is also T-periodic. Furthermore, we have the following results.

Lemma 11.4. *If the input to a stable single-input linear system (A, b) is T-periodic signal that is orthogonal to a frequency mode ω_k, for any positive integer k, the steady state is orthogonal to the frequency mode and the state variable is asymptotically orthogonal to the frequency mode. Furthermore, if the linear system (A, b, C) has no zero at $j\omega_k$, the steady-state output is orthogonal to the frequency mode ω_k if and only if the input to the system is orthogonal to the frequency mode.*

Proof. We denote the input as μ, and the state variable x satisfies

$$\dot{x} = Ax + b\mu.$$

Since μ is T-periodic, the steady-state solution of the above state equation, denoted by x_s, is also T-periodic and

$$\dot{x}_s = Ax_s + b\mu.$$

Let

$$J_k = \int_0^T x_s(\tau) \sin \omega_k \tau \, d\tau.$$

Using integration by part, we have

$$J_k = -\omega_k^{-1} \int_0^T x_s d \cos \omega_k \tau$$

$$= \omega_k^{-1} \int_0^T \dot{x}_s \cos \omega_k d\tau$$

$$= \omega_k^{-1} \int_0^T (Ax_s + b\mu(t)) \cos \omega_k \tau \, d\tau$$

$$= \omega_k^{-1} A \int_0^T x_s \cos \omega_k \tau \, d\tau$$

$$= \omega_k^{-2} A \int_0^T x_s d \sin \omega_k \tau$$

$$= -\omega_k^{-2} A \int_0^T (Ax_s + b\mu(t)) \sin \omega_k \tau \, d\tau$$

$$= -\omega_k^{-2} A_k^2 J_k.$$

Hence, we have

$$(\omega_k^2 I + A^2) J_k = 0.$$

Since A is a Hurwitz matrix which cannot have $\pm\omega_k j$ as its eigenvalues, we conclude $J_k = 0$. Similarly we can establish

$$\int_0^T x_s(\tau)\cos\omega_k\tau d\tau = 0$$

and therefore x_s is orthogonal to the frequency mode ω_k.

If we denote $e_x = x - x_s$, we have $\dot{e}_x = Ae_x$. It is clear that e_x exponentially converges to zero, and therefore we can conclude that x is asymptotically orthogonal to the frequency mode ω_k. This completes the proof of the first part.

For the second part of the lemma, the 'if' part follows directly from the first part of the lemma, and we now establish the 'only if' part by seeking a contradiction. Suppose that the output y is orthogonal to the frequency mode ω_k, and the input μ is not. In this case, it can be shown in a similar way as in the proof Lemma 11.3, that there exists a proper initial condition for $w_k(0)$ such that $\mu - g_k^T w_k$ is orthogonal to ω_k. Since the system is linear, we can write

$$y = y_\perp + y_w,$$

where $y_\perp$ denotes the steady-state output generated by $\mu - g_k^T w_k$ and y_w generated by $g_k^T w_k$. We have $y_\perp$ orthogonal to the frequency mode ω_k. However, y_w would be a sinusoidal function with frequency ω_k and it is definitely not orthogonal. Thus we conclude y is not orthogonal to ω_k, which is a contradiction. Therefore, μ must be orthogonal to ω_k if y is. This completes the proof.   □

To remove frequency modes in the input μ is to find an estimate $\hat{\mu}$ such that $\mu - \hat{\mu}$ does not contain those frequency modes asymptotically. For a single frequency mode with frequency ω_k, the task to obtain a $\hat{\mu}$ is accomplished by

$$\begin{aligned}
\dot{\hat{w}}_k &= S_k\hat{w}_k + l_k(y(t) - g^T\hat{w}_k) \\
\hat{\mu} &= (Q_k g)^T\hat{w}_k = g_k^T\hat{w}_k,
\end{aligned} \tag{11.18}$$

where l_k is chosen such that $S_k - l_k g^T$ is Hurwitz. For this observer, we have a useful result stated in the following lemma.

Lemma 11.5. *For any positive integer k, with the observer as designed in (11.18), $(\mu(\tau) - g_k^T\hat{w}_k)$ is asymptotically orthogonal to the frequency mode ω_k, i.e.,*

$$\lim_{t\to\infty}\int_t^{t+T}(\mu(\tau) - g_k^T\hat{w}_k)\sin\omega_k\tau d\tau = 0, \tag{11.19}$$

$$\lim_{t\to\infty}\int_t^{t+T}(\mu(\tau) - g_k^T\hat{w}_k)\cos\omega_k\tau d\tau = 0. \tag{11.20}$$

Proof. Consider the steady-state output y of input μ. From Lemma 11.3, there exists an initial state $w_k(0)$ such that

$$\int_0^T (y(\tau) - g^T w_k(\tau)) \sin \omega_k \tau d\tau = 0,$$

$$\int_0^T (y(\tau) - g^T w_k(\tau)) \cos \omega_k \tau d\tau = 0,$$

which implies that $\mu - g_k^T w_k$ is orthogonal to the frequency mode ω_k, again based on Lemma 11.3.

Let $\tilde{w}_k = w_k - \hat{w}_k$. The dynamics of $\tilde{w}_k$ can be obtained from (11.14) and (11.18) as

$$\dot{\tilde{w}}_k = \bar{S}_k \tilde{w}_k - l_k(y - g^T w_k), \tag{11.21}$$

where $\bar{S}_k = S_k - l_k g^T$. Note that $\bar{S}_k$ is a Hurwitz matrix and $(y - g^T w_k)$ is a T-periodic signal. There exists a periodic steady-state solution of (11.21) such that

$$\dot{\pi}_k = \bar{S}_k \pi_k - l_k(y - g^T w_k).$$

From Lemma 11.4, π_k is orthogonal to the frequency mode ω_k because $(y - g^T w_k)$ is. Let $e_k = \tilde{w}_k - \pi_k$. We have

$$\dot{e}_k = \bar{S}_k e_k,$$

which implies that e_k exponentially converges to zero. The observer state $\hat{w}_k$ can be expressed as

$$\hat{w}_k = w_k - \pi_k - e_k.$$

Therefore, (11.19) and (11.20) can be established, and this completes the proof. $\qquad\square$

The result in Lemma 11.5 shows how an individual frequency mode can be removed with the observer designed in the way as if the output would not contain other frequency modes. From the proof of Lemma 11.5, it can be seen that there is an asymptotic error, π_k, between the observer state and the actual state variables associated with the frequency mode ω_k. Although π_k is orthogonal to the frequency mode ω_k, it does in general contain components generated from all the other frequency modes. Because of this, a set of observers of the same form as shown in (11.18) would not be able to extract multiple frequency modes simultaneously. To remove multiple frequency modes, it is essential to find an estimate which is asymptotically orthogonal to the multiple frequency modes. For this, the interactions between the observers must be dealt with.

Suppose that we need to remove a number of frequency modes ω_k for all the k in a finite set of positive integers $\mathrm{K} = \{k_i\}$, for $= 1, \ldots, m$. To estimate the frequency modes for $\omega_{k,i}$, $i = 1, \ldots, m$, we propose a sequence of observers,

$$\dot{\hat{w}}_{k,1} = S_{k,1} \hat{w}_{k,1} + l_{k,1}(y - g^T \hat{w}_{k,1}) \tag{11.22}$$

and, for $i = 2, \ldots, m$,

$$\dot{\eta}_{k,i-1} = A\eta_{k,i-1} + bg^T_{k,i-1}\hat{w}_{k,i-1} \tag{11.23}$$

$$\dot{\hat{w}}_{k,i} = S_{k,i}\hat{w}_{k,i} + l_{k,i}\left(y - \sum_{j=1}^{i-1} C\eta_{k,j} - g^T\hat{w}_{k,i}\right), \tag{11.24}$$

where $l_{k,i}$, for $i = 1, \ldots, m$, are designed such that $\bar{S}_{k,i} := S_{k,i} - l_{k,i}g^T$ are Hurwitz, and

$$g_{k,i} = \frac{1}{m_{k,i}}\begin{bmatrix} \cos\phi_{k,i} & \sin\phi_{k,i} \\ -\sin\phi_{k,i} & \cos\phi_{k,i} \end{bmatrix} g := Q_{k,i}g$$

with $m_{k,i} = |C(jw_{k,i} - A)^{-1}b|$ and $\phi_{k,i} = \angle C(jw_{k,i} - A)^{-1}b$.

The estimate for the input disturbance which contains the required frequency modes for asymptotic rejection is given by

$$\hat{\mu}_m = \sum_{i=1}^{m} g^T_{k,i}\hat{w}_{k,i}. \tag{11.25}$$

The estimate $\hat{\mu}_m$ contains all the frequency modes $\omega_{k,i}$, for $i = 1, \ldots, m$. The useful property of the estimate is given in the following theorem.

Theorem 11.6. *For the estimate $\hat{\mu}_m$ given in (11.25), $\mu - \hat{\mu}_m$ is asymptotically orthogonal to the frequency modes $\omega_{k,i}$ for $i = 1, \ldots, m$.*

Proof. In the proof, we will show how to establish the asymptotic orthogonality in detail by induction.

We introduce the notations $\tilde{w}_{k,i} = w_{k,i} - \hat{w}_{k,i}$. We use $\pi_{k,i}$ to denote the steady-state solutions of $\tilde{w}_{k,i}$ and $e_{k,i} = \tilde{w}_{k,i} - \pi_{k,i}$, for $i = 1, \ldots, m$.

Lemma 11.5 shows that the results hold for $m = 1$. Let

$$\mu_1 = g^T_{k,1}(w_{k,1} - \pi_{k,1})$$

and $\mu - \mu_1$ is orthogonal to the frequency mode $\omega_{k,1}$.

We now establish the result for $m = 2$. From Lemma 11.3, there exists an initial state variable $w_{k,2}(0)$ for the dynamic system

$$\dot{w}_{k,2} = S_{k,2}w_{k,2} \tag{11.26}$$

such that $y - Cq_{k,1} - g^Tw_{k,2}$ is orthogonal to the frequency mode $\omega_{k,2}$ where $q_{k,i}$, for $i = 1, \ldots, m - 1$, denote the steady-state solution of

$$\dot{q}_{k,i} = Aq_{k,i} + bg^T_{k,i}(w_{k,i} - \pi_{k,i}).$$

Note that $g^Tw_{k,2}$ can be viewed as the output for the input $g^T_{k,2}w_{k,2}$ to the system (A, b, C), based on Lemma 11.3. Hence, $y - Cq_{k,1} - g^Tw_{k,2}$ is the output for the input $\mu - g^T_{k,1}(w_{k,1} - \pi_{k,1}) - g^T_{k,2}w_{k,2}$ to the system (A, b, C). Therefore, from Lemma 11.4, $\mu - g^T_{k,1}(w_{k,1} - \pi_{k,1}) - g^T_{k,2}w_{k,2}$ is orthogonal to $\omega_{k,2}$. Furthermore, from the previous step $\pi_{k,1}$ is orthogonal to $\omega_{k,1}$. Hence, $\mu - g_{k,1}^T(w_{k,1} - \pi_{k,1}) - g_{k,2}^T w_{k,2}$ is orthogonal to the frequency modes $\omega_{k,j}$ for $j = 1, 2$.

The dynamics of $\tilde{w}_{k,2}$ and $\pi_{k,2}$ are obtained as

$$\dot{\tilde{w}}_{k,2} = \bar{S}_{k,2}\tilde{w}_{k,2} - l_{k,2}(y - C\eta_{k,1} - g^T w_{k,2}), \tag{11.27}$$

$$\dot{\pi}_{k,2} = \bar{S}_{k,2}\pi_{k,2} - l_{k,2}(y - Cq_{k,1} - g^T w_{k,2}), \tag{11.28}$$

and the error $e_{k,2} = \tilde{w}_{k,2} - \pi_{k,2}$ satisfies

$$\dot{e}_{k,2} = \bar{S}_{k,2}e_{k,2} - l_{k,2}C(q_{k,1} - \eta_{k,1}).$$

Since $q_{k,1} - \eta_{k,1}$ exponentially converges to zero, so does $e_{k,2}$. From (11.28) and Lemma 11.4, $\pi_{k,2}$ is orthogonal to the frequency modes $\omega_{k,j}$ for $j = 1, 2$. Therefore, by letting

$$\mu_2 = g_{k,1}^T(w_{k,1} - \pi_{k,1}) + g_{k,2}^T(w_{k,2} - \pi_{k,2}),$$

$\mu - \mu_2$ is orthogonal to the frequency modes $\omega_{k,j}$ for $j = 1, 2$.

Notice that

$$\mu_2 - \hat{\mu}_2 = \sum_{j=1}^{2} g_{k,j}^T(w_{k,j} - \pi_{k,j} - \hat{w}_{k,j})$$

$$= \sum_{j=1}^{2} g_{k,1}^T e_{k,j}.$$

Hence $\mu_2 - \hat{\mu}_2$ converges to zero exponentially. Therefore, we conclude that $\mu - \hat{\mu}_2$ is asymptotically orthogonal to the frequency modes $\omega_{k,j}$ for $j = 1, 2$.

Now suppose that the result holds for $m = i$ and therefore $\mu - \mu_i$ is orthogonal to the frequency modes $\omega_{k,j}$ for $j = 1, \ldots, i$, with

$$\mu_i = \sum_{j=1}^{i} g_{k,j}^T(w_{k,j} - \pi_{k,j}).$$

We need to establish that the result holds for $m = i + 1$.

For the frequency mode $\omega_{k,i+1}$, there exists an initial state variable $w_{k,i}(0)$ for the dynamic system

$$\dot{w}_{k,i+1} = S_{k,i}w_{k,i+1}$$

such that $y - C\sum_{j=1}^{i} q_{kj} - g^T w_{k,i+1}$ is orthogonal to the frequency mode $\omega_{k,i+1}$. Note that $y - C\sum_{j=1}^{i} q_{kj} - g^T w_{k,i+1}$ is the output for the input $\mu - \mu_i - g_{k,i+1}^T w_{k,i+1}$ to the system (A, b, C). Therefore from Lemma 11.4, $\mu - \mu_i - g_{k,i+1}^T w_{k,i+1}$ is orthogonal to $\omega_{k,i+1}$. Since $\mu - \mu_i$ is orthogonal to the frequency modes $\omega_{k,j}$ for $j = 1, \ldots, i$, $\mu - \mu_i - g_{k,i+1}^T w_{k,i+1}$ is orthogonal to the frequency modes $\omega_{k,j}$ for $j = 1, \ldots, i + 1$ and so is $(y - C\sum_{j=1}^{i} q_{kj} - g^T w_{k,i+1})$.

The dynamics of $\tilde{w}_{k,i+1}$ and $\pi_{k,i+1}$ are obtained as

$$\dot{\tilde{w}}_{k,i+1} = \bar{S}_{k,i+1}\tilde{w}_{k,i+1}$$

$$- l_{k,i+1}\left(y - C\sum_{j=1}^{i}\eta_{k,j} - g^T w_{k,i+1}\right), \tag{11.29}$$

$$\dot{\pi}_{k,i+1} = \bar{S}_{k,i+1}\pi_{k,i+1}$$

$$- l_{k,i+1}\left(y - C\sum_{j=1}^{i}q_{k,j} - g^T w_{k,i+1}\right). \tag{11.30}$$

Since the input is orthogonal to the frequency modes $\omega_{k,j}$ for $j = 1,\ldots,i+1$, we conclude from Lemma 11.4 that π_{k+i+1} is orthogonal to $\omega_{k,j}$ for $j = 1,\ldots,i+1$, and therefore $\mu - \mu_{i+1}$ is orthogonal to the frequency modes $\omega_{k,j}$ for $j = 1,\ldots,i+1$ with

$$\mu_{i+1} = \mu_i + g_{k,i+1}^T(w_{k,i+1} - \pi_{k+i+1}).$$

The error $e_{k,i+1} = \tilde{w}_{k,i+1} - \pi_{k,i+1}$ satisfies

$$\dot{e}_{k,i+1} = \bar{S}_{k,i+1}e_{k,i+1} - l_{k,i+1}C\sum_{j=1}^{i}(q_{k,j} - \eta_{k,j}).$$

Since $q_{k,j} - \eta_{k,j}$, for $j = 1,\ldots,i$, exponentially converge to zero, so does $e_{k,i+1}$. With

$$\mu_{i+1} - \hat{\mu}_{i+1} = \sum_{j=1}^{i+1}g_{k,j}^T(w_{k,j} - \pi_{k,j} - \hat{w}_{k,j})$$

$$= \sum_{j=1}^{i}g_{k,j}^T e_{k,j},$$

$\mu_{i+1} - \hat{\mu}_{i+1}$ converges to zero exponentially. Therefore, we conclude that $\mu - \hat{\mu}_{i+1}$ is asymptotically orthogonal to the frequency modes $\omega_{k,j}$ for $j = 1,\ldots,i+1$.

Since i is an arbitrary integer between 1 and m, the result holds for $i = m - 1$, and this completes the proof. $\qquad\square$

11.1.3 Estimation of specific frequency modes in input

From the previous analysis on the equivalent input disturbance, we can convert the unmatched disturbance case to the matched one. Therefore, we can now carry on the disturbance rejection based on the matched case:

$$\dot{x} = A_c x + \phi(y) + b(u - \mu)$$

$$y = Cx. \tag{11.31}$$

If the disturbance does not exist in (11.31), the system (11.31) is in the linear observer error with output injection that is shown in Chapter 8. In that case, we can design a state observer as

$$\dot{p} = (A_c - LC)p + \phi(y) + bu + Ly, \tag{11.32}$$

where $p \in \mathbb{R}^n$, $L \in \mathbb{R}^n$ is chosen so that $A_c - LC$ is Hurwitz. The difficulty in the state estimation is due to the unknown disturbance μ. Consider

$$\dot{q} = (A_c - LC)q + b\mu, \tag{11.33}$$

where q denotes the steady-state solution. Such a solution exists, and an explicit solution is given earlier. Each element of q is a periodic function, as μ is periodic. We have an important property for p and q stated in the following lemma.

Lemma 11.7. *The state variable x can be expressed as*

$$x = p - q + \epsilon, \tag{11.34}$$

where p is generated from (11.32) with q satisfying (11.33) and ϵ satisfying

$$\dot{\epsilon} = (A_c - LC)\epsilon. \tag{11.35}$$

From (11.33), we have Cq as the steady-state output of the system $((A_c - kC), b, C)$ for the equivalent input disturbance μ. If Cq is available, we are ready to use the results presented in the previous section for disturbance estimation of the specified frequency modes. The result shown in Lemma 11.7 indicates that $Cq = Cp - y - C\epsilon$, and hence $Cp - y$ exponentially converges to Cq. Therefore, we now propose an observer for frequency modes in the input disturbances using $Cp - y$. To estimate the frequency modes for $\omega_{k,i}$ $i = 1, \ldots, m$, we have

$$\dot{\hat{w}}_{k,1} = S_{k,1}\hat{w}_{k,1} + l_{k,1}(Cp - y - g^T\hat{w}_{k,1}) \tag{11.36}$$

and, for $i = 2, \ldots, m$,

$$\dot{\eta}_{k,i-1} = (A_c - kC)\eta_{k,i-1} + bg_{k,i-1}^T\hat{w}_{k,i-1}, \tag{11.37}$$

$$\dot{\hat{w}}_{k,i} = S_{k,i}\hat{w}_{k,i} + l_{k,i}\left(Cp - y - \sum_{j=1}^{i-1}C\eta_{k,j} - g^T\hat{w}_{k,i}\right), \tag{11.38}$$

where $l_{k,i}$, for $i = 1, \ldots, m$, are designed such that $\bar{S}_{k,i} := S_{k,i} - l_{k,i}g^T$ are Hurwitz, and

$$g_{k,i} = \frac{1}{m_{k,i}}\begin{bmatrix} \cos\phi_{k,i} & \sin\phi_{k,i} \\ -\sin\phi_{k,i} & \cos\phi_{k,i} \end{bmatrix} g := Q_{k,i}g \tag{11.39}$$

with $m_{k,i} = |C(jw_{k,i} - (A - kC))^{-1}b|$ and $\phi_{k,i} = \angle C(jw_{k,i} - (A - kC))^{-1}b$.

The estimate for the input disturbance which contains the required frequency modes for asymptotic rejection is given by

$$\hat{\mu}_m = \sum_{i=1}^{m} g_{k,i}^T \hat{w}_{k,i}. \tag{11.40}$$

The estimate $\hat{\mu}_m$ contains all the frequency modes $\omega_{k,i}$, for $i = 1, \ldots, m$. The useful property of the estimate is given in the following theorem.

Theorem 11.8. *The estimate $\hat{\mu}_m$ given in (11.40) is bounded and satisfies the following:*

$$\lim_{t \to \infty} \int_t^{t+T} [\mu(\tau) - \hat{\mu}_m(\tau)] \sin \omega_{k,i} \tau d\tau = 0, \tag{11.41}$$

$$\lim_{t \to \infty} \int_t^{t+T} [\mu(\tau) - \hat{\mu}_m(\tau)] \cos \omega_{k,i} \tau d\tau = 0, \tag{11.42}$$

for $i = 1, \ldots, m$.

Proof. From Theorem 11.6, the results in (11.41) and (11.42) hold if the input $Cp - y$ in (11.36), (11.37) and (11.38) equals Cq. Since $Cp - y - Cq$ converges exponentially to 0, and (11.36)–(11.38) are stable linear systems, the errors in the estimation of $\hat{\mu}$ caused by replacing Cq by $Cp - y$ in (11.36)–(11.38) are also exponentially convergent to zero. Therefore, (11.41) and (11.42) can be established from Theorem 11.6. □

11.1.4 Rejection of frequency modes

With the estimated frequency modes $\hat{\mu}_m$, we set the control input as

$$u = v + \hat{\mu}_m,$$

where v is designed based on backstepping design which is shown in Chapter 9. Due to the involvements of disturbance in the system, we need certain details of stability analysis with explicit expression of Lyapunov functions. For the convenience of stability analysis, we adopt an approach using backstepping with filtered transformation shown in Section 9.4. The control input v can be designed as a function

$$v = v(y, \xi),$$

where ξ is the state variable of the input filter for the filtered transformation. The final control input is designed as

$$u = v(y, \xi) + \hat{\mu}. \tag{11.43}$$

For the stability of the closed-loop system with disturbance rejection, we have the following theorem.

Theorem 11.9. *The closed-loop system of (11.1) under the control input (11.43) ensures the boundedness of all the state variables, and the disturbance modes of specified frequencies $\omega_{k,i}$, for $i = 1, \ldots, m$, are asymptotically rejected from the system in the sense that all the variables of the closed-loop system is bounded and the system is asymptotically driven by the frequency modes other than the specified modes.*

Proof. For the standard backstepping design for the nonlinear systems, we establish the stability by considering the Lyapunov function

$$V = \beta z^T P z + \frac{1}{2} y^2 + \frac{1}{2} \sum_{i=1}^{\rho-1} \tilde{\xi}_i^2, \tag{11.44}$$

where β is a constant and $\tilde{\xi}_i = \xi_i - \hat{\xi}_i$ and P is a positive definite matrix satisfying $D^T P + P D = -I$. The standard backstepping design ensures that there exists a positive real constant γ_1 such that

$$\dot{V} \leq -\gamma_1 V - [2\beta z^T P b_z + y b_y]\tilde{\mu}$$

$$\leq -\gamma_1 V - \gamma_2 \sqrt{V} \tilde{\mu}$$

$$\leq -\gamma_3 V + \gamma_4 \tilde{\mu}^2 \tag{11.45}$$

for some positive real constants γ_i, $i = 2, 3, 4$. Hence, the boundedness of $\tilde{\mu}$ ensures the boundedness of all the state variables. It is clear from Theorem 11.8 that all the specified frequency modes are asymptotically removed from $\tilde{\mu}$. This completes the proof.

Remark 11.5. In the control design, we have used the control input obtained with the backstepping design for the proof of Theorem 11.9. In case that there exists a known control design for the disturbance-free system which ensures exponential stability of the disturbance-free case, we can pursue the proof of Theorem 11.9 in a similar way to obtain the result shown in (11.45). This is very useful in particular for the case of non-minimum phase systems, to which the standard backstepping presented earlier does not apply.   ◁

11.1.5 Example

In this section, we use a simple example to demonstrate estimation of harmonics using the method proposed earlier in this section.

We consider a voltage signal with diados as shown in Figure 11.1. We are going to use iterative observers to estimate harmonics in this signal. To simplify the presentation, we assume that the signal is directly measurable. In this case, there are no dynamics between the measurement and the point of estimation. Hence, we can simplify the algorithms presented in Subsection 11.1.2 ((11.22),

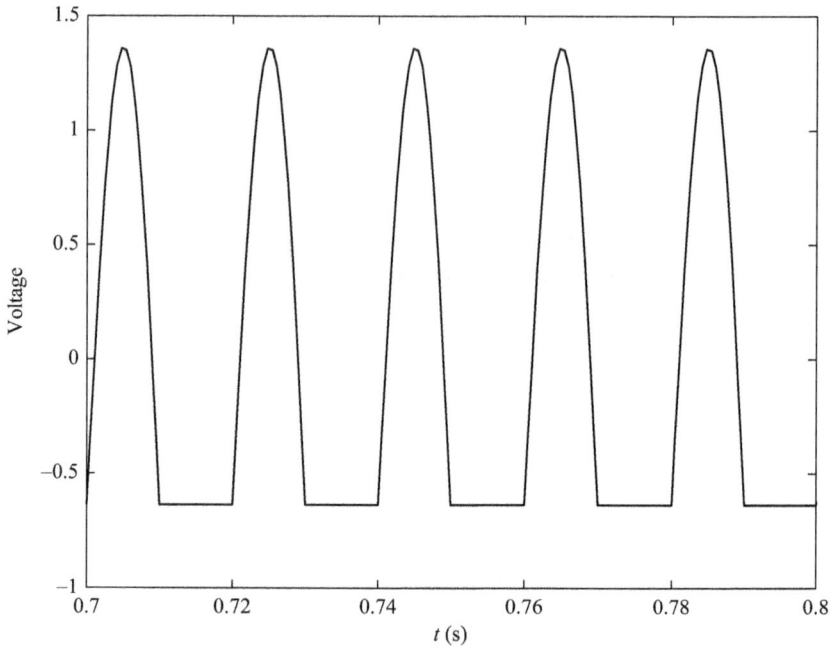

Figure 11.1   Voltage signal of a nonlinear load

(11.23) and (11.24)). In this case, there is no need to use the observer model in (11.23). The simplified algorithm for direct estimation of harmonics is presented below.

To estimate the frequency modes for $\omega_{k,i}$, $i = 1, \ldots, m$, the iterative observers are given by

$$\dot{\hat{w}}_{k,1} = S_{k,1}\hat{w}_{k,1} + l_{k,1}(y - g^T\hat{w}_{k,1}) \tag{11.46}$$

and, for $i = 2, \ldots, m$,

$$\dot{\hat{w}}_{k,i} = S_{k,i}\hat{w}_{k,i} + l_{k,i}\left(y - \sum_{j=1}^{i} g^T\hat{w}_{k,j}\right), \tag{11.47}$$

where $l_{k,i}$, for $i = 1, \ldots, m$, are designed such that $\bar{S}_{k,i} := S_{k,i} - l_{k,i}g^T$ are Hurwitz.

For the electricity, the base frequency is 50 Hz. We will show the estimates of harmonics from the base frequency, that is, we set $\omega_{k,1} = 100\pi$, $\omega_{k,2} = 200\pi$, $\omega_{k,3} = 300\pi$, etc. For these frequency modes, we have

$$S_{k,i} = \begin{bmatrix} 0 & i100\pi \\ -i100\pi & 0 \end{bmatrix}.$$

The observer gain $l_{k,i}$ can be easily designed to place the close-loop poles at any specified positions. For the simulation study, we place the poles at $\{-200, -200\}$. For such pole positions, we have

$$l_{k,i} = \begin{bmatrix} 400 \\ \frac{40000}{i100\pi} - 200\pi \end{bmatrix}.$$

The harmonics of second and third orders are plotted with the original signal and the estimated base frequency component in Figure 11.2. The approximations of accumulative harmonics to the original signal are shown in Figure 11.3.

In this example, we have shown the estimation of harmonics from direct measurements. The advantage of this estimation is that the estimation is on-line and can be implemented in real time. Also, the estimation using the proposed method can provide the phase information, which is very important for harmonic rejection. We did not show the rejection of harmonics in this example. However, it is not difficult to see that rejection can be simulated easily with the proposed method.

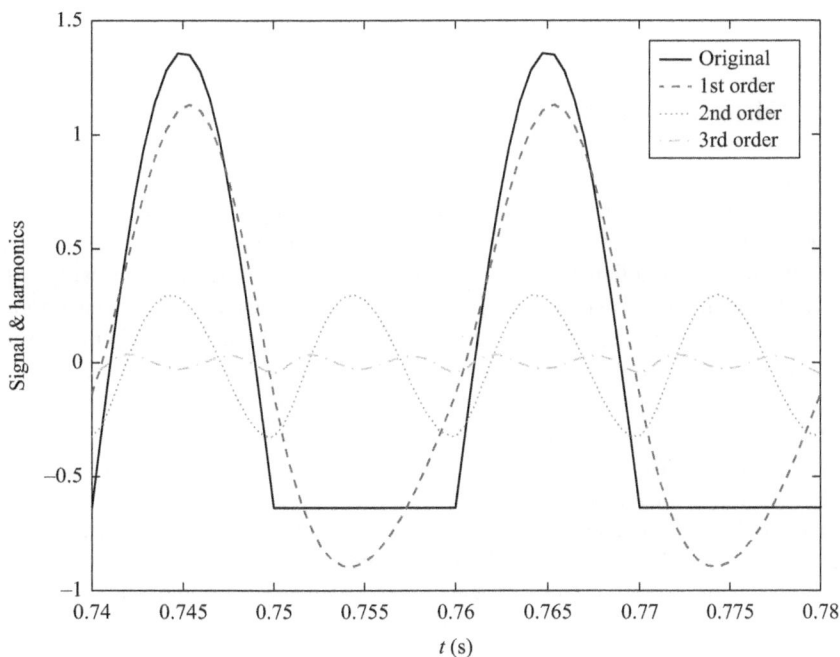

Figure 11.2   Estimated harmonics

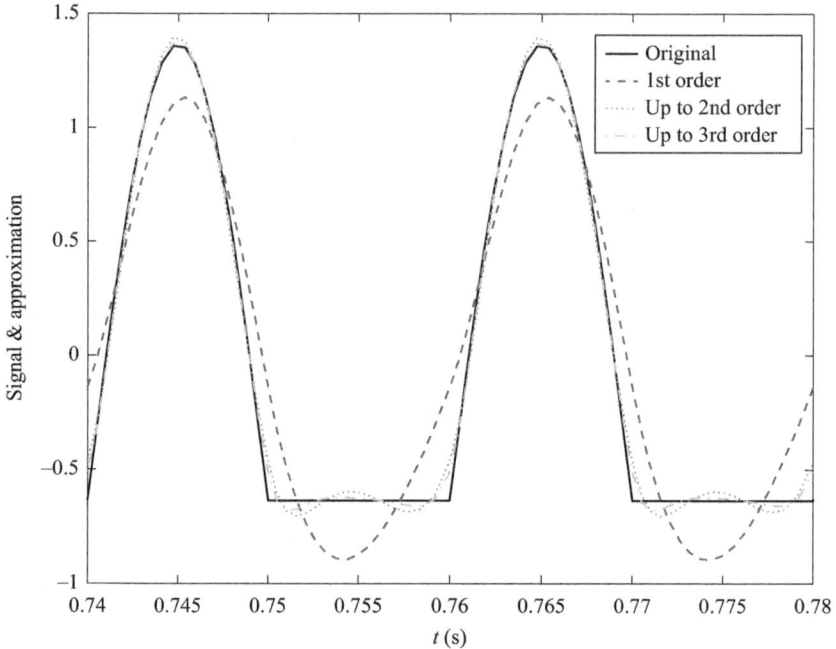

Figure 11.3　Approximation using harmonic components

11.2　Observer and control design for circadian rhythms

Circadian rhythms play an important role in daily biological activities of living species. Circadian rhythms, or biological clocks, normally have a period of 24 h, and changes to the rhythms can be referred to as changes in phases. Circadian disorder is a phenomenon of circadian rhythms which occurs when internal rhythms cannot keep up with the changes of external environmental rhythms. Changes of environmental rhythms, presented by the change of light/dark cycles or by irregular rhythms, result in phase shifts between internal and external rhythms. For example, jet lags are caused when people travel from one time zone to another. The existence of these phase shifts in longer term has negative effect to health. Therefore, in biological study of circadian rhythms, it is important to find methods to recover the shifted phases, such as jet lags, to their normal rhythms. These methods could then lead to treatments of circadian disorder, like sleep disorder and jet lag. Circadian rhythms can be accurately modelled as limit cycles of nonlinear dynamic systems. We will apply the observer and control design shown in earlier chapters for analysis and control of circadian rhythms. The phase restoration is carried out by the synchronisation of trajectories generated from a controlled model with the trajectories of a reference system via nonlinear control design. Both reference and controlled systems are based on a given third-order model of Neurospora circadian rhythms. An observer is also designed to estimate two unknown states from the measurement of a single state.

11.2.1 Circadian model

For almost every living organism on the Earth, their daily biological activities are governed by rhythms. These biological rhythms are called circadian rhythms. The circadian rhythms exist as self-sustained and periodic oscillations, and they are also known with their entrainment to 24-h day/night cycle. The entrainment is based on the activity of a circadian clock gene. For mammals, the circadian clock gene, also known as the circadian pacemaker, is found in the suprachiasmatic nuclei (SCN) of the anterior hypothalamus. The circadian pacemaker captures information sent from an external environment cue such as light, and then coordinates the timing of other slave clocks or slave oscillators in other parts of the body. Any changes of environment cues which cause the mismatch between external and internal rhythms can lead to disruption of circadian rhythms. This phenomenon is known as circadian disorders. Jet lags due to trans-continent flights and sleeping disorders due to irregular sleep–wake cycles are two typical examples of circadian disorders. In practice, one of the known medical treatments for circadian disorders is the application of light. Light is major external environmental cue, and with light input, the circadian phase can be adjusted to the light/dark rhythms at destination.

In this section, we aim for the restoration of circadian phase using nonlinear control. In order to achieve this objective, we propose an alternative control design method which synchronises trajectories generated from a controlled model with the trajectories generated from a reference model via backstepping approach. Both reference system and controlled system are based on a third-order mathematical model of Neurospora circadian rhythms. The trajectories generated by controlled system represent the altered rhythms. Meanwhile, the reference trajectories represent the desired rhythms which the trajectories of controlled system are adjusted to match. We also present an observer design for this circadian rhythm based on observer design presented in Chapter 8 for nonlinear systems with Lipschitz nonlinearities.

A third-order mathematical model is developed to describe molecular mechanism of circadian rhythms in Neurospora. Its molecular mechanism is based on the negative feedback exerted by FRQ proteins on the expression of *frq* gene. Transcription of *frq* gene yields messenger RNA (mRNA), and the translation of which synthesises FRQ protein. These synthesised FRQ proteins are then transferred back into nucleus where they inhibit the transcription of *frq* gene. A new activation of *frq* gene transcription will restart the cycle. Dynamics of these variables, *frq* mRNA, FRQ protein and nuclear FRQ protein, are by the following dynamic model:

$$
\begin{aligned}
\dot{x}_1 &= v_s \frac{K_i^n}{K_i^n + x_3^n} - v_m \frac{x_1}{K_M + x_1} \\
\dot{x}_2 &= k_s x_1 - v_d \frac{x_2}{K_d + x_2} - k_1 x_2 + k_2 x_3 \\
\dot{x}_3 &= k_1 x_2 - k_2 x_3,
\end{aligned}
\tag{11.48}
$$

where x_1, x_2 and x_3 denote concentration of *frq* mRNA, concentration of FRQ protein outside nucleus and concentration of nucleus FRQ protein respectively. Values of three state variables x_1, x_2 and x_3 are assumed to be positive values. In system (11.49), the parameter v_s denotes the transcription rate of *frq* gene. The other parameters involved in system (11.49) are $K_i, n, v_m, K_M, k_s, k_1, k_2, v_d$ and K_d. The parameters K_i, n, v_m, K_M represent the threshold constant beyond which nuclear FRQ protein inhibits the transcription of *frq*, the Hill coefficient showing the degree of co-operativity of the inhibition process, the maximum rate of *frq* mRNA degradation and the Michaelis constant related to the latter process respectively. The parameters k_s, k_1 and k_2 denote the rate constant measuring the rate of FRQ synthesis, the rate constants of the transport of FRQ into and out of the nucleus respectively. The parameter v_d denotes maximum rate of FRQ degradation and K_d is the Michaelis constant related to this process. For the third-order Neurospora model, the parameters have their typical values as $v_s = 1.6\,\text{nM h}^{-1}$, $K_i = 1\,\text{nM}$, $n = 4$, $v_m = 0.7\,\text{nM h}^{-1}$, $K_M = 0.4\,\text{nM}$, $k_s = 1\,\text{h}^{-1}$, $v_d = 4\,\text{nMh}^{-1}$, $K_d = 1.4\,\text{nM}$, $k_1 = 0.3\,\text{h}^{-1}$, $k_2 = 0.15\,\text{h}^{-1}$.

Circadian rhythms are self-sustained and periodic oscillations. With the set parameters as above, dynamics of state variables of (11.48) can sustain periodic oscillations, and in fact, a limit cycle. Figure 11.4 shows the plots of three state variables obtained in simulation with the model (11.48). For simulation study, we choose the initial values $x(0) = \begin{bmatrix} 5 & 1 & 1 \end{bmatrix}^T$. This particular initial value is actual a point on the limit cycle. If the initial value starts from a point outside the limit cycle, then there trajectory will converge to the limit cycle. There can be other selections for values of initial conditions.

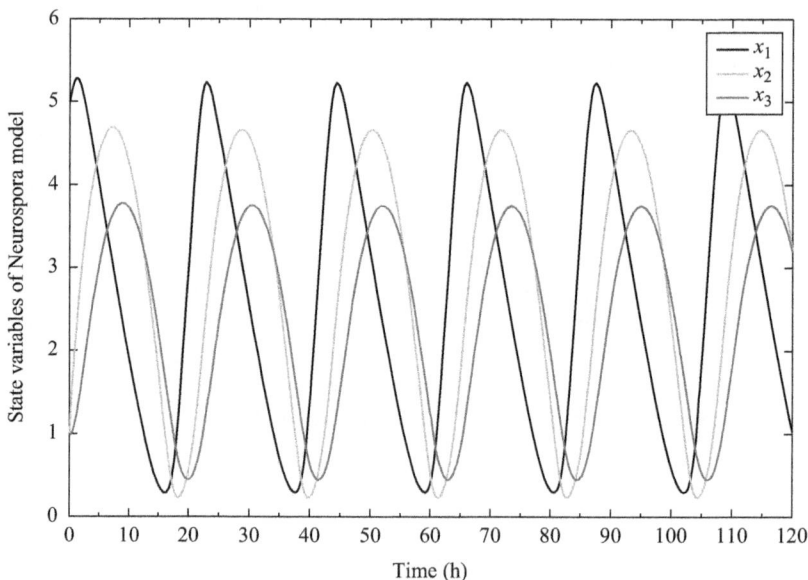

Figure 11.4   Circadian rhythm of Neurospora

The model (11.48) is not in the right format for control and observer design. We introduce a state transformation as

$$z = Tx \tag{11.49}$$

with

$$T = \begin{bmatrix} 0 & 0 & 1 \\ 0 & k_1 & 0 \\ k_1 k_s & 0 & 0 \end{bmatrix},$$

and the transformed system is described by

$$\dot{z}_1 = z_2 - k_2 z_1$$

$$\dot{z}_2 = z_3 - k_1 z_2 + k_1 k_2 z_1 - v_d \frac{k_1 z_2}{k_1 K_d + z_2} \tag{11.50}$$

$$\dot{z}_3 = v_s \frac{k_1 k_s K_i^n}{K_i^n + z_1^n} - v_m \frac{k_1 k_s z_3}{K_M k_1 k_s + z_3}.$$

Now the transformed model (11.50) is in the lower-triangular format.

11.2.2 Lipschitz observer design

We have presented observer design for nonlinear dynamic systems with Lipschitz nonlinearity in Section 8.4. Basically, for a nonlinear system described by (8.29)

$$\dot{x} = Ax + \phi(x, u)$$

$$y = Cx,$$

a nonlinear observer can be designed as in (8.30)

$$\dot{\hat{x}} = A\hat{x} + L(y - C\hat{x}) + \phi(\hat{x}, u), \tag{11.51}$$

provided that the conditions specified in Theorem 8.10 are satisfied. That suggests that we need find the Lipschitz constant v together with a positive definite matrix P for nonlinearities in the circadian model (11.48) such that the inequality (8.36)

$$A^T P + PA + 2vI - 2\sigma C^T C < 0$$

is satisfied, and then we obtain the observer gain as

$$L = \sigma P^{-1} C^T.$$

The one-sided Lipschitz condition shown in (8.36) can be checked using the Lipschitz constant for individual elements in the vector ϕ. Let us denote the Lipschitz constants of ϕ_i by γ_i. The condition shown in (8.36) can be guaranteed by

$$A^T P + PA + 2n \sum_{i=1}^{n} \gamma_i \lambda_i I - 2\sigma C^T C < 0, \tag{11.52}$$

where λ_i are positive real constants, and $P \in \cap_{i=1}^n \mathcal{P}''(i, \lambda_i)$. The definition of $P \in \cap_{i=1}^n \mathcal{P}''(i, \lambda_i)$ is given by

$$\mathcal{P}''(i, \lambda_i) = \{P : |p_{ji}| < \lambda_i, \text{ for } j = 1, 2, \dots, n\}.$$

For the system (11.50), we set

$$C_1 = \begin{bmatrix} 1 & 0 & 0 \end{bmatrix}$$

and therefore state variables z_2 and z_3 are unknown. Comparing with the structure in (8.29), we have

$$\phi(z, u) = \begin{bmatrix} 0 \\ \phi_2 \\ \phi_{3a} + \phi_{3b} \end{bmatrix},$$

where clearly $\phi_1 = 0$, and

$$\phi_2 = -v_d \frac{k_1 z_2}{k_1 K_d + z_2},$$

$$\phi_{3a} = v_s \frac{k_1 k_s K_i^n}{K_i^n + z_1^n},$$

$$\phi_{3b} = -v_m \frac{k_1 k_s z_3}{K_M k_1 k_s + z_3}.$$

From the definition of Lipschitz constant in Definition 8.2, we have

$$\left\| \frac{-v_d k_1 z_2}{k_1 K_d + z_2} + \frac{v_d k_1 \hat{z}_2}{k_1 K_d + \hat{z}_2} \right\| \le \frac{v_d}{K_d} \|z_2 - \hat{z}_2\|,$$

$$\left\| \frac{-v_m k_1 k_s z_3}{K_M k_1 k_s + z_3} + \frac{v_m k_1 k_s \hat{z}_3}{K_M k_1 k_s + \hat{z}_3} \right\| \le \frac{v_m}{K_M} \|z_3 - \hat{z}_3\|,$$

and therefore we obtain the Lipschitz constants as $\gamma_2 = \frac{v_d}{K_d} = 10.7962$ for $\varphi_2(z_2)$, and $\gamma_{3b} = \frac{v_m}{K_M} = 1.01$ for $\varphi_{3b}(z_3)$. For nonlinear function $\varphi_{3a}(z_1)$, its Lipschitz constant can be computed by using mean value theorem which is described by

$$|f'(\zeta)| = \left| \frac{f(x) - f(\hat{x})}{x - \hat{x}} \right|,$$

where $\zeta \in [x, \hat{x}]$. Setting $f = \phi_{3a}$, we have

$$|f'(\zeta)| = \left| -\frac{n v_s k_1 k_s K_i^n \zeta^{n-1}}{(K_i^n + \zeta^n)^2} \right| = \left| \frac{\phi_{3a}(z_1) - \phi_{3a}(\hat{z}_1)}{z_1 - \hat{z}_1} \right|,$$

where $\zeta \in [\min(z_1, \hat{z}_1), \max(z_1, \hat{z}_1)]$. We find maximum value of $|f'(\zeta)|$ by solving $|f''(\zeta)| = 0$, and obtain the result as 0.325. Since the value of Lipschitz constant is equivalent to maximum value of $f'(\zeta)$, the Lipschitz constant is given by $\gamma_{3a} = 0.325$. The Lipschitz constant λ_3 is given by $\lambda_3 = \lambda_{3a} + \lambda_{3b} = 1.326$.

After the Lipschitz constants are found, we solve (11.52) to obtain

$$\sigma = 67.8032, \quad L = \begin{bmatrix} 3.9887 \\ 2.6067 \\ 1.3469 \end{bmatrix}$$

for the observer (11.51). Simulation study for the observer has been carried out with the results shown in Figures 11.5 and 11.6.

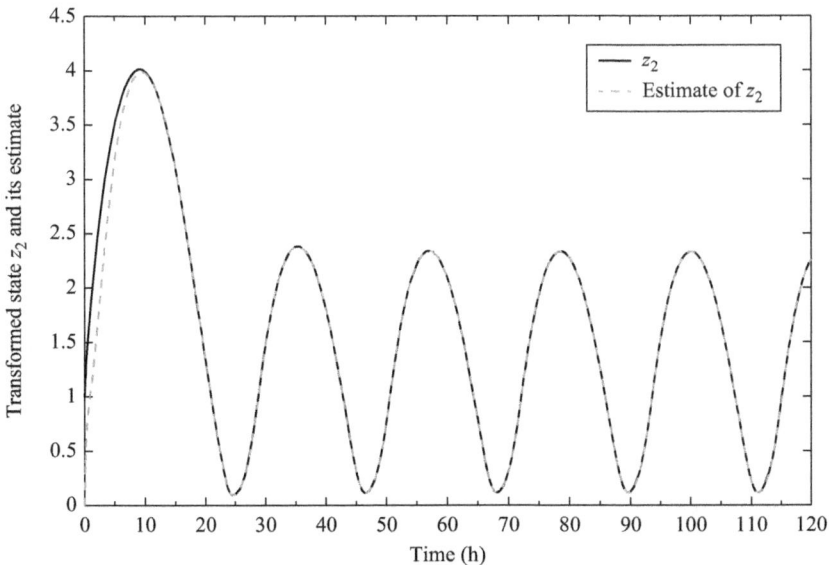

Figure 11.5   Observer state for circadian model

11.2.3   *Phase control of circadian rhythms*

The objective of phase control is to reset a distorted phase. Our strategy is to control the distorted circadian rhythm to follow a target rhythm. In order to design a control strategy, we need to define a sensible control input for the circadian model. Among the parameters appeared in (11.48), parameter v_s, which denotes the rate of *frq* mRNA transcription, is sensitive to light input. Therefore, for Neurospora circadian rhythms, this parameter is usually used as control input in many results which have been presented in literature. If v_s is taken as the control input, the circadian model (11.48) is rewritten as

$$\dot{x}_1 = (v_s + u) \frac{K_i^n}{K_i^n + x_3^n} - v_m \frac{x_1}{K_M + x_1}$$

$$\dot{x}_2 = k_s x_1 - v_d \frac{x_2}{K_d + x_2} - k_1 x_2 + k_2 x_3$$

$$\dot{x}_3 = k_1 x_2 - k_2 x_3,$$

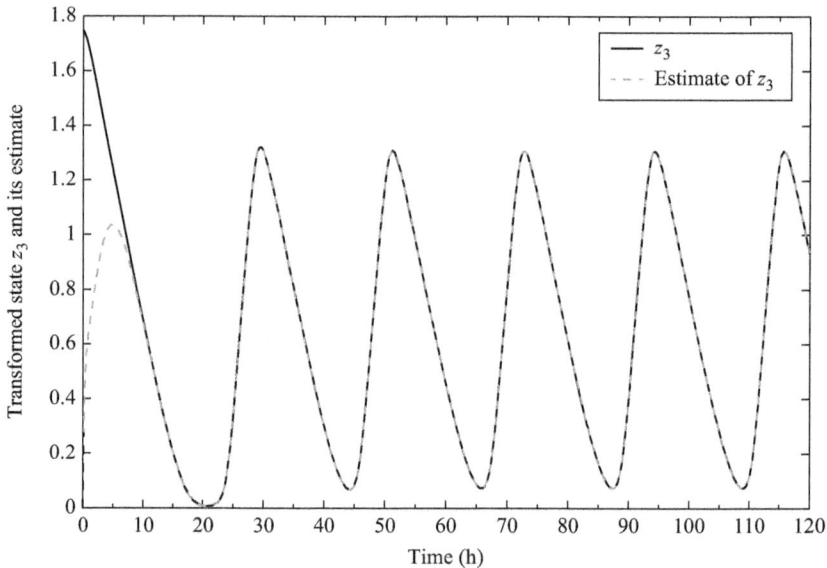

Figure 11.6   Observer state for circadian model

where u is the control input. The system model after the state transformation (11.49) is then obtained as

$$\dot{z}_1 = z_2 - k_2 z_1$$
$$\dot{z}_2 = z_3 - k_1 z_2 + k_1 k_2 z_1 - v_d \frac{k_1 z_2}{k_1 K_d + z_2}$$
$$\dot{z}_3 = v_s \frac{k_1 k_s K_i^n}{K_i^n + z_1^n} - v_m \frac{k_1 k_s z_3}{K_M k_1 k_s + z_3} + u \frac{k_1 k_s K_i^n}{K_i^n + z_1^n}. \tag{11.53}$$

This model (11.53) is then used for phase control design. We use q to denote the state variable for the target circadian model

$$\dot{q}_1 = q_2 - k_2 q_1$$
$$\dot{q}_2 = q_3 - k_1 q_2 + k_1 k_2 q_1 - v_d \frac{k_1 q_2}{k_1 K_d + q_2}$$
$$\dot{q}_3 = v_s \frac{k_1 k_s K_i^n}{K_i^n + q_1^n} - v_m \frac{k_1 k_s q_3}{K_M k_1 k_s + q_3}, \tag{11.54}$$

to which the variable z is controlled to follow.

The transformed dynamic models (11.53) and (11.54) are in the triangular form, and therefore the iterative backstepping method shown in Section 9.2 can be applied. There is a slight difference in the control objective here from the convergence to zero in Section 9.2. For this, we define

$$e_1 = z_1 - q_1,$$
$$e_2 = z_2 - q_2,$$
$$e_3 = z_3 - q_3,$$

(11.55)

where e_i, for $i = 1, 2$ and 3, are the tracking errors. The error dynamics are obtained as

$$\dot{e}_1 = e_2 - k_2 e_1$$

$$\dot{e}_2 = e_3 - k_1 e_2 + k_1 k_2 e_1 - \frac{v_d k_1^2 K_d e_2}{(k_1 K_d + z_2)(k_1 K_d + q_2)}$$

$$\dot{e}_3 = v_s \frac{k_1 k_s K_i^n}{K_i^n + z_1^n} - v_s \frac{k_1 k_s K_i^n}{K_i^n + q_1^n} - \frac{v_m K_M (k_1 k_s)^2 e_3}{(K_M k_1 k_s + z_3)(K_M k_1 k_s + q_3)} + u \frac{k_1 k_s K_i^n}{K_i^n + z_1^n}.$$

(11.56)

It is easy to see that the model (11.56) is still in the triangular form. The phase resetting is achieved if we can ensure that the errors e_i, $i = 1, 2, 3$, converge to zero. To the model (11.56), iterative backstepping can be applied. Following the procedures shown in Section 9.2, we define

$$w_1 = e_1,$$
$$w_2 = e_2 - \alpha_1,$$
$$w_3 = e_3 - \alpha_2,$$

(11.57)

where α_1 and α_2 are stabilising functions to be designed.

From the first equation of (11.56), we obtain dynamics of w_1 as

$$\dot{w}_1 = e_2 - k_2 e_1$$
$$= w_2 + \alpha_1 - k_2 e_1.$$

The stabilising function α_1 is then designed as

$$\alpha_1 = -c_1 w_1 + k_2 e_1,$$

(11.58)

where c_1 is a positive real constant. The resultant dynamics of w_1 are given by

$$\dot{w}_1 = -c_1 w_1 + w_2.$$

(11.59)

The dynamics of w_2 are obtained as

$$\dot{w}_2 = e_3 - k_1 e_2 + k_1 k_2 e_1 - \frac{v_d k_1^2 K_d e_2}{(k_1 K_d + z_2)(k_1 K_d + q_2)} - \frac{\partial \alpha_1}{\partial e_1} \dot{e}_1$$

$$= w_3 + \alpha_2 - k_1 e_2 + k_1 k_2 e_1 - \frac{v_d k_1^2 K_d e_2}{(k_1 K_d + z_2)(k_1 K_d + q_2)} - \frac{\partial \alpha_1}{\partial e_1} \dot{e}_1.$$

The stabilising function α_2 can be designed as

$$\alpha_2 = -w_1 - c_2 w_2 + k_1 e_2 - k_1 k_2 e_1 + v_d k_1 \frac{k_1 K_d e_2}{(k_1 K_d + z_2)(k_1 K_d + q_2)} + \frac{\partial \alpha_1}{\partial e_1} \dot{e}_1$$

with c_2 being a positive real constant, which results in the dynamics of w_2

$$\dot{w}_2 = -w_1 - c_2 w_2 + w_3. \tag{11.60}$$

From the dynamics of w_3

$$\dot{w}_3 = v_s \frac{k_1 k_s K_i^n}{K_i^n + z_1^n} - v_s \frac{k_1 k_s K_i^n}{K_i^n + q_1^n} + u \frac{k_1 k_s K_i^n}{K_i^n + z_1^n}$$
$$- \frac{v_m K_M (k_1 k_s)^2 e_3}{(K_M k_1 k_s + z_3)(K_M k_1 k_s + q_3)} - \frac{\partial \alpha_2}{\partial e_1} \dot{e}_1 - \frac{\partial \alpha_2}{\partial e_2} \dot{e}_2,$$

we design the control input u as

$$u = \frac{K_i^n + z_1^n}{k_1 k_s K_i^n} \left[-w_2 - c_3 w_3 - v_s \frac{k_1 k_s K_i^n}{K_i^n + z_1^n} + v_s \frac{k_1 k_s K_i^n}{K_i^n + q_1^n} + \frac{\partial \alpha_2}{\partial e_1} (e_2 - k_2 e_1) \right.$$
$$\left. + \frac{\partial \alpha_2}{\partial e_2} (e_3 - k_1 e_2 + k_1 k_2 e_1) - \frac{\partial \alpha_2}{\partial e_2} \frac{v_d k_1^2 K_d e_2}{(k_1 K_d + z_2)(k_1 K_d + q_2)} \right], \tag{11.61}$$

with c_3 as a positive real constant. The resultant dynamics of w_3 are then obtained as

$$\dot{w}_3 = -w_2 - c_3 w_3. \tag{11.62}$$

The stability analysis of the proposed control design can be established in the same way as the proof shown in Section 9.2 for iterative backstepping control design. Indeed, consider a Lyapunov function candidate

$$V = \frac{1}{2} \left(w_1^2 + w_2^2 + w_3^2 \right). \tag{11.63}$$

Using the dynamics of w_i, $i = 1, 2, 3$, in (11.59), (11.60) and (11.62), we obtain

$$\begin{aligned} \dot{V} &= w_1 \dot{w}_1 + w_2 \dot{w}_2 + w_3 \dot{w}_3 \\ &= w_1 \left(-c_1 w_1 + w_2 \right) + w_2 \left(-w_1 - c_2 w_2 + w_3 \right) + w_3 \left(-w_2 - c_3 w_3 \right) \\ &= -c_1 w_1^2 - c_2 w_2^2 - c_3 w_3^2. \end{aligned}$$

Therefore, we can conclude that the closed-loop system under the control of u in (11.61) is exponentially stable with respect to the variables w_i, $i = 1, 2, 3$, which suggests that the controlled circadian rhythm asymptotically tracks the targeted one.

Simulation study of the control design proposed above has been carried with the control parameters $c_1 = c_2 = c_3 = 0.1$. The states z_2 and z_3 are shown in Figures 11.7 and 11.8 when the control input is applied at $t = 50$ h, and the control input is shown in Figure 11.9. The plot of z_1 is very similar to the plot of z_2, and it is omitted. It can be seen from the figures that there is a phase difference before the control input is applied, and the control input resets the phase of the control circadian rhythm to the targeted one.

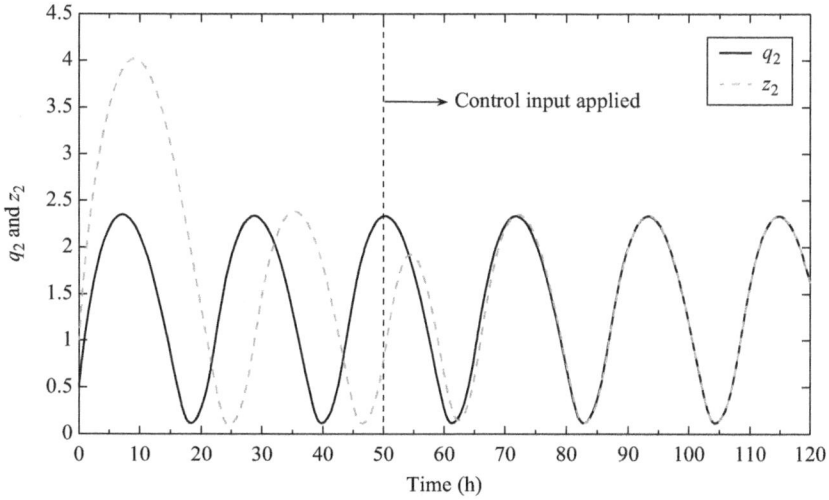

Figure 11.7   Circadian phase reset for state 2

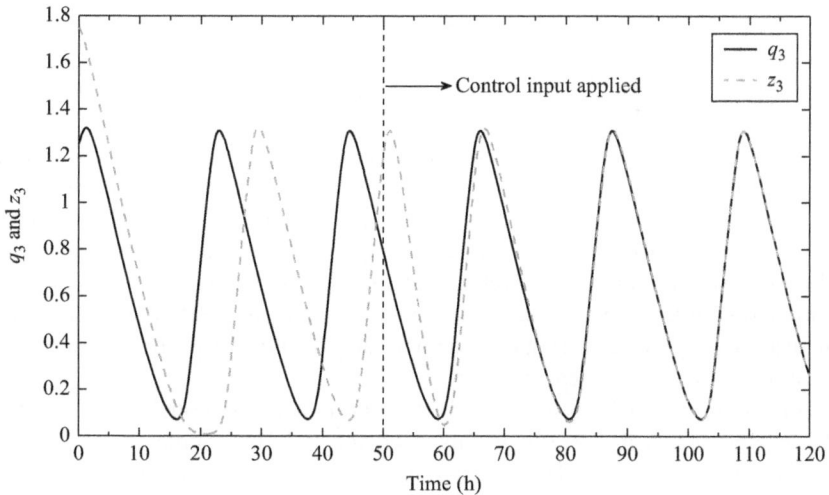

Figure 11.8   Circadian phase reset for state 3

11.3   Sampled-data control of nonlinear systems

A practical issue of applying nonlinear control strategies is to implement them in computers or microprocessors which are discrete-time in nature. For a continuous-time system with a discrete-time controller, inevitably we end with a sampled-data

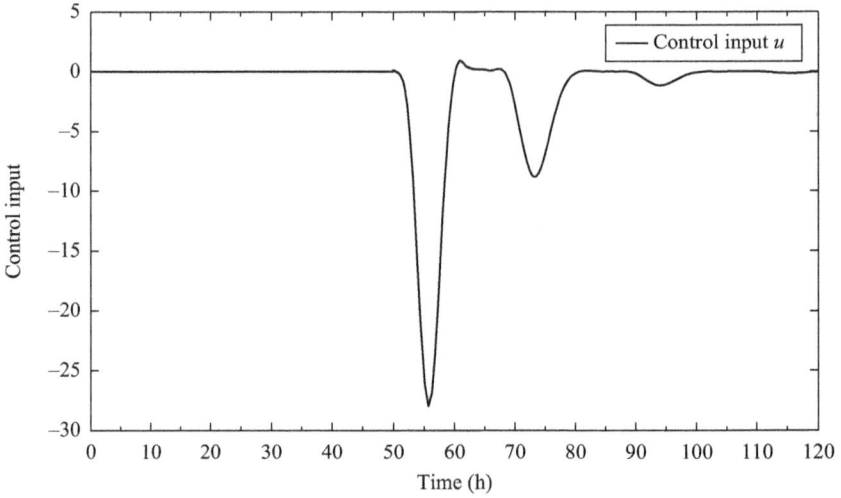

Figure 11.9   Control input for circadian phase reset

control system. It is well known for linear systems, we can use emulation method, that is, to design a controller in continuous-time and then implement its discrete-time version. Alternative to emulation method is the direct design in discrete-time. Stability analysis for sampled-data control of linear dynamic systems can then be carried out in the framework of linear systems in discrete-time. Sampled-data control of nonlinear systems is a much more challenging task.

For a nonlinear system, it is difficult or even impossible to obtain a nonlinear discrete-time model after sampling. Even with a sampled-data model in discrete-time, the nonlinear model structure cannot be preserved in general. It is well known that nonlinear control design methods do require certain structures. For example backstepping can be applied to lower-triangular systems, but a lower-triangular system in continuous-time will not have its discrete-time model in the triangular structure in general after sampling. Hence, it is difficult to carry out directly design in discrete-time for continuous-time nonlinear systems.

For a control input designed based on the continuous-time model, it can be sampled and implemented in discrete-time. However, the stability cannot be automatically guaranteed for the sampled-data system resulted from emulation method. The stability analysis is challenging because there is no corresponding method in discrete-time for nonlinear systems after sampling, unlike linear systems. One would expect that if the sampling is fast enough, the stability of the sampled-data system might be guaranteed by the stability of the continuous-time system. In fact, there is a counter example to this claim. Therefore, it is important to analyse the stability of sampled-data system in addition to the stability of the continuous-time control, for which the stability is guaranteed in control design.

In this section, we will show the stability analysis of sampled-data control for a class of nonlinear systems in the output feedback form. A link between the sampling period and initial condition is established, which suggests that for a given domain of initial conditions, there always exists a sampling time such that if the sampling is faster than that time, the stability of the sampled-data system is guaranteed.

11.3.1 System model and sampled-data control

In this section, we consider sampled-data control for a class of nonlinear systems in the output feedback form. This class of nonlinear systems (9.17) has been considered for control design in continuous-time in Chapter 9, using observer backstepping in Section 9.3 and using backstepping with filtered transformation in Section 9.4. We will present sampled-data control based on the control input in continuous-time obtained using backstepping with filtered transformation. For the convenience of presentation, we will show the system again and outline the key steps in the continuous-time here. The system, as shown in (9.17) and (9.32), is described as

$$\dot{x} = A_c x + bu + \phi(y)$$
$$y = Cx,$$

$$(11.64)$$

with

$$A_c = \begin{bmatrix} 0 & 1 & 0 & \cdots & 0 \\ 0 & 0 & 1 & \cdots & 0 \\ \vdots & \vdots & \vdots & \ddots & \vdots \\ 0 & 0 & 0 & \cdots & 1 \\ 0 & 0 & 0 & \cdots & 0 \end{bmatrix}, \quad C = \begin{bmatrix} 1 \\ 0 \\ \vdots \\ 0 \end{bmatrix}^T, \quad b = \begin{bmatrix} 0 \\ \vdots \\ 0 \\ b_\rho \\ \vdots \\ b_n \end{bmatrix},$$

where $x \in \mathbb{R}^n$ is the state vector; $u \in \mathbb{R}$ is the control input; $\phi : \mathbb{R} \to \mathbb{R}^n$ with $\phi(0) = 0$ is a nonlinear function with element ϕ_i being differentiable up to the $(n - i)$th order; and $b \in \mathbb{R}^n$ is a known constant Hurwitz vector with $b_\rho \neq 0$, which implies the relative of the system is ρ.

We design an output feedback control for the system in (11.64). As shown in Chapter 9, the continuous-time control input can be designed using the backstepping with filtered transformation in Section 9.4. In this section, we carry out the stability analysis based on the control input obtained using backstepping with filtered transformation.

In this section, we need to deal with control in continuous-time and discrete-time. For notation, we use u_c for the control input obtained in continuous-time and u_d for the resultant sampled-data control input. For notational convenience, we re-write the filter (9.33) as

$$\dot{\xi} = \Lambda \xi + b_f u$$

$$(11.65)$$

$$u = u_c(y, \xi),$$

$$(11.66)$$

where

$$
\Lambda = \begin{bmatrix} -\lambda_1 & 1 & 0 & \cdots & 0 \\ 0 & -\lambda_2 & 1 & \cdots & 0 \\ \vdots & \vdots & \vdots & \vdots & \vdots \\ 0 & 0 & 0 & \cdots & -\lambda_{\rho-1} \end{bmatrix}, \quad b_f = \begin{bmatrix} 0 \\ \vdots \\ 0 \\ 1 \end{bmatrix}.
$$

For a given u_c, the sampled-data controller based on emulation method is given as

$$
u_d(t) = u_c(y(mT), \xi(mT)), \quad \forall t \in [mT, mT + T), \tag{11.67}
$$

$$
\xi(mT) = e^{\Lambda T}\xi((m-1)T) + b_f u_c(y((m-1)T), \xi((m-1)T)) \int_0^T e^{\Lambda \tau} d\tau, \tag{11.68}
$$

where $y(mT)$ is obtained by sampling $y(t)$ at each sampling instant; $\xi(mT)$ is the discrete-time implementation of the filter shown in (11.65); T is the fixed sampling period; and m is the discrete-time index, starting from 0.

Before the analysis of the sampled-data control, we briefly review the control design in continuous-time.

As shown in Section 9.4, through the state transformations, we obtain

$$
\begin{aligned}
\dot\zeta &= D\zeta + \psi(y) \\
\dot y &= \zeta_1 + \psi_y(y) + b_\rho \xi_1.
\end{aligned} \tag{11.69}
$$

For the system with relative degree $\rho = 1$, as shown in Lemma 9.5 and its proof, the continuous-time control u_{c1} can be designed as

$$
u_{c1} = -\left(c_0 + \frac{1}{4}\right)y - \psi_y(y) - \frac{\|P\|^2 \|\psi(y)\|^2}{y}, \tag{11.70}
$$

where c_0 is a positive real constant and P is a positive real constant that satisfies

$$
D^T P + PD = -3I.
$$

For the control (11.70), the stability of the continuous-time system can be established with the Lyapunov function

$$
V = \zeta^T P \zeta + \frac{1}{2}y^2
$$

and its derivative

$$
\dot V \le -c_0 y^2 - \|\zeta\|^2.
$$

For the case of $\rho > 1$, backstepping is used to obtain the final control input u_{c2}. We introduce the same notations z_i, for $i = 1, \ldots, \rho$, as in Section 9.4 for backstepping

$$
z_1 = y, \tag{11.71}
$$

$$
z_i = \xi_{i-1} - \alpha_{i-1}, \quad \text{for } i = 2, \ldots, \rho, \tag{11.72}
$$

$$
z_{\rho+1} = u - \alpha_\rho, \tag{11.73}
$$

where α_i for $i = 2, \ldots, \rho$ are stabilising functions to be designed. We also use the positive real design parameters c_i and k_i for $i = 1, \ldots, \rho$ and $\gamma > 0$.

The control design has been shown in Section 9.4. We list a few key steps here for the convenience of the stability analysis. The stabilising functions are designed as

$$\alpha_1 = -c_1 z_1 - k_1 z_1 + \psi_y(y) - \frac{\gamma \|P\|^2 \|\psi(y)\|^2}{y},$$

$$\alpha_2 = -z_1 - c_2 z_2 - k_2 \left(\frac{\partial \alpha_1}{\partial y}\right)^2 z_2 + \frac{\partial \alpha_1}{\partial y}\psi(y) + \lambda_1 \xi_1$$

$$\alpha_i = -z_{i-1} - c_i z_i - k_i \left(\frac{\partial \alpha_{i-1}}{\partial y}\right)^2 z_i + \frac{\partial \alpha_{i-1}}{\partial y}\psi(y) + \lambda_{i-1}\xi_{i-1}$$

$$+ \sum_{j=1}^{i-2} \frac{\partial \alpha_{i-1}}{\partial \xi_j}(-\lambda_j \xi_j + \xi_{j+1}) \quad \text{for } i = 3, \ldots, \rho.$$
(11.74)

The continuous-time control input is given by $u_{c2} = \alpha_\rho$, that is

$$u_{c2} = -z_{\rho-1} - c_\rho z_\rho - k_\rho \left(\frac{\partial \alpha_{\rho-1}}{\partial y}\right)^2 z_\rho + \frac{\partial \alpha_{\rho-1}}{\partial y}\psi(y)$$

$$+ \sum_{j=1}^{i-2} \frac{\partial \alpha_{\rho-1}}{\partial \xi_j}(-\lambda_j \xi_j + \xi_{j+1}) + \lambda_{\rho-1}\xi_{\rho-1}.$$
(11.75)

For the control input (11.75) in the continuous-time, the stability result has been shown in Theorem 9.6. In the stability analysis, the Lyapunov function candidate is chosen as

$$V = \sum_{i=1}^{\rho} z_i^2 + \gamma \zeta^T P \zeta$$

and its derivative is shown to satisfy

$$\dot{V} \le -\sum_{i=1}^{\rho} c_i z_i^2 - \gamma \|\zeta\|^2.$$
(11.76)

11.3.2 Stability analysis of sampled-data systems

The following lemma is needed for stability analysis in sampled-data case.

Lemma 11.10. *Let* $V : \mathbb{R}^n \to \mathbb{R}^+$ *be a continuously differentiable, radially unbounded, positive definite function. Define* $\mathcal{D} := \{\chi \in \mathbb{R}^n | V(\chi) \le r\}$ *with* $r > 0$. *Suppose*

$$\dot{V} \le -\mu V + \beta V_m, \quad \forall t \in (mT, (m+1)T],$$
(11.77)

hold for all $\chi(mT) \in \mathcal{D}$, *where* μ, β *are any given positive reals with* $\mu > \beta$, $T > 0$ *the fixed sampling period and* $V_m := V(\chi(mT))$. *If* $\chi(0) \in \mathcal{D}$, *then the following holds:*

$$\lim_{t \to \infty} \chi(t) = 0.$$

Proof. Since $\chi(0) \in \mathcal{D}$, then (11.77) holds for $t \in (0, T]$ with the following form:

$$\dot{V} \le -\mu V + \beta V(\chi(0)).$$

Using the comparison lemma (Lemma 4.5), it is easy to obtain from the above that for $t \in (0, T]$

$$V(\chi(t)) \le e^{-\mu t} V_0 + \frac{1 - e^{-\mu t}}{\mu} \beta V_0 = q(t) V_0, \tag{11.78}$$

where $q(t) := (e^{-\mu t} + \frac{\beta}{\mu}(1 - e^{-\mu t}))$. Since $\mu > \beta > 0$, then $q(t) \in (0, 1)$, $\forall t \in (0, T]$. Then we have

$$V(\chi(t)) < V_0, \quad \forall t \in (0, T]. \tag{11.79}$$

Particularly, setting $t = T$ in (11.78) leads to

$$V_1 \le q(T) V_0, \tag{11.80}$$

which means that $\chi(T) \in \mathcal{D}$. Therefore, (11.77) holds for $t \in (T, 2T]$. By induction, we have

$$V(\chi(t)) < V_m, \quad \forall t \in (mT, (m+1)T]. \tag{11.81}$$

which states inter-sample behaviour of the sampled-data system concerned, and in particular

$$V_{m+1} \le q(T) V_m \tag{11.82}$$

indicating that V decreases at two consecutive sampling points with a fixed ratio. From (11.82),

$$V_m \le q(T) V_{m-1} \le q^m(T) V_0, \tag{11.83}$$

which implies that $\lim_{m \to \infty} V_m = 0$. The conclusion then follows from (11.81), which completes the proof.   □

Remark 11.6. Lemma 11.10 plays an important role in stability analysis for the sampled-data system considered in this section. When sampled-data systems are analysed in discrete-time in literature, only the performances at sampling instances are considered. In this section, Lemma 11.10 provides an integrated analysis framework where the behaviour of the sampled-data system both at and between sampling instants can be characterised. In particular, the system's behaviour at sampling instants is portrayed in (11.82), and (11.81) shows that the inter-sample behaviour of the system is well bounded by a compact set defined by V_m.   ◁

For the case of relative degree 1, the sampled-data system takes the following form:

$$\dot{\zeta} = D\zeta + \psi(y)$$
$$\dot{y} = \zeta_1 + \psi_y(y) + u_{d1}, \tag{11.84}$$

where u_{d1} is the sampled-data controller and can be simply implemented via a zero-order hold device as the following:

$$u_{d1}(t) = u_{c1}(y(m)), \quad \forall t \in [mT, mT + T). \tag{11.85}$$

Define $\chi := [\zeta^T, y^T]^T$ and we have the following result.

Theorem 11.11. *For system (11.84) and the sampled-data controller u_{d1} shown in (11.85), and a given neighbourhood of the origin $B_r := \{\chi \in \mathbb{R}^n |\ \|\chi\| \le r\}$ with r any given positive real, there exists a constant $T_1 > 0$ such that, for all $0 < T < T_1$ and for all $\chi(0) \in B_r$, the system is asymptotically stable.*

Proof. We choose

$$V(\chi) = \zeta^T P\zeta + \frac{1}{2}y^2$$

as the Lyapunov function candidate for the sampled-data system. We start with some sets used throughout the proof. Define

$$c := \max_{\chi \in B_r} V(\chi)$$

and the level set

$$\Omega_c := \{x \in \mathbb{R}^n | V(\chi) \le c\}.$$

There exist two class $\mathcal{K}$ functions ϱ_1 and ϱ_2 such that

$$\varrho_1(\|\chi\|) \le V(\chi) \le \varrho_2(\|\chi\|).$$

Let $l > \varrho_1^{-1}(c)$, and define $B_l := \{\chi \in \mathbb{R}^n |\ \|\chi\| \le l\}$. Then we have

$$B_r \subset \Omega_c \subset B_l.$$

The constants L_{u1}, L_1 and L_2 are Lipschitz constants of the functions u_{c1} and ψ_y and ψ with respect to B_l.

These local Lipschitz conditions establish that for the overall sampled-data system with $\chi(0) \in \Omega_c$, there exists a unique solution $\chi(t)$ over some interval $[0, t_1)$. Notice that t_1 might be finite. However, later analysis shows that the solution can be extended one sampling interval after another, and thus exists for all $t \ge 0$ with the property that $\lim_{t \to \infty} \chi(t) = 0$. Particularly, we intend to formulate the time derivative of the Lyapunov function V into the form shown in (11.77) or (11.93), which is shown below.

Consider the case when $t = 0$, $\chi(0) \in B_r \subset \Omega_c$. First, choose a sufficiently large l such that there exists a $T_1^* > 0$, and for all $T \in (0, T_1^*)$, the following holds:

$$\chi(t) \in B_l, \quad \forall t \in [0, T], \quad \chi(0) \in \Omega_c. \tag{11.86}$$

The existence of T_1^* is ensured by continuous dependency of the solution $\chi(t)$ on the initial conditions.

Next, calculate the estimate of $|y(t) - y(0)|$ forced by the sampled-data control u_{d1} during the interval $[0, T]$, provided that $\chi(0) \in B_r \subset \Omega_c$ and $T \in (0, T_1^*)$. From the second equation of (11.84), the dynamics of y are given by

$$\dot{y} = \zeta_1 + \psi_y(y) + u_{d1}.$$

It follows that

$$y(t) = y(0) + \int_0^t \zeta_1(\tau)d\tau + \int_0^t u_{d1}(\tau)d\tau$$

$$+ \int_0^t \left(\psi_y(y) - \psi_y(y(0)) \right) d\tau + \int_0^t \psi_y(y(0))\, d\tau.$$

Then we have

$$|y(t) - y(0)| \leq \underbrace{\int_0^t \|\zeta(\tau)\|d\tau + \int_0^t L_{u1}|y(0)|d\tau}_{\Delta_1}$$

$$+ \int_0^t L_1|y(\tau) - y(0)|d\tau + \int_0^t L_1|y(0)|d\tau. \tag{11.87}$$

We first calculate the integral Δ_1. From the first equation of system (11.69), we obtain

$$\zeta(t) = e^{Dt}\zeta(0) + \int_0^t e^{Dt}\psi(y(\tau))d\tau. \tag{11.88}$$

Since D is a Hurwitz matrix, there exist positive reals κ_1 and σ_1 such that $\|e^{Dt}\| \leq \kappa_1 e^{-\sigma_1 t}$. Thus, from (11.88)

$$\|\zeta(t)\| \leq \kappa_1 e^{-\sigma_1 t}\|\zeta(0)\| + \int_0^t \kappa_1 e^{-\sigma_1(t-\tau)}\|\psi(y(\tau)) - \psi(y(0))\|d\tau$$

$$+ \int_0^t \kappa_1 e^{-\sigma_1(t-\tau)}\|\psi(y(0))\|d\tau$$

$$\leq \kappa_1 e^{-\sigma_1 t}\|\zeta(0)\| + L_2 \int_0^t \kappa_1 e^{-\sigma_1(t-\tau)}|y(\tau) - y(0)|d\tau$$

$$+ L_2 \int_0^t \kappa_1 e^{-\sigma_1(t-\tau)}|y(0)|d\tau.$$

Then the following inequality holds:

$$\Delta_1 \leq \frac{\kappa_1\|\zeta(0)\|}{\sigma_1}(1 - e^{-\sigma_1 t}) + \frac{\kappa_1 L_2}{\sigma_1}|y(0)|t$$

$$+ \frac{\kappa_1 L_2}{\sigma_1} \int_0^t |y(\tau) - y(0)|d\tau. \tag{11.89}$$

Now we are ready to compute $|y(t) - y(0)|$. In fact, we have from (11.87) and (11.89)

$$|y(t) - y(0)| \le A_1(1 - e^{-\sigma_1 t}) + B_1 t + H \int_0^t |y(\tau) - y(0)| d\tau, \qquad (11.90)$$

where

$$A_1 = \sigma_1^{-1} \kappa_1 \|\zeta(0)\|,$$

$$B_1 = L_{u1}|y(0)| + L_1|y(0)| + \sigma_1^{-1}\kappa_1 L_2|y(0)|,$$

$$H = \sigma_1^{-1}\kappa_1 L_2 + L_1.$$

Applying Gronwall–Bellman inequality to (11.90) produces

$$|y(t) - y(0)| \le A_1(1 - e^{-\sigma_1 t}) + \frac{B_1}{H}(e^{Ht} - 1)$$

$$+ A_1(\sigma_1 e^{Ht} + He^{-\sigma_1 t} - (H + \sigma_1))(H + \sigma_1)^{-1} \qquad (11.91)$$

Setting $t = T$ on the right side of (11.91) leads to

$$|y(t) - y(0)| \le \delta_1(T)|y(0)| + \delta_2(T)\|\zeta(0)\|,$$

where

$$\delta_1(T) = H^{-1}(L_{u1} + L_1 + \sigma_1^{-1}\kappa_1 L_2)(e^{HT} - 1)$$
$$\delta_2(T) = \sigma_1^{-1}\kappa_1(\sigma_1 e^{HT} + He^{-\sigma_1 T} - (H + \sigma_1))(H + \sigma_1)^{-1}$$
$$+ \sigma_1^{-1}\kappa_1(1 - e^{-\sigma_1 T}). \qquad (11.92)$$

Note that $\delta_1(T)$ and $\delta_2(T)$ only depend on the sampling period T once the Lipschitz constants and the control parameters are chosen.

Next we shall study the behaviour of the sampled-data system during each interval using a Lyapunov function candidate $V(y, \zeta) = \zeta^T P \zeta + \frac{1}{2}y^2$. Consider $\chi(0) = [\zeta(0), y(0)]^T \in B_r$. When $t \in (0, T]$, its time derivative satisfies

$$\dot{V} = -3\|\zeta\|^2 + 2\zeta^T P\psi + y(\zeta_1 + \psi_y + u_{d1})$$

$$\le -c_0 y^2 - \|\zeta\|^2 + |y||u_{d1}(y(0)) - u_{c1}|$$

$$\le -c_0 y^2 - \|\zeta\|^2 + L_{u1}|y - y(0)||y|$$

$$\le -\left(c_0 - \frac{L_{u1}}{2}(\delta_1(T) + \delta_2(T))\right)y^2 - \|\zeta\|^2$$

$$+ \frac{L_{u1}}{2}\delta_1(T)|y(0)|^2 + \frac{L_{u1}}{2}\delta_2(T)\|\zeta(0)\|^2$$

$$\le -\mu_1(T)V + \beta_1(T)V(\zeta(0), y(0)), \qquad (11.93)$$

where

$$\mu_1 = \min \left\{ 2c_0 - L_{u1}\delta_1 - L_{u1}\delta_2(T), \frac{1}{\lambda_{max}(P)} \right\},$$

$$\beta_1 = \max \left\{ L_{u1}\delta_1(T), \frac{L_{u1}\delta_2(T)}{2\lambda_{min}(P)} \right\},$$

(11.94)

with $\lambda_{max}(\cdot)$ and $\lambda_{min}(\cdot)$ denoting the maximum and minimum eigenvalues of a matrix respectively.

Next we shall show that there exists a constant $T_2^* > 0$ such that the condition $\mu_1(T) > \beta_1(T) > 0$ is satisfied for all $0 < T < T_2^*$. Note from (11.92) that both $\delta_1(T)$ and $\delta_2(T)$ are actually the continuous functions of T with $\delta_1(0) = \delta_2(0)$. Define $e_1(T) := \mu_1(T) - \beta_1(T)$ and we have $e_1(0) > 0$. It can also be established from (11.94) that $e_1(T)$ is a decreasing and continuous function of T, which asserts by the continuity of $e_1(T)$ the existence of T_2^* so that for $0 < T < T_2^*$, $e_1(T) > 0$, that is $0 < \beta_1(T) < \mu_1(T)$.

Finally, set $T_1 = \min(T_1^*, T_2^*)$, and from Lemma 11.10, it is known that $V_1 \leq c$, which means $\chi(T) \in \Omega_c$, and subsequently, all the above analysis can be repeated for every interval $[mT, mT + T]$. Applying Lemma 11.10 completes the proof.   □

For systems with relative degree $\rho > 1$, the implementation of the sampled-data controller u_{d2} is given in (11.67) and (11.68). It is easy to see from (11.68) that $\xi(mT)$ is the exact, discrete-time model of the filter

$$\dot{\xi} = -\Lambda\xi + b_f u_{d2}$$

(11.95)

due to the fact that u_d remains constant during each interval and the dynamics of ξ shown in (11.95) is linear. Then (11.68) and (11.95) are virtually equivalent at each sampling instant. This indicates that we can use (11.95) instead of (11.68) for stability analysis of the sampled-data system.

Let $\chi := [\zeta^T, z^T]^T$ and we have the following result.

Theorem 11.12. *For the extended system consisting (11.65), (11.69) and the sampled-data controller u_{d2} shown in (11.67) and (11.68), and a given neighbourhood of the origin $B_r := \{\chi \in R^n | \|\chi\| \leq r\}$ with r any given positive real, there exists a constant $T_2 > 0$ such that, for all $0 < T < T_2$ and for all $\chi(0) \in B_r$, the system is asymptotically stable.*

Proof. The proof can be carried out in a similar way to that for the case $\rho = 1$, except that the effect of the dynamic filter of ξ has to be dealt with.

For the overall sampled-data system, a Lyapunov function candidate is chosen the same as for continuous-time case, that is

$$V = \gamma\zeta^T P\zeta + \frac{1}{2}\sum_{i=1}^{\rho} z_i^2.$$

Similar to the case of $\rho = 1$, the sets B_r, Ω_c and B_l can also be defined such that $B_r \subset \Omega_c \subset B_l$, and there exists a $T_3^* > 0$ such that for all $T \in (0, T_3^*)$, the following holds:

$$\chi(t) \in B_l, \quad \forall t \in (0, T], \quad \chi(0) \in \Omega_c.$$

As in the proof for Theorem 11.11, we also aim to formulate the time derivative of $V(\chi)$ into form (11.77). Next we shall derive the bounds for $\|\xi(t) - \xi(0)\|$ and $|y(t) - y(0)|$ during $t \in [0, T]$ with $0 < T < T_3^*$. Consider the case where $\chi(0) \in B_r$. We have from (11.95)

$$\xi(t) = e^{\Lambda t} \xi(0) + \int_0^t e^{\Lambda(t-\tau)} b_f u_{d2} d\tau. \tag{11.96}$$

Since Λ is a Hurwitz matrix, there exist positive reals κ_2, κ_3 and σ_2 such, that

$$\|e^{\Lambda t}\| \leq \kappa_2 e^{-\sigma_2 t},$$

$$\|e^{\Lambda t} - I\| \leq \kappa_3 (1 - e^{-\sigma_2 t}),$$

where I is the identity matrix. Then, using the Lipschitz property of u_{c2} with respect to the set B_l and the fact that $u_{d2}(0,0) = 0$, it can be obtained from (11.96) that

$$\int_0^t \|\xi(\tau)\| d\tau \leq \frac{\kappa_2 \|\xi(0)\|}{\sigma_2} (1 - e^{-\sigma_2 t}) + \frac{\kappa_2 L_{u2}}{\sigma_2} (|y(0)| + \|\xi(0)\|) t \tag{11.97}$$

and

$$\|\xi(t) - \xi(0)\| \leq \kappa_3 \|\xi(0)\| (1 - e^{-\sigma_2 t}) + \|u_{d2}(y(0), \xi(0))\| \int_0^t \kappa_2 e^{-\sigma_2(t-\tau)} d\tau$$

$$\leq \delta_3(T) |y(0)| + \delta_4(T) \|\xi(0)\|, \tag{11.98}$$

where

$$\delta_3(T) = \sigma_2^{-1} \kappa_2 L_{u2} (1 - e^{-\sigma_2 T}),$$

$$\delta_4(T) = (\kappa_3 + \sigma_2^{-1} \kappa_2 L_{u2})(1 - e^{-\sigma_2 T}),$$

and L_{u2} is a Lipschitz constant of u_{c2}. As for $|y - y(0)|$, we have from (11.69)

$$y(t) = y(0) + \int_0^t \zeta_1(\tau) d\tau + \int_0^t \xi(\tau) d\tau$$

$$+ \int_0^t (\psi_y(y) - \psi_y(y(0))) d\tau + \int_0^t \psi_y(y(0)) d\tau.$$

It can then be shown that

$$|y(t) - y(0)| \leq \underbrace{\int_0^t \|z(\tau)\| d\tau}_{\Delta_1} + \underbrace{\int_0^t \|\xi(\tau)\| d\tau}_{\Delta_2}$$

$$+ \int_0^t L_1 |y(\tau) - y(0)| d\tau + \int_0^t L_1 |y(0)| d\tau, \tag{11.99}$$

where Δ_1 is already shown in (11.89) and Δ_2 in (11.97). With (11.89), (11.97) and (11.99), it follows that

$$|y(t) - y(0)| \leq A_1(1 - e^{-\sigma_1 t}) + A_2(1 - e^{-\sigma_2 t}) + B_2 t$$
$$+ H \int_0^t |y(\tau) - y(0)| d\tau \qquad (11.100)$$

where A_1 and H are defined in (11.90)

$$A_2 = \sigma_2^{-1} \kappa_2 \|\xi(0)\|,$$
$$B_2 = L_1 |y(0)| + \sigma_1^{-1} \kappa_1 L_2 |y(0)| + \sigma_2^{-1} \kappa_2 L_{u2} |y(0)| + \sigma_2^{-1} \kappa_2 L_{u2} \|\xi(0)\|.$$

Defining $A_3 := A_1 + A_2$ and $\sigma_0 := \max(\sigma_1, \sigma_2)$, we have

$$|y(t) - y(0)| \leq A_3(1 - e^{-\sigma_0 t}) + B_2 t + H \int_0^t |y(\tau) - y(0)| d\tau.$$

Applying the Gronwall–Bellman lemma produces

$$|y(t) - y(0)| \leq \delta_5(T)|y(0)| + \delta_6(T)\|z(0)\| + \delta_7(T)\|\xi(0)\|, \qquad (11.101)$$

where

$$\delta_5(T) = H^{-1}(L_1 + \sigma_1^{-1}\kappa_1 L_2 + \sigma_2 \kappa_2^{-1} L_{u2})(e^{HT} - 1),$$
$$\delta_6(T) = \sigma_1^{-1}\kappa_1(\sigma_0 e^{HT} + H e^{-\sigma_0 T} - (H + \sigma_0))(H + \sigma_0)^{-1} + \sigma_1^{-1}\kappa_1(1 - e^{-\sigma_0 T}),$$
$$\delta_7(T) = \sigma_2^{-1}\kappa_2(1 - e^{-\sigma_0 T}) + \sigma_2^{-1}\kappa_2 L_{u2}(e^{HT} - 1)$$
$$+ \sigma_2^{-1}\kappa_2 L_{u2}(\sigma_0 e^{HT} + H e^{-\sigma_0 T} - (H + \sigma_0))(H + \sigma_0)^{-1}.$$

Note that $\|\xi(0)\|$ appears in (11.98) and (11.101) while for the analysis using Lyapunov function to carry on, $\|z(0)\|$ is needed. Therefore, it is necessary to find out an expression of $\|\xi(0)\|$ that exclusively involves $\|z(0)\|$, which is shown below.

Notice that due to the special structure of the filter (11.65) and the backstepping technique, each stabilising function has the property that $\alpha_1 = \alpha_1(y)$, $\alpha_1(0) = 0$, and $\alpha_i = \alpha_i(y, \xi_1, \ldots, \xi_{i-1})$ and $\alpha_i(0, \ldots, 0) = 0$, $i = 2, \ldots, \rho - 1$. From (11.74) we have

$$|\xi_1(0)| \leq |z_2(0)| + |\alpha_1(0)| \leq z_2(0) + \mathcal{L}_1|y(0)|$$

$$|\xi_2(0)| \leq |z_3(0)| + |\alpha_2(0)| \leq |z_3(0)| + \mathcal{L}_2|y(0)| + \mathcal{L}_2|\xi_1(0)|$$

$$\vdots$$

$$|\xi_{\rho-1}(0)| \leq |z_\rho(0)| + |\alpha_{\rho-1}(0)| \leq |z_\rho(0)| + \mathcal{L}_{\rho-1}|y(0)| + \mathcal{L}_{\rho-1}\sum_{i=1}^{\rho-2}|\xi_i(0)|$$

where with a bit abuse of notation, $\mathcal{L}_i$ is the Lipschitz constant of $\hat{\xi}_i$ with respect to the set B_l. Thus, a constant $\mathcal{L}_0$ can be found such that the following holds:

$$\|\xi(0)\| \leq \mathcal{L}_0(\|\bar{z}(0)\| + |y(0)|). \qquad (11.102)$$

with $\bar{z} = [z_2, \ldots, z_\rho]^T$, which implies that if $[\zeta(0)^T, y(0)^T, \bar{z}(0)^T]^T \in B_l$, then $[\zeta(0)^T, y(0)^T, \xi(0)^T]^T$ will be confined in a bounded set, denoted by B'_l.

Then the time derivative of

$$V_{d2} = \gamma \zeta^T P \zeta + \frac{1}{2}y^2 + \frac{1}{2}\sum_{i=2}^{\rho} z_i^2$$

during the interval $(0, T]$ satisfies

$$\dot{V}_{d2} \leq -\gamma \|\zeta\|^2 - \sum_{i=1}^{\rho} c_i z_i^2 + z_\rho(u_{d2} - u_{c2})$$

$$\leq -c_1 y^2 - \gamma \|\zeta\|^2 - \lambda_0 \sum_{i=2}^{\rho} z_i^2 + \|\bar{z}\| |u_{d2} - u_{c2}|, \tag{11.103}$$

where $\lambda_0 = \min\{c_2, \ldots, c_\rho\}$. In addition, we have

$$\|\xi\| |u_{d2} - u_{c2}| \leq L_{u2} \|\xi\| (|y - y(0)| + \|\xi - \xi(0)\|)$$

$$\leq L_{u2} \|\xi\| (\delta_3(T) + \delta_5(T)) |y(0)| + L_{u2} \|\xi\| \delta_6(T) \|\zeta(0)\|$$

$$+ L_{u2} \|\xi\| (\delta_4(T) + \delta_7(T)) \|\xi(0)\|$$

$$\leq \varepsilon_1(T) |y(0)|^2 + \varepsilon_2(T) \|\zeta(0)\|^2 + \varepsilon_3(T) \|\xi(0)\|^2$$

$$+ \varepsilon_4(T) \|\bar{z}\|^2, \tag{11.104}$$

where $\varepsilon_1(T) = \frac{L_{u2}}{2}(\delta_3(T) + \delta_5(T))$, $\varepsilon_2 = \frac{L_{u2}}{2}\delta_6(T)$, $\varepsilon_3 = \frac{L_{u2}}{2}(\delta_4(T) + \delta_7(T))$, $\varepsilon_4 = \frac{L_{u2}}{2}\sum_{i=3}^{7}\delta_i(T)$ and L_{u2} is a Lipschitz constant of u_{c2} with respect to the set B'_l dependent on B_l.

From (11.102) to (11.103), we then have

$$\dot{V}_{d2} \leq -c_1 y^2 - \gamma \|\zeta\|^2 - (\lambda_0 - \varepsilon_4)\bar{z}^2$$

$$+ (\varepsilon_1(T) + 2\mathcal{L}_0^2 \varepsilon_3(T)) |y(0)|^2 + \varepsilon_2(T) \|\zeta(0)\|^2 + 2\mathcal{L}_0^2 \varepsilon_3(T) \|\bar{z}(0)\|^2$$

$$= -\alpha_2(T) V_{d2} + \beta_2(T) V_{d2}(\zeta(0), y(0), \bar{z}(0)),$$

where

$$\alpha_2(T) = \min\left\{2c_1, \frac{\gamma}{\lambda_{max}(P)}, 2(\lambda_0 - \varepsilon_4(T))\right\},$$

$$\beta_2(T) = \max\left\{2(\varepsilon_1(T) + 2\mathcal{L}_0^2 \varepsilon_3(T)), \frac{\varepsilon_2(T)}{\lambda_{min}(P)}, 4\mathcal{L}_0^2 \varepsilon_3(T)\right\}.$$

Note from (11.98), (11.101) and (11.104) that each ε_i $(1 \le i \le 4)$ is a continuous function of T with $\varepsilon_i(0) = 0$. Then, as shown in the analysis for the case $\rho = 1$, it can be claimed that given arbitrary positive reals c_1, γ and λ, there exists a constant $T_4^* > 0$ such that for all $0 < T < T_4^*$, $\alpha_2(T) > \beta_2(T) > 0$.

Finally, letting $T_2 := \min(T_3^*, T_4^*)$ proves the theorem. □

Remark 11.7. In the results shown in Theorems 11.11 and 11.12, the radius r of B_r can be any positive value. It means that we can set the stability region as large as we like, and the stability can still be guaranteed by determining a fast enough sampling time. Therefore, Theorems 11.11 and 11.12 establish the semi-global stability of the sampled-data control of nonlinear systems in the output feedback form. ◁

11.3.3 Simulation

The following example is a simplified model of a jet engine without stall:

$$\dot{\phi}_m = \psi + \frac{3}{2}\phi_m^2 - \frac{1}{2}\phi_m^3$$

$$\dot{\psi} = u,$$

where ϕ_m is the mass flow and ψ the pressure rise. Take ϕ_m as the output and then the above system is in the output feedback form. The filter $\dot{\xi} = -\lambda\xi + u$ is introduced so that the filtered transformation $y = x_1$ and $\bar{\zeta} = x_2 - \xi$, and the state transformation $\zeta = \bar{\zeta} - \lambda y$ can render the system into the following form:

$$\dot{\zeta} = -\lambda\zeta - \lambda\left(\frac{3}{2}y^2 - \frac{1}{2}y^3\right) - \lambda^2 y$$

$$\dot{y} = \zeta + \lambda y + \left(\frac{3}{2}y^2 - \frac{1}{2}y^3\right) + \xi.$$

Finally, the stabilising function

$$\alpha_1 = -y - \left(\frac{3}{2}y^2 - \frac{1}{2}y^3\right) - \lambda^2 y\left(\frac{3}{2}y - \frac{1}{2}y^2 + \lambda\right)^2$$

with $P = 1$, and the control u_c can be obtained using α_2 as shown in (11.74). For simulation, we choose $\lambda = 0.5$.

Simulations are carried out with the initial values set as $x_1(0) = 1$ and $x_2(0) = 1$, which are for simulation purpose only and have no physical meaning. Results shown in Figures 11.10 and 11.11 indicate that the sampled-data system is asymptotically stable when $T = 0.01$ s, which is confirmed by a closer look at the convergence of V shown in Figure 11.12. Further simulations show that the overall system is unstable if $T = 0.5$ s. In summary, the example illustrates that for a range of sampling period T, the sampled-data control design presented earlier in this section can asymptotically stabilise the sampled-data system.

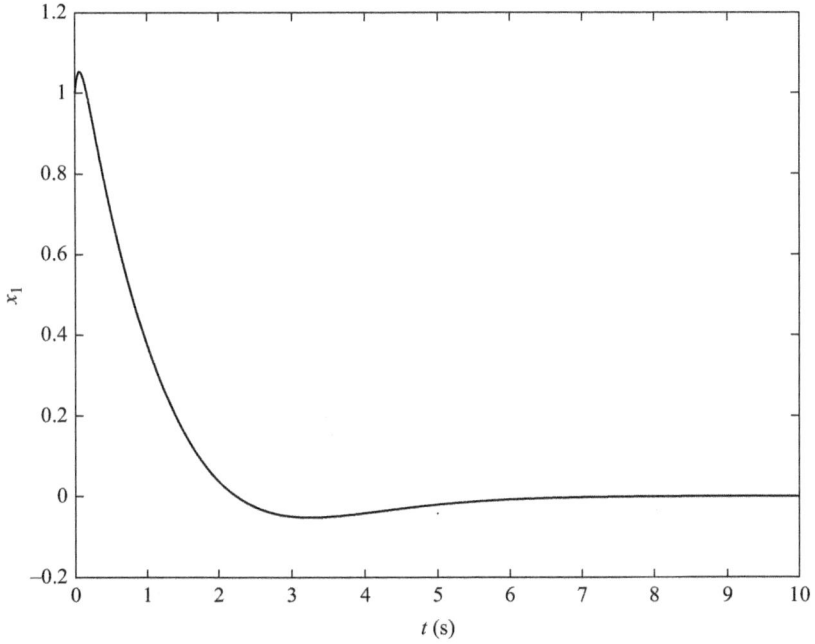

Figure 11.10   The time response of x_1 for $T = 0.01$ s

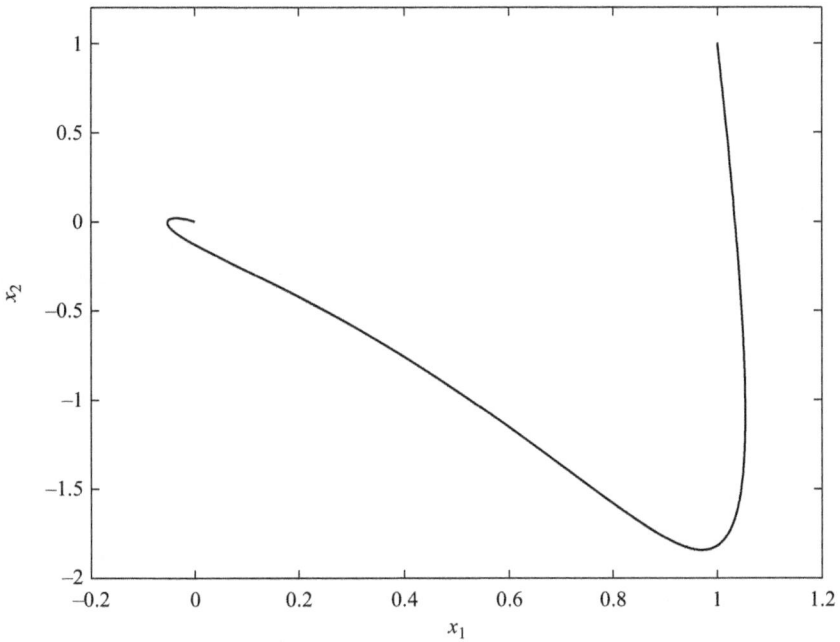

Figure 11.11   Phase portrait of the system for $T = 0.01$ s

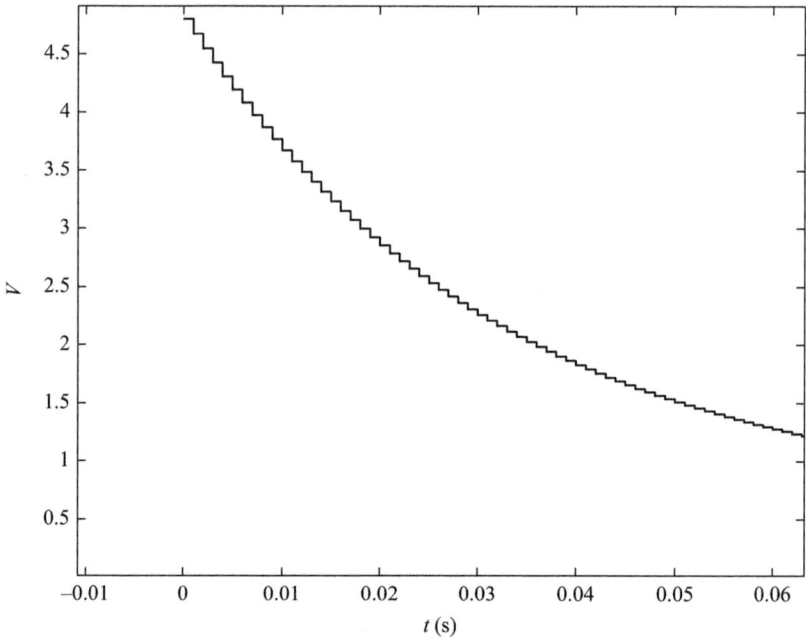

Figure 11.12   The convergence of V for T = 0.01 s

Bibliographical Notes

Chapter 1. The basic concepts of systems and states of nonlinear systems discussed in this chapter can be found in many books on nonlinear systems and nonlinear control systems, for example Cook [1], Slotine and Li [2], Vidyasagar [3], Khalil [4] and Verhulst [5]. For the background knowledge of linear systems, readers can consult Ogata [6], Kailth [7]. Antsaklis and Michel [8], Zheng [9], and Chen, Lin and Shamash [10]. The existence of unique solutions of differential equations is discussed in the references such as Arnold [11] and Borrelli and Coleman [12]. Many concepts such as stability and backstepping control design are covered in detail in later chapters of the book. For system with saturation, readers may consult Hu and Lin [13]. Sliding mode control is covered in detail in Edwards and Spurgeon [14] and feedforward control in Isidori [15]. Limit cycles and chaos are further discussed in Chapter 2. Semi-global stability is often related to systems with saturation, as discussed in Hu and Lin [13], and Isidori [15].

Chapter 2. Lipschitz conditions and the existence of solutions for differential equations are covered in detail in Arnold [11] and Borrelli and Coleman [12]. Classifications of singular points of second-order nonlinear systems are covered in almost all the introductory texts on nonlinear control systems, Cook [1], Slotine and Li [2], Vidyasagar [3] and Khalil [4]. More phase portraits of systems similar to the one in Example 2.1 can be found in Slotine and Li [2], Khalil [4] and Verhulst [5]. The swing machine equation in Example 2.2 was a simplified model of the system shown in Elgerd [16] and a similar treatment is shown in Cook [1]. More examples of similar systems can also be found in Verhulst [5]. The van der Pol oscillator was first discussed in his publication van der Pol [17], and perhaps it is one of the best known systems with limit cycles. Detailed coverage of van der Pol oscillator can be found in several books, including Verhulst [5] and Borrelli and Coleman [12]. Lorenz attractor was first shown in Lorenz [18] for his research on heat transfer. The boundedness of the trajectory shown in the book was taken from Borrelli and Coleman [12].

Chapter 3. Describing function analysis is covered in many classic texts on nonlinear control, such as Cook [1], Slotine and Li [2], Vidyasagar [3], and Khalil [4]. The stability criterion of limit cycles with describing functions is influenced by the treatment in Slotine and Li [2]. The describing function analysis for van der Pol systems is adapted from Cook [1]. More results on describing functions are to be found in Atherton [19] and Gelb and Velde [20].

Chapter 4. Lyapunov's work on stability was published in Russian in 1892 and was translated in French in 1907 [21]. Lyapunov stability is covered in almost all the books on nonlinear systems and control. One of the early books on motion stability is published by Hahn [22]. Example 4.4 is adapted from Slotine and Li [2], and a more general form is shown in Reference [5]. More discussions on radially unbounded functions and global stability can be found in Khalil [4] and in Slotine and Li [2]. Further results on comparison lemma and its applications are to be found in Khalil [4] and Vidyasagar [3].

Chapter 5. The definition of positive real systems are common in many books such as Slotine and Li [2], Khalil [4] and Vidyasagar [3]. For strictly positive real systems, there may have some variations in definitions. A discussion on the variations can be found in Narendra and Annaswamy [23]. The definition in this book follows Slotine and Li [2], Khalil [4]. Lemma 5.3 was adapted from Khalil [4], where a poof of necessity can also be found. Kalman–Yakubovich Lemma can appear in different variations. A proof of it can be found in Khalil [4] and Marino and Tomei [24]. A book on absolute stability was published by Narendra and Taylor [25]. Circle criterion is covered in a number of books such as Cook [1], Slotine and Li [2], Vidyasagar [3] and Khalil [4]. The treatment of loop-transformation to obtain the condition of circle criterion is similar to Cook [1]. The interpretation based on Figure 5.3 is adapted from Khalil [4]. Complex bilinear mapping is also used to justify the circle interpretation. Basic concepts of complex bilinear mapping can be found from text books on complex analysis, such as Brown and Churchill [26]. Input-to-state stability was first introduced by Sontag [27]. For the use of comparison functions of classes $\mathcal{K}$ and $\mathcal{K}_\infty$, more are to be found in Khalil [4]. The definitions for ISS are adapted from Isidori [15]. Theorem 5.8 is a simplified version, by using powers of the state norm, rather than $\mathcal{K}_\infty$ to characterise the stability for the convenience of presentation. The original result of Theorem 5.8 was shown by Sontag and Wang [28]. Using ISS pair (α, σ) for interconnected systems is based on a technique, changing supply functions, which is shown in Sontag and Teel [29]. The small gain theorem for ISS was shown in Jiang, Teel and Praly [30]. A systematic presentation of ISS stability issues can be found in Isidori [15]. Differential stability describes the stability issue of nonlinear systems for observer design. It is introduced in Ding [31], and the presentation in Section 5.4 is adapted from that paper.

Chapter 6. Fundamentals of Lie derivatives and differential manifolds can be found in Boothby [32]. A brief description of these concepts, sufficient for the concepts used in this book, is presented in Isidori [33]. Early results on exact linearisation in state space started from Brockett [34], and the presentation in this chapter was greatly influenced by the work of Isidori [33] and Marino and Tomei [24]. A proof of Frobenius Theorem can be found in Isidori [33].

Chapter 7. For self-tuning control, readers can refer to the textbooks such as Astrom and Wittenmark [35] and Wellstead and Zarrop [36]. The approach to obtain the model reference control is based on Ding [37]. Early work of using Lyapunov function for stability analysis can be found in Parks [38]. A complete treatment of MRAC of linear systems with high relative degrees can be found in Narendra and Annaswamy [23] and Ioannou and Sun [39]. The example to show the divergence of parameter estimation under a bounded disturbance is adapted from Ioannou and Sun [39], where more robust adaptive laws for linear systems are to be found. Results on robust adaptive control and on relaxing assumptions of the sign of high-frequency gain, minimum-phase, etc., for nonlinear systems can be found in Ding [37, 40–45].

Chapter 8. Linear observer design is covered in many books on linear systems. A proof of Lemma 8.2 can be found in Zheng [9]. Early results on nonlinear observer design can be traced back to 1970s, for example in Thau [46], and Kou, Ellitt and Tarn [47]. The result on the output injection form was shown in Krener and Isidori [48]. The geometric conditions for the existence of a state transformation to the output injection form can also be found in Isidori [33] and Marino and Tomei [24]. Basic knowledge of differential manifolds can be found in Boothby [32]. Linear observer error dynamics via direct state transformation was initially shown by Kazantzis and Kravaris [49]. Further results on this topic can be found in Xiao [50] and Ding [51]. In the latter one, polynomial approximation to nonlinear functions for a solution to a nonlinear state transformation is discussed and an explicit region of convergence of such an approximation is given. The basic idea of observer design for systems with Lipschitz nonlinearities is to use the linear part of the observer dynamics to dominate the nonlinear terms. This idea was shown in some of the early results on nonlinear observers, but the systematic introduction to the topic was due to Rajamani [52], in which the condition (8.31) was shown. Some later results on this topic are shown in Zhu and Han [53]. One-sided Lipschtiz condition for nonlinear observer design was shown in Zhao, Tao and Shi [54]. The result on observer design for systems with Lipschitz output nonlinearities is adapted from Ding [55]. The result on reduced-order observer of linear systems can be traced back to the very early stage of observer design in Luenberger [56]. Early results on this topic were discussed in Kailath [7]. The reduced-order observer design is closely related to observer design for systems with unknown inputs, of which some results may be found in Hou and Muller [57]. For nonlinear systems, a reduced-order observer was used for estimation of unknown states for control design of a class of nonlinear systems with nonlinearities of unmeasured states by Ding [58]. The result shown here is based on Ding [31]. An early result on adaptive observer for nonlinear systems is reported in Bastin and Gevers [59]. Results on adaptive observers with output nonlinearities are covered in Marino and Tomei [24]. The results shown in Section 8.6 is mainly based on Cho and Rajamani [60]. The method to construct a first-order positive real system in Remark 8.9 is based on Ding [40].

Chapter 9. Early ideas of backstepping appeared in a number of papers, such as Tsinias [61]. Iterative backstepping for state feedback is shown in Kanellakopoulos, Kokotovic and Morse [62] with adaptive laws for unknown parameters.

Backstepping designs without adaptive laws shown in this chapter are basically simplified versions of their adaptive counterparts. Filtered transformations were used for backstepping in Marino and Tomei [63], and the presentation in this chapter is different from the forms used in [63]. Initially, multiple estimation parameters are used for adaptive control with backstepping, for example the adaptive laws in [63]. Tuning function method was first introduced in Krstic, Kanellakopoulos and Kokotovic [64] to remove multiple adaptive parameters for one unknown parameter vector with backstepping. Adaptive backstepping with filtered transformation was based on Ding [65], without the disturbance rejection part. Adaptive observer backstepping is largely based on the presentation in Krstic, Kanellakopoulos and Kokotovic [66]. Nonlinear adaptive control with backstepping is also discussed in details in Marino and Tomei [24]. Some further developments of nonlinear adaptive control using backstepping can be found in Ding [40] for robust adaptive control with dead-zone modification, robust adaptive control with σ-modification in Ding [67], with unknown control directions and unknown high-frequency gains in Ding [41, 42]. Adaptive backstepping was also applied to nonlinear systems with nonlinear parameterisation in Ding [68].

Chapter 10.   Early results on rejection of sinusoidal disturbances can be found in Bodson, Sacks and Khosla [69] and Bodson and Douglas [70]. Asymptotic rejection of sinusoidal disturbances for nonlinear systems in the output feedback form is reported in Ding [71] when the frequencies are known. The result shown in Section 10.1 is adapted mainly from Ding [72]. Asymptotic rejection of periodic disturbances can be formulated as an output regulation problem when the periodic disturbances are generated from a dynamic system, which is known as an exosystem. Output regulation for linear systems is shown in Davison [73] and Francis [74], and for nonlinear systems with local stability in Huang and Rugh [75], and Isidori and Byrnes [76]. More results on output regulation with local stability can be found in the books by Isidori [33] and Huang [77]. The result shown in Section 10.2 is based on Ding [65]. The format of the exosystem for internal model design was inspired by Nikiforov [78]. The Nussbaum gain was first used by Nussbaum [79], and the treatment here is similar to Ye and Ding [80] used in the control design to tackle the unknown sign of high frequency gain. Early results on output regulation with nonlinear exosystems can be found in Ding [81, 82], and Chen and Huang [83]. The result shown in Section 10.3 is based on Xi and Ding [84]. Asymptotic rejection of general disturbances was studied as an extension of output regulation of nonlinear exosystems. Waveform profiles of the general periodic disturbances were used to estimate the equivalent input disturbance in some early results in Ding [85–87]. The approach for asymptotic rejection with an observer-like internal model is adapted from Ding [55, 88].

Chapter 11.   Harmonic estimation is treated as a special case of state estimation. The recursive estimation and rejection method in Section 11.1 is based on Ding [89]. Practical issues on harmonics estimation and rejections are covered in Arrillaga, Watson and Chen [90] and Wakileh [91].

Clock gene was reported in 1994 by Takahashi *et al.* [92]. Reports on key genes in circadian oscillators are in Albrecht *et al.* [93], van der Horst *et al.* [94] and Bunger *et al.* [95]. The effects of jet lags and sleep disorders to circadian rhythms are reported in Sack *et al.* [96] and Sack *et al.* [97]. The effects of light and light treatments to disorders of circadian rhythms are shown by Boulos *et al.* [98], Kurosawa and Goldbeter [99] and Geier *et al.* [100]. The circadian model (11.48) is adapted from Gonze, Leloup and Goldbeter [101]. The condition (11.52) is shown by Zhao, Tao and Shi [54]. The observer design is adapted from That and Ding [102], and the control design is different from the version shown in that paper. More results on observer design for a seventh-order circadian model are published by That and Ding [103].

There are many results in literature on sampled-data control of linear systems, for example see Chen and Francis [104] and the references therein. The difficulty of obtaining exact discrete-time model for nonlinear systems is shown by Nesic and Teel [105]. Structures in nonlinear systems cannot be preserved as shown in Grizzle and Kokotovic [106]. Approximation to nonlinear discrete-time models is shown by several results, for example Nesic, Teel and Kokotovic [107] and Nesic and Laila [108]. The effect on fast-sampling of nonlinear static controllers is reported in Owens, Zheng and Billings [109]. Several other results on emulation method for nonlinear systems are shown in Laila, Nesic and Teel [110], Shim and Teel [111] and Bian and French [112]. The results presented in Section 10.3 are mainly based on Wu and Ding [113]. Sampled-data control for disturbance rejection of nonlinear systems is shown by Wu and Ding [114]. A result on sampled-data adaptive control of a class of nonlinear systems is to be found in Wu and Ding [115].

References

[1]   Cook, P. A. *Nonlinear Dynamic Systems*. London: Prentice-Hall, 1986.
[2]   Slotine, J.-J. E. and Li, W. *Applied Nonlinear Control*. London: Prentice-Hall, 1991.
[3]   Vidyasagar, M. *Nonlinear Systems Analysis*, 2nd ed. Englewood Cliffs, New Jersey: Prentice-Hall International, 1993.
[4]   Khalil, H. K. *Nonlinear Systems*, 3rd ed. Upper Saddle River, New Jersey: Prentice Hall, 2002.
[5]   Verhulst, F. *Nonlinear Differential Equations and Dynamic Systems*. Berlin: Springer-Verlag, 1990.
[6]   Ogata, K. *Modern Control Engineering*, 3rd ed. London: Prentice-Hall, 1997.
[7]   Kailath, T. *Linear Systems*. London: Prentice-Hall, 1980.
[8]   Antsaklis, P. J. and Michel, A. N. *Linear Systems*. New York: McGraw-Hill, 1997.
[9]   Zheng, D. Z. *Linear Systems Theory (in Chinese)*. Beijing: Tsinghua University Press, 1990.
[10] Chen, B. M., Lin, Z. and Shamash, Y. *Linear Systems Theory*. Boston: Birkhauser, 2004.
[11] Arnold, V. I. *Ordinary Differential Equations*. Berlin: Springer International, 1994.
[12] Borrelli, R. L. and Coleman, C. S. *Differential Equations*. New York: John Wiley & Sons, 1998.
[13] Hu, T. and Lin, Z. *Control Systems With Actuator Saturation: Analysis And Design*. Boston, MA: Birhäuser, 2001.
[14] Edwards, C. and Spurgeon, S. *Sliding Mode Control*. London: Taylor & Francis, 1998.
[15] Isidori, A. *Nonlinear Control Systems II*. London: Springer-Verlag, 1999.
[16] Elgerd, O. I. *Electric Energy Systems Theory*. New York: McGraw-Hill, 1971.
[17] van der Pol, B. 'On relaxation oscillations,' *Phil. Mag.*, vol. 2, pp. 978–992, 1926.
[18] Lorenz, E. N. 'Deterministic nonperiodic flow,' *J. Atmos. Sci.*, vol. 20, pp. 130–141, 1963.
[19] Atherton, D. P. *Nonlinear Control Engineering*. Workingham: Van Nostraind Reinhold, 1975.

[20] Gelb, A. and Velde, W. E. V. *Multi-Input Describing Functions and Nonlinear System Design*. New York: McGraw-Hill, 1963.

[21] Lyapunov, A. M. 'The general problem of motion stability (translated in french),' *Ann. Fac. Sci. Toulouse*, vol. 9, pp. 203–474, 1907.

[22] Hahn, W. *Stability of Motion*. Berlin: Springer, 1967.

[23] Narendra, K. S. and Annaswamy, A. M. *Stable Adaptive Systems*. Englewood Cliffs, New Jersey: Prentice-Hall, 1989.

[24] Marino, R. and Tomei, P. *Nonlinear Control Design: Geometric, Adaptive, and Robust*. London: Prentice-Hall, 1995.

[25] Narendra, K. S. and Taylor, J. H. *Frequency Domain Stability for Absolute Stability*. New York: Academic Press, 1973.

[26] Brown, J.W. and Churchill, R.V. *Complex Variables and Applications*, 7th ed. New York: McGraw-Hill, 2004.

[27] Sontag, E. D. 'Smooth stabilization implies coprime factorization,' *IEEE Transaction on Automatic Control*, vol. 34, pp. 435–443, 1989.

[28] Sontag, E.D. and Wang, Y. 'On characterization of the input-to-state stability properties,' *Syst. Contr. Lett.*, vol. 24, pp. 351–359, 1995.

[29] Sontag, E. and Teel, A. 'Changing supply functions input/state stable systems,' *IEEE Trans. Automat. Control*, vol. 40, no. 8, pp. 1476–1478, 1995.

[30] Jiang, Z. -P., Teel, A. R. and Praly, L. 'Small-gain theorem for iss systems and applications,' *Math. Contr. Sign. Syst.*, vol. 7, pp. 95–120, 1994.

[31] Ding, Z. 'Differential stability and design of reduced-order observers for nonlinear systems,' *IET Control Theory and Applications*, vol. 5, no. 2, pp. 315–322, 2011.

[32] Boothby, W. A. *An Introduction to Differential Manifolds and Riemaniann Geometry*. London: Academic Press, 1975.

[33] Isidori, A. *Nonlinear Control Systems*, 3rd ed. Berlin: Springer-Verlag, 1995.

[34] Brockett, R. W. 'Feedback invariants for non-linear systems,' in *The Proceedings of 7th IFAC Congress*, vol. 6, Helsinki, Finland.

[35] Astrom, K. J. and Wittenmark, B. *Adaptive Control*, 2nd ed. Reading, MA: Addison-Wesley, 1995.

[36] Wellstead, P. E. and Zarrop, M. B. *Self-Tuning Systems: Control and Signal Processing*. New York: John Wiley & Sons, 1991.

[37] Ding, Z. 'Model reference adaptive control of dynamic feedback linearisable systems with unknown high frequency gain,' *IEE Proceedings Control Theory and Applications*, vol. 144, pp. 427–434, 1997.

[38] Parks, P. C. 'Lyapunov redesign of model reference adaptive control systems,' *IEEE Trans. Automat. Contr.*, vol. 11, pp. 362–367, 1966.

[39] Ioannou, P. A. and Sun, J. *Robust Adaptive Control*. Upper Saddle River, New Jersey: Prentice Hall, 1996.

[40] Ding, Z. 'Robust adaptive control of nonlinear output-feedback systems under bounded disturbances,' *IEE Proc. Control Theory Appl.*, vol. 145, pp. 323–329, 1998.

[41]   Ding, Z. 'Global adaptive output feedback stabilization of nonlinear systems of any relative degree with unknown high frequency gain,' *IEEE Trans. Automat. Control*, vol. 43, pp. 1442–1446, 1998.

[42]   Ding, Z. 'Adaptive control of nonlinear systems with unknown virtual control coefficients,' *Int. J. Adaptive Control Signal Proc.*, vol. 14, no. 5, pp. 505–517, 2000.

[43]   Ding, Z. 'Analysis and design of robust adaptive control for nonlinear output feedback systems under disturbances with unknown bounds,' *IEE Proc. Control Theory Appl.*, vol. 147, no. 6, pp. 655–663, 2000.

[44]   Ding, Z. and Ye, X. 'A flat-zone modification for robust adaptive control of nonlinear output feedback systems with unknown high frequency gains,' *IEEE Trans. Automat. Control*, vol. 47, no. 2, pp. 358–363, 2002.

[45]   Ding, Z. 'Adaptive stabilization of a class of nonlinear systems with unstable internal dynamics,' *IEEE Trans. Automat. Control*, vol. 48, no. 10, pp. 1788–1792, 2003.

[46]   Thau, F. 'Observing the states of nonlinear dynamical systems,' *Int. J. Control*, vol. 18, pp. 471–479, 1973.

[47]   Kou, S., Ellitt, D. and Tarn, T. 'Exponential observers for nonlinear dynamical systems,' *Inf. Control*, vol. 29, pp. 204–216, 1975.

[48]   Krener, A. J. and Isidori, A. 'Linearization by output injection and nonlinear observers,' *Syst. Control Lett.*, vol. 3, pp. 47–52, 1983.

[49]   Kazantzis, M. and Kravaris, C. 'Nonlinear observer design using lyapunov's auxiliary theorem,' *Syst. Control Lett.*, vol. 34, pp. 241–247, 1998.

[50]   Xiao, M. 'The global existence of nonlinear observers with linear error dynamics: A topological point of view,' *Syst. Control Lett.*, vol. 55, no. 10, pp. 849–858, 2006.

[51]   Ding, Z. 'Observer design in convergent series for a class of nonlinear systems,' *IEEE Trans. Automat. Control*, vol. 57, no. 7, pp. 1849–1854, 2012.

[52]   Rajamani, R. 'Observers for Lipschitz nonlinear systems,' *IEEE Trans. Automat. Control*, vol. 43, no. 3, pp. 397–401, 1998.

[53]   Zhu, F. and Han, Z. 'A note on observer design for Lipschitz nonlinear systems,' *IEEE Trans. Automat. Control*, vol. 47, no. 10, pp. 1751–1754, 2002.

[54]   Zhao, Y., Tao, J. and Shi, N. 'A note on observer design for one-sided Lipschitz nonlinear systems,' *Syst. Control Lett.*, vol. 59, pp. 66–71, 2010.

[55]   Ding, Z. 'Asymptotic rejection of unmatched general periodic disturbances with nonlinear Lipschitz internal model,' *Int. J. Control*, vol. 86, no. 2, pp. 210–221, 2013.

[56]   Luenberger, D. G. 'Observing the state of a linear system,' *IEEE Trans. Mil. Electron.*, vol. 8, pp. 74–80, 1964.

[57]   Hou, M. and Muller, P. 'Design of observers for linear systems with unknown inputs,' *IEEE Trans. Automat. Control*, vol. 37, no. 6, pp. 871–875, 1992.

[58]   Ding, Z. 'Global output feedback stabilization of nonlinear systems with nonlinearity of unmeasured states,' *IEEE Trans. Automat. Control*, vol. 54, no. 5, pp. 1117–1122, 2009.

[59] Bastin, G. and Gevers, M. 'Stable adaptive observers for nonlinear time varying systems,' *IEEE Trans. Automat. Control*, vol. 33, no. 7, pp. 650–658, 1988.

[60] Cho, Y. M. and Rajamani, R. 'A systematic approach to adaptive observer synthesis for nonlinear systems,' *IEEE Trans. Automat. Control*, vol. 42, no. 4, pp. 534–537, 1997.

[61] Tsinias, J., 'Sufficient lyapunov-like conditions for stabilization,' *Math. Control Signals Systems*, vol. 2, pp. 343–357, 1989.

[62] Kanellakopoulos, I., Kokotovic, P. V. and Morse, A. S. 'Systematic design of adaptive controllers for feedback linearizable systems,' *IEEE Trans. Automat. Control*, vol. 36, pp. 1241–1253, 1991.

[63] Marino, R. and Tomei, P. 'Global adaptive output feedback control of nonlinear systems, part i: Linear parameterization,' *IEEE Trans. Automat. Control*, vol. 38, pp. 17–32, 1993.

[64] Krstic, M., Kanellakopoulos, I. and Kokotovic, P. V. 'Adaptive nonlinear control without overparametrization,' *Syst. Control Lett.*, vol. 19, pp. 177–185, 1992.

[65] Ding, Z. 'Adaptive output regulation of class of nonlinear systems with completely unknown parameters,' in *Proceedings of 2003 American Control Conference*, Denver, CO, 2003, pp. 1566–1571.

[66] Krstic, M., Kanellakopoulos, I. and Kokotovic, P. V. *Nonlinear and Adaptive Control Design*. New York: John Wiley & Sons, 1995.

[67] Ding, Z. 'Almost disturbance decoupling of uncertain output feedback systems,' *IEE Proc. Control Theory Appl.*, vol. 146, pp. 220–226, 1999.

[68] Ding, Z. 'Adaptive control of triangular systems with nonlinear parameterization,' *IEEE Trans. Automat. Control*, vol. 46, no. 12, pp. 1963–1968, 2001.

[69] Bodson, M., Sacks, A. and Khosla, P. 'Harmonic generation in adaptive feedforward cancellation schemes,' *IEEE Trans. Automat. Control*, vol. 39, no. 9, pp. 1939–1944, 1994.

[70] Bodson, M. and Douglas, S. C. 'Adaptive algorithms for the rejection of sinusoidal disturbances with unknown frequencies,' *Automatica*, vol. 33, no. 10, pp. 2213–2221, 1997.

[71] Ding, Z. 'Global output regulation of uncertain nonlinear systems with exogenous signals,' *Automatica*, vol. 37, pp. 113–119, 2001.

[72] Ding, Z. 'Global stabilization and disturbance suppression of a class of nonlinear systems with uncertain internal model,' *Automatica*, vol. 39, no. 3, pp. 471–479, 2003.

[73] Davison, E. J. 'The robust control of a servomechanism problem for linear time-invariant multivariable systems,' *IEEE Trans. Automat. Control*, vol. 21, no. 1, pp. 25–34, 1976.

[74] Francis, B. A. 'The linear multivariable regulator problem,' *SIAM J. Control Optimiz.*, vol. 15, pp. 486–505, 1977.

[75] Huang, J. and Rugh, W. J. 'On a nonlinear multivariable servomechanism problem,' *Automatica*, vol. 26, no. 6, pp. 963–972, 1990.

[76]   Isidori, A. and Byrnes, C. I. 'Output regulation of nonlinear systems,' *IEEE Trans. Automat. Control*, vol. 35, no. 2, pp. 131–140, 1990.

[77]   Huang, J. *Nonlinear Output Regulation Theory and Applications*. Philadelphia, PA: SIAM, 2004.

[78]   Nikiforov, V. O. 'Adaptive non-linear tracking with complete compensation of unknown disturbances,' *Eur. J. Control*, vol. 4, pp. 132–139, 1998.

[79]   Nussbaum, R. D. 'Some remarks on a conjecture in parameter adaptive control,' *Syst. Control Lett.*, vol. 3, pp. 243–246, 1983.

[80]   Ye, X. and Ding, Z. 'Robust tracking control of uncertain nonlinear systems with unknown control directions,' *Syst. Control Lett.*, vol. 42, pp. 1–10, 2001.

[81]   Ding, Z. 'Output regulation of uncertain nonlinear systems with nonlinear exosystems,' in *Proceeding of the 2004 American Control Conference*, Boston, MA, 2004, pp. 3677–3682.

[82]   Ding, Z. 'Output regulation of uncertain nonlinear systems with nonlinear exosystems,' *IEEE Trans. Automat. Control*, vol. 51, no. 3, pp. 498–503, 2006.

[83]   Chen, Z. and Huang, J. 'Robust output regulation with nonlinear exosystems,' *Automatica*, vol. 41, pp. 1447–1454, 2005.

[84]   Xi, Z. and Ding, Z. 'Global adaptive output regulation of a class of nonlinear systems with nonlinear exosystems,' *Automatica*, vol. 43, no. 1, pp. 143–149, 2007.

[85]   Ding, Z. 'Asymptotic rejection of general periodic disturbances in output-feedback nonlinear systems,' *IEEE Trans. Automat. Control*, vol. 51, no. 2, pp. 303–308, 2006.

[86]   Ding, Z. 'Asymptotic rejection of asymmetric periodic disturbances in output-feedback nonlinear systems,' *Automatica*, vol. 43, no. 3, pp. 555–561, 2007.

[87]   Ding, Z. 'Asymptotic rejection of unmatched general periodic disturbances in a class of nonlinear systems,' *IET Control Theory Appl.*, vol. 2, no. 4, pp. 269–276, 2008.

[88]   Ding, Z. 'Observer design of Lipschitz output nonlinearity with application to asymptotic rejection of general periodic disturbances,' in *The Proceedings of The 18th IFAC Congress*, vol. 8, Milan, Italy, 2011.

[89]   Ding, Z. 'Asymptotic rejection of finite frequency modes of general periodic disturbances in output-feedback nonlinear systems,' *Automatica*, vol. 44, no. 9, pp. 2317–2325, 2008.

[90]   Arrillaga, J., Watson, N. R. and Chen, S. *Power System Quality Assessment*. Chichester: John Wiley & Sons, 2000.

[91]   Wakileh, G. J. *Power Systems Harmonics*. Berlin: Springer-Verlag, 2010.

[92]   Takahashi, J., Vitaterna, M., King, D., Chang, A., Kornhauser, J., Lowrey, P., McDonald, J., Dove, W., Pinto, L., and Turek, F. 'Mutagenesis and mapping of a mouse gene, clock, essential for circadian behaviour,' *Science*, vol. 264, pp. 719–725, 1994.

[93]   Albrecht, U., Sun, Z., Eichele, G. and Lee, C. 'A differential response of two putative mammalian circadian regulators, mper1 and mper2 to light,' *Cell Press*, vol. 91, pp. 1055–1064, 1997.

[94]   van der Horst, G., Muijtjens, M., Kobayashi, K., Takano, R., Kanno, S., Takao, M., de Wit, J., Verkerk, A., Eker, A., van Leenen, D., Buijs, R., Bootsma, D., Hoeijmakers, J., and Yasui, A. 'Mammalian cry1 and cry2 are essential for maintenance of circadian rhythms,' *Nature*, vol. 398, pp. 627–630, 1999.

[95]   Bunger, M., Wilsbacher, L., Moran, S., Clendenin, C., Radcliffe, L., Hogenesch, J., Simon, M., Takahashi, J., and Bradfield, C., 'Mop3 is an essential component of the master circadian pacemaker in mammals,' *Cell*, vol. 103, pp. 1009–1017, 2000.

[96]   Sack, R., Auckley, D., Auger, R., Carskadon, M., Wright, K., Vitiello, M., and Zhdanova, I. 'Circadian rhythm sleep disorders: part i, basic principles, shift work and jet lag disorders,' *Sleep*, vol. 30, pp. 1460–1483, 2007.

[97]   Sack, R., Auckley, D., Auger, R., Carskadon, M., Wright, K., Vitiello, M., and Zhdanova, I. 'Circadian rhythm sleep disorders: Part ii, advanced sleep phase disorder, delayed sleep phase disorder, free-running disorder, and irregular sleep-wake rhythm,' *Sleep*, vol. 30, pp. 1484–1501, 2007.

[98]   Boulos, Z., Macchi, M., Sturchler, M., Stewart, K., Brainard, G., Suhner, A., Wallace, G., and Steffen, R., 'Light visor treatment for jet lag after west ward travel across six time zones,' *Aviat. Space Environ. Med.*, vol. 73, pp. 953–963, 2002.

[99]   Kurosawa, G. and Goldbeter, A. 'Amplitude of circadian oscillations entrained by 24-h light-dark cycles,' *J. Theor. Biol.*, vol. 242, pp. 478–488, 2006.

[100]   Geier, F., Becker-Weimann, S., Kramer, K., and Herzel, H. 'Entrainment in a model of the mammalian circadian oscillator,' *J. Biol. Rhythms*, vol. 20, pp. 83–93, 2005.

[101]   Gonze, D., Leloup, J., and Goldbeter, A., 'Theoretical models for circadian rhythms in neurospora and drosophila,' *Comptes Rendus de l'Academie des Sciences -Series III -Sciences de la Vie*, vol. 323, pp. 57–67, 2000.

[102]   That, L. T. and Ding, Z. 'Circadian phase re-setting using nonlinear output-feedback control,' *J. Biol. Syst.*, vol. 20, no. 1, pp. 1–19, 2012.

[103]   That, L. T. and Ding, Z. 'Reduced-order observer design of multi-output nonlinear systems with application to a circadian model,' *Trans. Inst. Meas. Control*, to appear, 2013.

[104]   Chen, T. and Francis, B. *Optimal Sampled-Data Control Systems*. London: Springer-Verlag, 1995.

[105]   Nesic, D. and Teel, A. 'A framework for stabilization of nonlinear sampled-data systems based on their approximate discrete-time models,' *IEEE Trans. Automat. Control*, vol. 49, pp. 1103–1122, 2004.

[106]   Grizzle, J. W. and Kokotovic, P. V. 'Feedback linearization of sampled-data systems,' *IEEE Trans. Automat. Control*, vol. 33, pp. 857–859, 1988.

[107]   Nesic, D., Teel, A., and Kokotovic, P.V. 'Sufficient conditions for stabiliza-
        tion of sampled-data nonlinear systems via discrete-time approximations,'
        Syst. Control Lett., vol. 38, pp. 259–270, 1999.

[108]   Nesic, D. and Laila, D. 'A note on input-to-state stabilization for nonlinear
        sampled-data systems,' *IEEE Trans. Automat. Control*, vol. 47, pp. 1153–
        1158, 2002.

[109]   Owens, D. H., Zheng, Y. and Billings, S. A. 'Fast sampling and stability of
        nonlinear sampled-data systems: Part 1. Existing theorems,' *IMA J. Math.
        Control Inf.*, vol. 7, pp. 1–11, 1990.

[110]   Laila, D. S., Nesic, D. and Teel, A. R. 'Open-and closed-loop dissipa-
        tion inequalities under sampling and controller emulation,' *Eur. J. Control*,
        vol. 18, pp. 109–125, 2002.

[111]   Shim, H. and Teel, A. R. 'Asymptotic controllability and observability
        imply semiglobal practical asymptotic stabilizability by sampled-data output
        feedback,' *Automatica*, vol. 39, pp. 441–454, 2003.

[112]   Bian, W. and French, M. 'General fast sampling theorems for nonlinear
        systems,' *Syst. Control Lett.*, vol. 54, pp. 1037–1050, 2005.

[113]   Wu, B. and Ding, Z. 'Asymptotic stablilisation of a class of nonlinear systems
        via sampled-data output feedback control,' *Int. J. Control*, vol. 82, no. 9,
        pp. 1738–1746, 2009.

[114]   Wu, B. and Ding, Z. 'Practical disturbance rejection of a class of nonlin-
        ear systems via sampled output,' *J. Control Theory Appl.*, vol. 8, no. 3,
        pp. 382–389, 2010.

[115]   Wu, B. and Ding, Z. 'Sampled-data adaptive control of a class of nonlinear
        systems,' *Int. J. Adapt. Control Signal Process.*, vol. 25, pp. 1050–1060,
        2011.

Index